Emerging Contaminants and Associated Treatment Technologies

Series Editors
Muhammad Zaffar Hashmi, Institute of Biology and Biotechnology
The University of Lahore
Lahore, Pakistan

Vladimir Strezov, Department of Environmental Sciences
Macquarie University
Sydney, NSW, Australia

Emerging Contaminants and Associated Treatment Technologies focuses on contaminant matrices (air, land, water, soil, sediment), the nature of pollutants (emerging, well-known, persistent, e-waste, nanomaterials, etc.), health effects (e.g., toxicology, occupational health, infectious diseases, cancer), treatment technologies (bioremediation, sustainable waste management, low cost technologies), and issues related to economic development and policy. The book series includes current, comprehensive texts on critical national and regional environmental issues of emerging contaminants useful to scientists in academia, industry, planners, policy makers and governments from diverse disciplines. The knowledge captured in this series will assist in understanding, maintaining and improving the biosphere in which we live. The scope of the series includes monographs, professional books and graduate textbooks, edited volumes and books devoted to supporting education on environmental pollution at the graduate and post-graduate levels.

Kanchan Kumari

Pollutants of Global Concern

A Comprehensive Overview of Persistent Organic Pollutants

Springer

Kanchan Kumari
CSIR-National Environmental Engineering Research Institute
(NEERI), Kolkata Zonal Centre
Kolkata, West Bengal, India

ISSN 2524-6402 ISSN 2524-6410 (electronic)
Emerging Contaminants and Associated Treatment Technologies
ISBN 978-3-031-50995-7 ISBN 978-3-031-50996-4 (eBook)
https://doi.org/10.1007/978-3-031-50996-4

This Springer imprint is published by the registered company Springer Nature Switzerland AG
The registered company address is: Gewerbestrasse 11, 6330 Cham, Switzerland

Preface

Persistent organic pollutants (POPs) have piqued the interest of the scientific community, the public, and policymakers across the globe. However, when it comes to their status in India, accessing comprehensive information has often proven elusive. This book provides a comprehensive compilation of crucial insights concerning POPs, systematically organized in accordance with the principles outlined by the Stockholm Convention. It spans an array of information, ranging from the technical characteristics of POPs to their national usage, remedial approaches adopted within India in comparison to international practices, their ecological impacts, and future perspectives. Significantly, in-house scientific research on each of these chemicals is presented succinctly in paragraphs and tabulations. The book distinguishes itself by dedicating each chapter to a specific POP, enabling a thorough exploration of its usage, effects, and fate.

In essence, this book serves as an invaluable guide for stakeholders, academicians, and anyone dealing with these chemicals, seeking a deeper understanding of the current state of POPs in India. Furthermore, there's an urgent need for raising awareness about these internationally recognized toxic pollutants, especially among the research community. This book, through its latest compiled data and information, contributes to clarifying India's standing in the global landscape.

Titled *Pollutants of Global Concern – A Comprehensive Overview of Persistent Organic Pollutants*, this book aims to shed light on various facets of POPs. The insights gathered from various scientific events, with the valuable contributions of national and international experts from diverse backgrounds, have been thoughtfully integrated into this book as individual chapters, forming a crucial part of the comprehensive exploration of POPs in our contemporary world. In the process of compiling this book, we've drawn from a wide spectrum of scientific literature and expert opinions, spanning research organizations, industries, academia, civil society, and non-governmental organizations. Our primary audience includes researchers, industry professionals, regulatory bodies, NGOs, MSMEs, industry associations, and other related stakeholders. Secondary audiences encompass academicians, students, and legal practitioners. The book provides an up-to-date compilation of data derived from various nationally and internationally conducted studies and offers an acute comparison with internationally available figures to position India within the global context.

The book addresses 24 chemicals currently listed under the Stockholm Convention placing a significant emphasis on POPs requiring greater attention such as brominated flame-retardants (polybrominated diphenyl ethers), perfluorooctane sulfonate (PFOS), perfluorooctanoic acid (PFOA), hexabromocyclododecane (HBCD), hexachlorobutadiene (HCBD), hexabromobiphenyl (HBB), pentachlorobenzene (PeCB), polychlorinated biphenyls (PCBs), hexabromobiphenyl (HBB), chlordecone, chlordane, dioxins/furans, and short-chain chlorinated paraffin (SCCPs), among others. These chemicals are not only of relevance to developed countries but also of significant concern in developing nations.

This book serves as a highly anticipated reference source, tailored to a wide-ranging audience, encompassing academicians, industrial scientists, and regulatory bodies. All the chapters are meticulously structured to provide a comprehensive understanding of diverse categories of persistent organic pollutants. This includes their technical intricacies, the present landscape of their usage in India, a comparative analysis of international versus Indian data, the extensive ramifications they pose, effective management and remedial strategies, and a concluding synthesis of insights. Within the pages of this book, our authors aim to provide a comprehensive understanding of persistent organic pollutants and their impact on both humans and ecosystems. Each chapter is dedicated to a particular persistent organic pollutant, allowing for a focused examination of its attributes.

As you delve into the pages that follow, we invite you on a journey of understanding, discovery, and collaboration. Our sincere hope is that this book will stand as a valuable resource, a beacon of knowledge, and a catalyst for positive change in our collective effort to combat the persistent threats posed by POPs and to safeguard the health of our planet and its inhabitants.

Kolkata, West Bengal, India Kanchan Kumari

Acknowledgement

We, the authors, are thankful to Director, CSIR-NEERI and Head, KZC, CSIR-NEERI for granting permission to undertake this challenging work. We also wish to express our sincere thanks to the Ministry of Environment, Forest and Climate Change (MoEF&CC), Government of India, for guiding us to take forward this work. All the chapters have been checked by iThenticate plagiarism checker software and approved to be published vide following KRC Number: CSIR-NEERI/KRC/2023/AUG/KZC/2-3; NIIST/2023/Aug/537; CSIR-NEERI/KRC/2023/SEP/KZC/1-12; CSIR-NEERI/KRC/2023/OCT/KZC/1-8, KZC-EBGD/1, KZC-APC/1.

Contents

List of Figures

Aldrin

Chlordane

Dichlorodiphenyltrichloroethane (DDT)

Dieldrin

Endrin

Heptachlor

Hexachlorobenzene

Mirex

Toxaphene

Polychlorinated Biphenyls (PCBs)

Dioxins and Furans

Hexachlorocyclohexane (α, β-HCH) and Lindane (γ-HCH)

Chlordecone

Hexabromobiphenyl (HBB)

Hexabromocyclododecane (HBCD)

Tetra-, Penta-, Hexa-, and Heptabromodiphenyl Ethers

Hexachlorobutadiene (HCBD)

Penta-Chlorobenzene (PeCB)

Pentachlorophenol (PCP): Its Salts and Esters

Perfluorooctane Sulphonic Acid (PFOS), Its Salts, and Perfluorooctane Sulphonyl Fluoride (PFOS-F)

Polychlorinated Naphthalenes (PCNs)

Endosulfan

Decabromodiphenyl Ether (decaBDE)

Short-Chain Chlorinated Paraffins (SCCPs)

List of Tables

Endrin

Heptachlor

Hexachlorobenzene

Mirex

Toxaphene

Polychlorinated Biphenyls (PCBs)

Dioxins and Furans

Hexachlorocyclohexane (α, β-HCH) and Lindane (γ-HCH)

Chlordecone

Hexabromobiphenyl (HBB)

Hexabromocyclododecane (HBCD)

Tetra-, Penta-, Hexa-, and Heptabromodiphenyl Ethers

Hexachlorobutadiene (HCBD)

Penta-Chlorobenzene (PeCB)

Pentachlorophenol (PCP): Its Salts and Esters

Perfluorooctane Sulphonic Acid (PFOS), Its Salts, and Perfluorooctane Sulphonyl Fluoride (PFOS-F)

Polychlorinated Naphthalenes (PCNs)

Endosulfan

Decabromodiphenyl Ether (decaBDE)

Short-Chain Chlorinated Paraffins (SCCPs)

About the Author

Kanchan Kumari is a Principal Scientist at CSIR-NEERI, Kolkata, Associate Professor at the Academy of Scientific and Innovative Research (AcSIR) and Former Contact Person of Stockholm Convention Regional Centre for Asia Region on Persistent Organic Pollutants (POPs). She is Member of (1) Program Advisory Committee (CEE & Power Scheme), Science and Engineering Research Board, Delhi; (2) Technical Expert Committee, NGT (OA 18/2020/EZ) and (3) Research Council, Central University of Gujarat and IIT Dhanbad. She has completed her PhD in the area of Environmental Toxicology. She is working in NEERI since 2011 and is actively involved in research and development in the area of Chemical Wastes, particularly POPs, along with Ecology and Biodiversity Study (QCI-NABET approved). She has organized more than 40 All India Training Programme related to waste management and water quality monitoring and also delivered several lectures in National and International Workshops/Training Programmes. She has published more than 50 research publications in peer reviewed journals, conference, proceedings and technical reports. Book entitled *Persistent Organic Pollutants – Gaps in Management and Associated Challenges* has been edited by her with 8 contributed chapters. She has guided 3 PhD Students and also supervised >25 MSc students for their dissertation work and has received various national awards (>13). She has been awarded the Young Scientist Award, 2005, 2007 and 2009; Scientist of Year 2013 Award; DBT Women Scientist Fellowship, 2012; Best paper presentation award, 2006,

2007 and 2009; and nominated for CSIR Young Scientist Award, 2013. She has successfully executed >35 major research projects; funding support mostly received from UNEP, UNDP, World Bank, DBT, MoEF&CC, CPCB, Industries, etc. in the capacity of Principal Investigator. Because of her initiative, one dedicated Centre on Chemical and Waste with special focus on POPs supported by UNEP has been established in CSIR-NEERI to provide a common platform where all the stakeholders dealing with chemical and wastes can share and exchange the information. She also executed a national study on the development of soil quality at ten different sites of the country along with the National Inventorization of Mercury in the waste sector. Because of her active involvement, two MoUs have been signed with NEERI – one with MoEF&CC and the other with RECETOX, Czech Republic, in order to work to fulfil the obligations of the Stockholm Convention.

Aldrin

Kanchan Kumari and Shaheen Akhtar

1 Introduction

An important member of the classic organochlorine pesticides, aldrin was extensively used until the 1970s when it was thoroughly banned in most countries. It was included to the Stockholm Convention's Annex A in 2001 as a persistent organic pollutant (POP) (Stockholm Convention, 2001). Aldrin is primarily prepared via the Diels-Alder reaction of hexachlorocyclopentadiene with norbornadiene and contains 95% of 1,2,3,4,10,10-hexachloro-1,4,4a,5,8,8a-hexa-hydro-exo-1,4-endo-5,8-dimethanonaphthalene (HHDN). Prior to its banning, it was majorly used to treat seed and soil but was later labelled a persistent organic pollutant because of its bio accumulative property. But due to its rapid conversion into dieldrin and also due to relatively rapid volatization, it is seldom found in nature (Metcalf, 2000).

As an effective soil and seed pesticide, aldrin is applied to kill grasshoppers, corn rootworms, seed corn beetle, rice water weevil, and other pests. Its emulsified concentrate or wettable powder was also used to treat wooden structures from termites (Jorgenson, 2001). It was discovered that aldrin might also harm people, fish, and birds in the early 1960s. The lethal dose in humans, where it mostly affects the central nervous system, is thought to be around 5 g for an adult male (Burt, 1975). Aldrin was included to the list of persistent chemicals by the Governing Council of the United Nations Environment Programme (UNEP) to decrease and/or eliminate persistent organic compounds emission (WHO, 2003). It is thus banned in many

K. Kumari (✉) · S. Akhtar
CSIR-National Environmental Engineering Research Institute (NEERI),
Kolkata Zonal Centre, Kolkata, West Bengal, India
e-mail: k_kumari@neeri.res.in

K. Kumari, *Pollutants of Global Concern*, Emerging Contaminants and Associated Treatment Technologies, https://doi.org/10.1007/978-3-031-50996-4_1

countries including Singapore, Bulgaria, Hungary, and Switzerland and its manufacture and application is severely restricted in countries like New Zealand, the USA, Austria, Japan, Canada, and many more (Ritter et al., 1995).

2 Technical Details of Aldrin

The physicochemical characteristics of aldrin are given in Table 1.

Table 1 Physico-chemical properties of aldrin

Details	Description (aldrin)
Structure	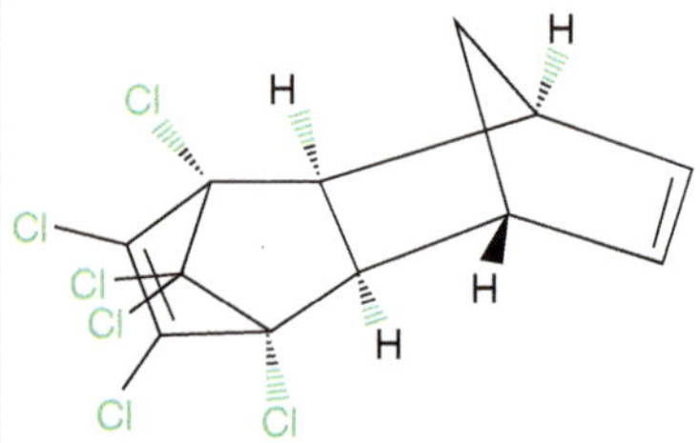
IUPAC name	(1*R*,2*R*,3*R*,6*S*,7*S*,8*S*)-1,8,9,10,11,11-Hexachlorotetracyclo[6.2.1.$1^{3,6}$.$0^{2,7}$]dodeca-4,9-diene
Molecular formula	$C_{12}H_8Cl_6$
CAS No.	309-00-2
Molecular weight	364.9 g/mol
Color/form	Colorless crystal
Odor	Mild chemical odor
Boiling point	145 °C at 2 mm Hg
Melting point	104 °C
Solubility	In water, 170 mg/l at 25 °C
Density	1.6 g/cm^3
Vapor pressure	8e−05 mm Hg
Octanol/water partition coefficient	6.50
Octanol-air partition coefficient	8.08
Henry's law constant	4.40e−05 atm-m^3/mole
Stability/shelf life	Stable in the heat up to 200 °C and between pH 4 and pH 8
Decomposition	When heated to the point of breakdown, it releases harmful hydrogen chloride vapors
Corrosivity	Corrosive to metals, owing to the slow formation of hydrogen chloride during storage
Safety and hazards	Acute toxic, health hazard and environmental hazard

3 Status of Works and Usage in India

Aldrin was majorly used in agricultural fields and farms in India as a pesticide in soil. It has been banned from manufacture, use, and import in India since 20 September 1996 (Kumari & Kate, 2022). Despite this, there have been reports of aldrin stockpiling in several regions of the nation (NIP, 2011) (Fig. 1).

India exported 439,980,000 kg of aldrin in 2014–2015, according to statistics from the Ministry of Commerce and Industry and the Government of India. In 2015–2016, 57.68 thousand kg was exported to Bangladesh, Denmark, Taiwan, and other countries, and 73.17 thousand kg in 2016–2017 to Hong Kong, Nepal, Iran, Mauritius, Israel, and other countries. Aldrin's export has progressively declined over the years. In 2014–2015, India exported the most aldrin from India, totaling 175,96,000 kg to Vietnam. The greatest shipment of aldrin, 218 kg, was made in 2014–2015 from Italy (Toxics Link, 2018).

Owing to its broad spectrum pesticidal and insecticidal effects as well as its low cost, aldrin was one of the most extensively used organochlorine pesticide in India. As residual quantities of aldrin have been periodically recorded in various parts of the country (Fig. 2), one of the prevalent environmental matrices where evident traces of aldrin were noted is water and the various riverine systems. Reasons that majorly contributed to their contamination are agricultural runoff, leaching from landfills, sewage, leakage of lagoons etc. Motekar (2011) reported that aldrin residues were detected in large number of samples in the Godavari Plain of Nanded district (Maharashtra). Similarly, groundwater samples collected from bore wells in Uttar Pradesh (Unnao) and Delhi showed considerable amount of aldrin residues,

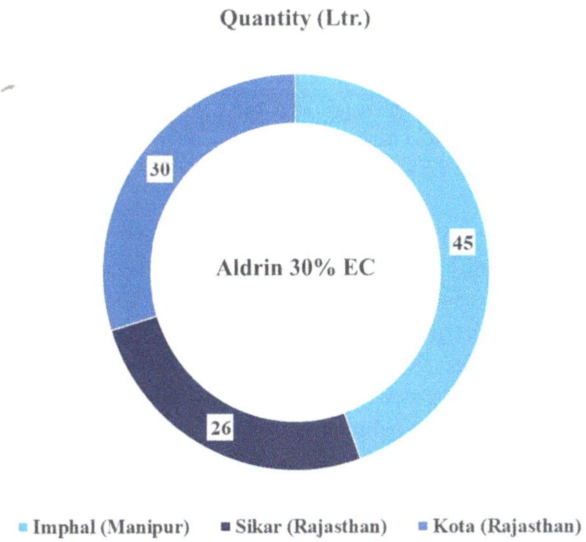

Fig. 1 Stockpiles of aldrin in India (NIP, 2011). *EC-Emulsifiable Concentrate

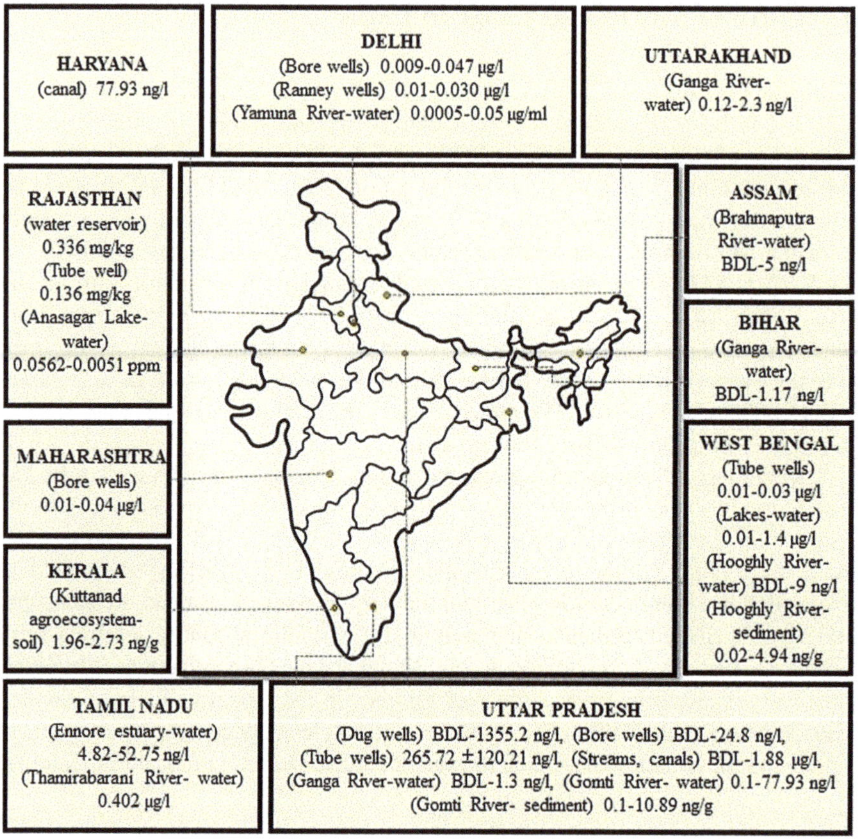

Fig. 2 Residual aldrin in groundwater, surface water, riverine systems, and soil of different states of India. (Source: Bakore et al., 2004; Singh et al., 2005; Ghose et al., 2009; Motekar, 2011; Mutiyar et al., 2011; Singh, 2001; Nair et al., 1991; Chakraborty et al., 2016; Mutiyar and Mittal, 2013; Leena et al., 2011; Malik et al., 2009; Arisekar et al., 2021; Khuman et al., 2020; Sruthi et al., 2017; Singh et al., 2007; Kaushik et al., 2008; Charan et al., 2007; Sundararajan et al., 2015). *BDL-Below Detection Limit

i.e., 24.8 ng/l and 0.04 µg/l, respectively (Singh et al., 2005; Mutiyar et al., 2011). Organochlorine pesticides (OCPs) were found in the Ganga River basin, which encompassed the three states of Uttarakhand, Uttar Pradesh, and Bihar and accounted for 72% of the river's length. Pesticide that dominated the Bihar stretch was aldrin, with a 34% detection rate. According to the source apportionment of the OCP, intense agriculture and industrial activity may be a substantial source of these pesticides in the lower reaches (Mutiyar & Mittal, 2013).

Since rivers are marked as the major repositories of these OCPs, it has been recorded that approximately 2500 tonnes of pesticides and 1.2 million tonnes of fertilizer are used annually in the Ganges' catchment area, which substantially pollutes the river (Kumari et al., 2001). However, there is no designated database of the quantities of harmful micro pollutants like organochlorines and heavy metals in the Ganges ecosystem or any such other riverine or surface water system. The upper trophic levels of the food chain, including fish of these waterbodies, are possibly harmed by pollution by the OCPs as they can acquire toxic pesticide residues at concentrations that are several times higher than in the surrounding water.

According to Mani and Konar (1975), stressed fish exhibit abnormal behavior, survival, growth, reproduction, and respiration. Not only aquatic species but also breeding issues affect birds, reptiles, mammals, and other species living in contaminated settings that contain a variety of identified as well as unidentified pollutants. The need for measuring bioaccumulation in fish of riverine systems has been evident by the presence of significant quantities of these pollutants in ecosystems. There are numerous reports available for the analysis of aldrin residues in fish (Singh & Singh, 2008). The quantities, as well as the distribution trends of some of these pesticides in samples of fish obtained from the Gomti River in India, were reported in research, with aldrin being the most frequent OCP with a value of 0.30–7.84 ng/g wet weight (Malik et al., 2007). Another study was conducted by Kumari et al. (2001) where quantities of persistent organochlorine contamination were detected in economically significant Ganges River fish from various Bihar locales. Aldrin was present in fish muscles in average wet weight values ranging from 3.1 to 86.1 ng/g (Table 2). Therefore, in a nation like India, where the use of organochlorine pesticides was prominent and where fish is one of the main food sources, monitoring organochlorine contaminants in river aquatic species is critical for conservation. Eating fish tainted with organochlorine pesticides poses a serious harm to human health.

Despite the fact that there are insufficient data on aldrin concentration in India's environmental matrices and atmosphere, these can be found in human diet, and it is well recognized that aldrin can concentrate up the food chain and consequently be found in human tissues. A study conducted in Delhi showed the average aldrin concentrations to be 0.048 ppb for adipose tissue, 0.004 ppb for blood serum, and 0.003 ppb for breast milk respectively. The adipose tissue from the older donors had larger concentrations of aldrin. These compounds were present in breast milk at higher concentrations in pregnant women. The aldrin concentration in adipose tissue and blood serum, as well as that in adipose tissue and breast milk, was found to be positively correlated (Nair et al., 1992). Another study showed levels of aldrin to be significantly higher in patients with chronic kidney disease (Siddarth et al., 2014). The findings of the study by Mustafa et al. (2010) made it evident that age and dietary habits affected the levels of aldrin in maternal and cord blood of pregnant women. Compared to women in the age range of 19–24 years, women in the 25–30 age group had higher levels of aldrin in their maternal blood. Similarly, aldrin levels in maternal blood were considerably greater in non-vegetarian women than in vegetarian women (Table 3).

Table 2 Residual aldrin in biota

Sample	Species	Study site	Location	Aldrin concentration	References
Fishes	*Rita rita* *Mystus tengara* *Cyprinus carpio* *Labeo rohita*	Ponds in Gujartal	Uttar Pradesh (Jaunpur)	0.004 ppm 0.003 ppm 0.005 ppm 0.002 ppm	Singh and Singh (2008)
Fishes	*Rita rita* *Cyprinus carpio*	Gomti River	Uttar Pradesh (Jaunpur)	0.02 ppm 0.01 ppm	Singh and Singh (2008)
Fishes	*Mystus tengara* *Labeo rohita*	Ganga River	Uttar Pradesh (Varanasi)	0.08 ppm 0.04 ppm	Singh and Singh (2008)
Birds	*Grus antigone* *Streptopelia decaocto*	Keoladeo National Park	Rajasthan (Bharatpur)	89.75 ppm 140 ppm	Muralidharan (1993)
Earthworms	–	Yamuna River	Delhi	0.0004–0.15 μg/g	Nair et al. (1991)
Fish	–	Yamuna River	Delhi	0.0001–0.03 μg/g	Nair et al. (1991)
Clams	–	Yamuna River	Delhi	0.0002–0.01 μg/g	Nair et al. (1991)
Fish	*Labeo rohita*	Thamirabarani River	Tamil nadu	0.8975 μg/kg	Arisekar et al. (2021)
Zooplanktons	–	Kaveri River and Kollidam River	Southeast Bay of Bengal	0.36–0.78 μg/g	Kannan and SenGupta (1987)
River dolphins	*Platanista gangetica*	Ganga river	Patna, Bihar	4.0–9.2 ng/g	Kumari et al. (2002)
Fishes	*Channa punctatus*	Gomti river	Uttar Pradesh	0.49–5.15 ng/g	Malik et al. (2007)
Fishes	*Labeo rohita, Labeo bata, Cirrhinus mrigala, Rita rita, Eutropichthys vacha, Clupisoma garua, Catla catla, Wallago attu, Silonia silondia, Aoriichthys seenghala* and *Aoriichthys aor*	Ganga river	Bihar	3.1–86.1 mg/g	Kumari et al. (2001)

4 Global Scenario

Around two million tonnes of OCPs were consumed annually throughout the world between 1950 and 2000 (Fang et al., 2017). Though their application was banned in the majority of countries, OCPs are still found in many environmental

Table 3 Residual aldrin found in human (breast/serum)

Study site	Gender	Medical condition	Aldrin concentration (ppb)	References
Delhi	Women	Pre term delivery (maternal)	3.32 ± 1.85	Mustafa et al. (2010)
Delhi	Women	Pre term delivery (cord)	2.41 ± 0.93	Mustafa et al. (2010)
Punjab	Women	Healthy	6.3	Mathur et al. (2005)
Ahmedabad	Men	Healthy	0.200± 0.045	Bhatnagar et al. (1992)
Delhi	Women	Healthy	0.004	Nair et al. (1992)
North India	Both	Parkinson's Disease	1.4–6.8	Chhillar et al. (2013)
North India	Both	Alzeimer's Disease	2.56 ± 1.38	Singh et al. (2013)
North India	Both	Chronic Kidney Disease	3.80 ± 2.5	Siddarth et al. (2014)

compartments, such as soil, sediment, water, etc., due to their prior over usage. Studies to assess the concentrations of OCPs including that of aldrin have been significantly carried out around the world where 0.06 ng/g of aldrin in soil was found in Uzbekistan while aldrin concentration in soil of Shanghai (China) was 0.32 ± 1.13 ng/g (Jorfi et al., 2019). This organochlorine pesticide residue was also reported in various water systems such as that of Ovia River in Nigeria, Buffalo River in South Africa, Nile River in Egypt, Vistule River in Poland etc. (Fig. 3).

Moreover, evidences of residual aldrin in higher organisms and eventually in humans have also been recorded in various parts of the globe. Serum samples collected from Canary Island in Spain showed detectable amounts of aldrin where the younger generation in the age group of below 18 years had higher concentrations of aldrin as compared to older generation of 65–75 years (Luzardo et al., 2006). Residues of this OCP were also detected in breast milk of rural Mayan women of Mexico (Polanco Rodríguez et al., 2017) (Table 4).

5 Adverse Effects of Aldrin

Aldrin residues that readily metabolize to dieldrin (UNEP/POPS/INC.1/INF/10, 2023) are either rarely or in very small amounts found in the bodies of animals. In studies conducted by Abdullah et al. (1990) and Mugambi et al. (1989), out of 100 samples, only 21 of Egyptian fish and 0.5% of the total sampled eggs of *Gallus domesticus* (domestic fowl from central Kenya) showed the presence of aldrin residues. Furthermore, aldrin binds firmly to the soil, has resistance to groundwater leaching, is very toxic, and is highly species-specific, making it a possible endocrine disruptor (Jorgenson, 2001). As with other organochlorine contaminants, most animal studies show that liver damage is the predominant consequence. In animal

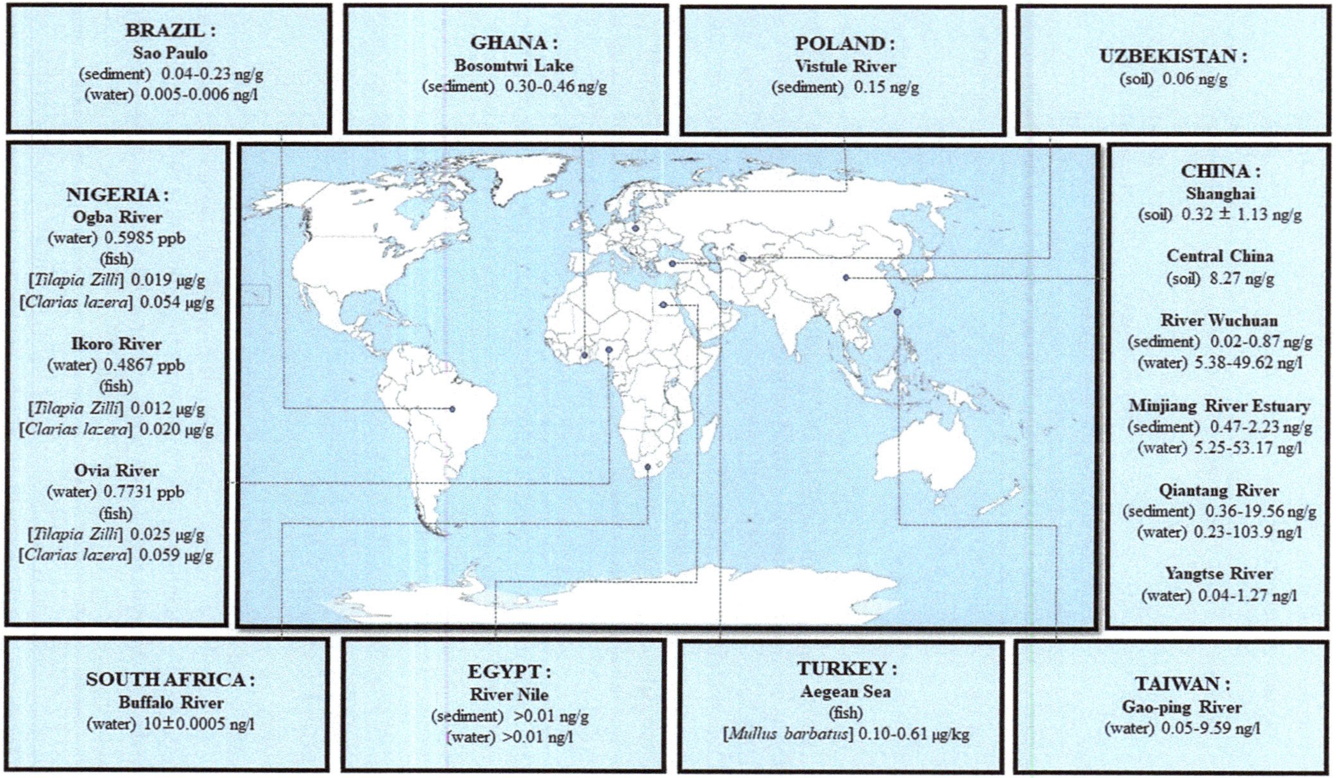

Fig. 3 Residual aldrin concentration in soil, sediment, and surface water around the world. (Source: Sruthi et al., 2017; Jorfi et al., 2019; Zhang et al., 2002, 2003; Zhou et al., 2006; Rissato et al., 2006; Darko et al., 2008; Ahmed et al., 2008; Falandysz & Strandberg, 2004; Xu et al., 2000; Awofolu & Fatoki, 2003; Doong et al., 2008; Ize-Iyamu et al., 2007; Kucuksezgin et al., 2001)

Table 4 Residual aldrin found in human (breast milk/serum)

Country	Gender	Medical condition	Aldrin concentration (ppb)	References
Mexico	Women (Rural Maya community)	Exposed to contaminated water	122	Polanco Rodríguez et al. (2017)
Spain	Women	Breast cancer	72.5 ± 107.2	Boada et al. (2012)
Spain	Both	Healthy population of different age groups	45.8 ±35.9	Luzardo et al. (2006)

populations, these herbicides produce hypersensitivity, convulsions, neural degeneration, temporary hypothermia, tremors, and anorexia (ATSDR, 1993). These cause negative enzymatic and hormonal alterations in fish, resulting in decreased reproductive capabilities (Environmental Health Criteria 91, 1989). Many ducks and shorebirds are suspected to have died as a result of ingestion of aldrin-treated rice along the Texas Gulf Coast of the United States (Ritter et al., 1995). Studies have indicated that mice exposed to aldrin on a long-term basis develop either benign or malignant liver tumors (Kolaja et al., 1996) while no substantial effects on the liver were seen when identical experiments on rats were performed. Large, single doses of aldrin in mice and hamsters during mid-gestation led in morphological abnormalities in the fetus, including foot webbing, cleft lip, and open-eye, as well as an elevated death rate (ATSDR, 1993). In rats exposed to high doses, extensive kidney impairment has been recorded while chronic aldrin exposure caused hepatic effects, such as increased liver weight, hepatocyte enlargement, increased levels and activity of cellular smooth endoplasmic reticulum and microsomal enzymes, as well as a rise in vacuolization (ATSDR, 1993).

Aldrin has comparable negative effects on people when exposed acutely. The most likely route of exposure for people is through food and drinking water, and it has been discovered mostly in food (ATDSR, 2022). Maximum Residue Limit (MRL) of the Food and Agricultural Organisation (FAO) and WHO vary from 0.006 mg/kg milk fat to 0.2 mg/kg beef fat and 0.1 to 180 g/l for water quality (Toxics Link, 2006). Aldrin exposure symptoms include headache, dizziness, nausea, loss of appetite, irritability, muscular twitching, loss of consciousness, and convulsions, and even death (ATSDR, 1993). These symptoms may last for a few weeks after exposure, but they have not been proven to be permanent. Although the study was constrained by a lack of quantitative data on exposure, a statistical increase in liver and biliary tract malignancies was reported among a sample of employees occupationally exposed to aldrin, dieldrin, and endrin (Ritter et al., 1995). The United States Environmental Protection Agency lists aldrin as a potential human carcinogen. Headaches, nausea and vomiting, muscular twitching, myoclonic jerks, anorexia, and, in the most severe cases, aberrant EEG readings have been documented among workers involved in the production or use of aldrin. Furthermore, two investigations of people exposed to aldrin at work revealed unusually high incidence of psychiatric disease. One study found aberrant electromyograms (EMGs), indicating peripheral neuropathy (ATSDR, 1993). Aldrin has also been reported in maternal blood, placenta, and umbilical cord blood of mother and fetuses, indicating its potential to traverse via placenta (Saxena et al., 1981). Mutagenic effects of aldrin have been reported by Ahmed et al., 1977, where 1–1000 μmol/l of aldrin has shown to induce unscheduled DNA synthesis transforming the human fibroblast cell. In addition, fluoroculturists exposed to aldrin exhibited increased exchange of sister chromatid and exchange type chromatic aberrations (Dulout et al., 1985).

Low water stability and semi-volatility properties of aldrin have primarily aided its transport over the polar regions where its metabolized form dieldrin has been detected in artic air, water, and organisms (Barrie et al., 1992; Fellin et al., 1996; Muir et al., 1992).

6 Management and Remedial Measures

According to the Ministry of Agriculture and Farmers' Welfare, the order published on 17th July 2001 vide S.O. 682 (E) sub-section (2) of section 27 read with Section 28 of the Insecticides Act, 1968, the central government formed an expert panel to look into the usage of aldrin and examine its effect on various biota.

After concluding that the import, production, formulation, and use of aldrin pose health risks to humans, the Ministry of Agriculture (Government of India) issued an order prohibiting the manufacture, usage, export, and import of aldrin beginning on July 17, 2001. The Food Safety and Standards (Contaminants, Toxins, and Residues) Regulations, 2011, may be referred to as these regulations, which state that no pesticides may be applied directly on food products and the amount of aldrin on the foods (Fig. 4), shall not exceed the tolerance limit prescribed.

Removal of aldrin from environmental compartments is a challenging task that has been studied for years implementing various methods of combustion, xenon lamps, ion conversion etc (Intisar et al., 2023; Saranya & Kanaga, 2019). Microbial degradation has thus far been validated to be promising removal method for aldrins especially from soil and sediments and it has been discussed in detail in review work of Raghunathan et al., 2022. Numerous microbial species have been identified for this purpose that include *Pseudomonas fluorescens, Pleurotus ostreatus, Bacillus* sp., *Burkholderia* sp., *Cupriavidus* sp., *Pseudonocardia* sp., *Trichoderma viride, Mucor racemosus,* and Arthrobacter. Oxidation, hydroxylation, and reduction pathways are the major ones adopted for aldrin degradation (Ritter et al., 1995; Pang et al., 2022)

7 Conclusion

Aldrin, a well-known persistent organic pollutant (POP), serves as a cautionary story of the unforeseen environmental repercussions of human activities. Its previous broad usage as an agricultural pesticide resulted in long-term pollution of soil,

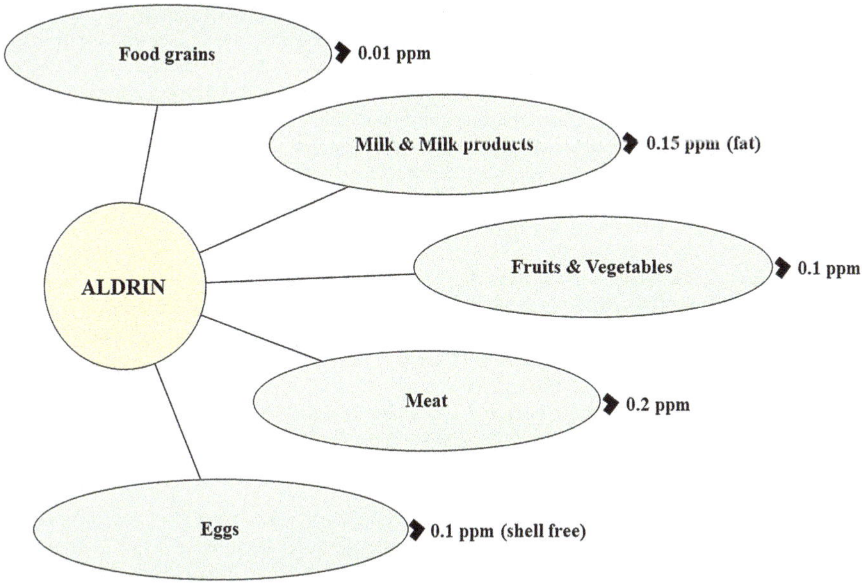

Fig. 4 Prescribed tolerance limit in types of food (Toxics Link, 2018)

water, and wildlife. Concerns have been expressed regarding aldrin's long-term persistence, bioaccumulation, and potential for long-distance transmission. Furthermore, prior aldrin use may continue to have an impact on species and ecosystems. Since it has been identified as a pollutant of concern in India owing to its hazardous nature, aldrin was prohibited as an agricultural pesticide in India in 1996, in accordance with the Stockholm Convention on Persistent Organic Pollutants. Inspite of that aldrin concentrations can still be discovered in several parts of India. According to the standard criteria, the permissible value of the aldrin in India is very less. Hence, no further evaluation of this OCP is needed.

Resolving aldrin's legacy in India would need coordinated actions, including monitoring methods to limit its prevalence in the environment. Public awareness efforts on the concerns of POPs can also help to promote appropriate farming practises and discourage the use of prohibited chemicals.

As we face environmental issues, aldrin's legacy serves as a reminder of the significance of responsible chemical usage and proactive actions to protect our world and future generations from the negative impacts of persistent organic pollutants.

References

Abdallah, M. A., Hassan, I. M., Naguib, M. M., & Abodonia, M. A. (1990). Survey of residues of organochlorine pesticides in some marketable Egyptian fish. *Journal of the Association of Official Analytical Chemists, 73*(4), 502–508.

Ahmed, F. E., Hart, R. W., & Lewis, N. J. (1977). Pesticide induced DNA damage and its repair in cultured human cells. *Mutation Research/Fundamental and Molecular Mechanisms of Mutagenesis, 42*, 161–173.

Ahmed, A., Barbary, E., Yehia, M. M., Mohamed, M., & Bouraie, E. (2008). Evaluation of organochlorine pesticides (OCPs) in surface water and bed sediment samples from the river Nile at Rosetta Branch, Egypt. *Journal of Applied Sciences Research, 4*(12), 1985–1993.

Arisekar, U., Shakila, R. J., Shalini, R., & Jeyasekaran, G. (2021). Pesticides contamination in the Thamirabarani, a perennial river in peninsular India: The first report on ecotoxicological and human health risk assessment. *Chemosphere, 267*, 129251.

ATSDR, Agency for Toxic Substances and Disease Registry. (1993). *Cancer policy framework.* U.S. Department of Health and Human Services, Public Health Service.

ATSDR, Agency for Toxic Substances and Disease Registry. (2022). *Toxicological profile for aldrin and dieldrin.* U.S. Department of Health and Human Services, Public Health Service, Agency for Toxic Substances and Disease Registry.

Awofolu, R. O., & Fatoki, O. S. (2003). Persistent organochlorine pesticide residues in freshwater systems and sediments from the Eastern Cape, South Africa. *Water SA, 29*(3), 323–330.

Bakore, N., John, P. J., & Bhatnagar, P. (2004). Organochlorine pesticide residues in wheat and drinking water samples from Jaipur, Rajasthan, India. *Environmental Monitoring and Assessment, 98*, 381–389.

Barrie, L. A., Gregor, D., Hargrave, B., Lake, R., Muir, D., Shearer, R., et al. (1992). Arctic contaminants: Sources, occurrence and pathways. *Science of the Total Environment, 122*(1–2), 1–74.

Bhatnagar, V. K., Patel, J. S., Variya, M. R., Venkaiah, K., Shah, M. P., & Kashyap, S. K. (1992). Levels of organochlorine insecticides in human blood from Ahmedabad (rural), India. *Bulletin of Environmental Contamination and Toxicology;(United States), 48*(2), 302–307.

Boada, L. D., Zumbado, M., Henríquez-Hernández, L. A., Almeida-González, M., Álvarez-León, E. E., Serra-Majem, L., & Luzardo, O. P. (2012). Complex organochlorine pesticide mixtures as determinant factor for breast cancer risk: A population-based case–control study in the Canary Islands (Spain). *Environmental Health, 11*, 1–9.

Burt, G. S. (1975). Use of behavioral techniques in the assessment of environmental contaminants. In *Behavioral toxicology* (pp. 241–263). Springer US.

Chakraborty, P., Khuman, S. N., Selvaraj, S., Sampath, S., Devi, N. L., Bang, J. J., & Katsoyiannis, A. (2016). Polychlorinated biphenyls and organochlorine pesticides in River Brahmaputra from the outer Himalayan Range and River Hooghly emptying into the Bay of Bengal: Occurrence, sources and ecotoxicological risk assessment. *Environmental Pollution, 219*, 998–1006.

Charan, P. D., Patni, A., & Sharma, K. C. (2007). Pesticide contamination in lake anasagar of Ajmer, Rajasthan. In *Proceedings of Taal2007: The 12th World Lake Conference* (Vol. 360, p. 365). B.M. Birla Auditorium and Convention Centre.

Chhillar, N., Singh, N. K., Banerjee, B. D., Bala, K., Mustafa, M., Sharma, D., & Chhillar, M. (2013). Organochlorine pesticide levels and risk of Parkinson's disease in north Indian population. *International Scholarly Research Notices, 2013*, 371034.

Darko, G., Akoto, O., & Oppong, C. (2008). Persistent organochlorine pesticide residues in fish, sediments and water from Lake Bosomtwi, Ghana. *Chemosphere, 72*(1), 21–24.

Doong, R. A., Lee, S. H., Lee, C. C., Sun, Y. C., & Wu, S. C. (2008). Characterization and composition of heavy metals and persistent organic pollutants in water and estuarine sediments from Gao-ping River, Taiwan. *Marine Pollution Bulletin, 57*(6–12), 846–857.

Dulout, F. N., Pastori, M. C., Olivero, O. A., Cid, M. G., Loria, D., Matos, E., et al. (1985). Sister-chromatid exchanges and chromosomal aberrations in a population exposed to pesticides. *Mutation Research Letters, 143*(4), 237–244.

Environmental Health Criteria 91. (1989). *International programme on chemical safety, environmental health criteria 91, aldrin and dieldrin.* World Health Organization. Retrieved on 2 Mar 2023 from https://apps.who.int/iris/bitstream/handle/10665/37536/9241542918eng.pdf?sequence=1&isAllowed=y

Falandysz, J., & Strandberg, B. (2004). Persistent organochlorine compounds in sludge and sediments from the Gdańsk region, Baltic Sea. *Polish Journal of Environmental Studies, 13*(2), 133–138.

Fang, Y., Nie, Z., Die, Q., Tian, Y., Liu, F., He, J., Huang, Q. (2017). Organochlorine pesticides in soil, air, and vegetation at and around a contaminated site in southwestern China: concentration, transmission, and risk evaluation. *Chemosphere 178:340–349.*

Fellin, P., Dougherty, D., Barrie, L. A., Toom, D., Muir, D., Grift, N., et al. (1996). Air monitoring in the Arctic: Results for selected persistent organic pollutants for 1992. *Environmental Toxicology and Chemistry: An International Journal, 15*(3), 253–261.

Food Safety and Standards (Contaminants, Toxins and Residues) Regulations, 2011. https://www.fssai.gov.in/upload/uploadfiles/files/Contaminants_Regulations.pdf

Ghose, N. C., Saha, D., & Gupta, A. (2009). Synthetic detergents (surfactants) and organochlorine pesticide signatures in surface water and groundwater of greater Kolkata, India. *Journal of Water Resource and Protection, 2009*(4), 227–298, 290.

Intisar, A., Ramzan, A., Hafeez, S., Hussain, N., Irfan, M., Shakeel, N., et al. (2023). Adsorptive and photocatalytic degradation potential of porous polymeric materials for removal of pesticides, pharmaceuticals, and dyes-based emerging contaminants from water. *Chemosphere, 336*, 139203.

Ize-Iyamu, O. K., Asia, I. O., & Egwakhide, P. A. (2007). Concentrations of residues from organochlorine pesticide in water and fish from some rivers in Edo State Nigeria. *International Journal of Physical Sciences, 2*(9), 237–241.

Jorfi, S., Atashi, Z., Akhbarizadeh, R., Khorasgani, Z. N., & Ahmadi, M. (2019). Distribution and health risk assessment of organochlorine pesticides in agricultural soils of the Aghili plain, Southwest Iran. *Environmental Earth Sciences, 78*, 1–9.

Jorgenson, J. L. (2001). Aldrin and dieldrin: A review of research on their production, environmental deposition and fate, bioaccumulation, toxicology, and epidemiology in the United States. *Environmental Health Perspectives, 109*(Suppl 1), 113–139.

Kannan, S. T., & SenGupta, R. (1987). Organochlorine residues in zooplankton off the Saurashtra coast, India. *Marine Pollution Bulletin, 18*, 92.

Kaushik, C. P., Sharma, H. R., Jain, S., Dawra, J., & Kaushik, A. (2008). Pesticide residues in river Yamuna and its canals in Haryana and Delhi, India. *Environmental Monitoring and Assessment, 144*, 329–340.

Khuman, S. N., Vinod, P. G., Bharat, G., Kumar, Y. M., & Chakraborty, P. (2020). Spatial distribution and compositional profiles of organochlorine pesticides in the surface soil from the agricultural, coastal and backwater transects along the south-west coast of India. *Chemosphere, 254*, 126699.

Kolaja, K. L., Stevenson, D. E., Johnson, J. T., Walborg, E. F., Jr., & Klaunig, J. E. (1996). Subchronic effects of dieldrin and phenobarbital on hepatic DNA synthesis in mice and rats. *Fundamental and Applied Toxicology, 29*(2), 219–228.

Kucuksezgin, F., Altay, O., Uluturhan, E. S. İ. N., & Kontas, A. (2001). Trace metal and organochlorine residue levels in red mullet (Mullus barbatus) from the eastern Aegean, Turkey. *Water Research, 35*(9), 2327–2332.

Kumari, K., & Kate, P. (2022). Gaps and challenges associated with management of POPs. In *Persistent organic pollutants: Gaps in management and associated challenges* (1st ed., pp. 253–270). CRC Press.

Kumari, A., Sinha, R. K., & Gopal, K. (2001). Organochlorine contamination in the fish of the River Ganges, India. *Aquatic Ecosystem Health & Management, 4*(4), 505–510.

Kumari, A., Sinha, R. K., Gopal, K., & Lata, S. (2002). Concentration of organochlorines in Ganges River dolphins from Patna, Bihar. *Journal of Environmental Biology, 23*(3), 279–281.

Leena, S., Choudhary, S. K., & Singh, P. K. (2011). Organochlorine and organophosphorous pesticides residues in water of river Ganga at Bhagalpur, Bihar, India. *International Journal of Research in Chemistry and Environment, 1*, 77–84.

Luzardo, O. P., Goethals, M., Zumbado, M., Alvarez-Leon, E. E., Cabrera, F., Serra-Majem, L., & Boada, L. D. (2006). Increasing serum levels of non-DDT-derivative organochlorine pesticides in the younger population of the Canary Islands (Spain). *Science of the Total Environment, 367*(1), 129–138.

Malik, A., Singh, K. P., & Ojha, P. (2007). Residues of organochlorine pesticides in fish from the Gomti River, India. *Bulletin of Environmental Contamination and Toxicology, 78*, 335–340.
Malik, A., Ojha, P., & Singh, K. P. (2009). Levels and distribution of persistent organochlorine pesticide residues in water and sediments of Gomti River (India)—A tributary of the Ganges River. *Environmental Monitoring and Assessment, 148*, 421–435.
Mani, V. G. T., & Konar, S. K. (1975). Chronic effect of malathion on feeding behaviour, survival, growth and reproduction of fish. *Environment and Ecology, 4*(4), 517–520.
Mathur, H. B., Agarwal, H. C., Johnson, S., & Saikia, N. (2005). Analysis of pesticide residues in blood samples from villages of Punjab. *CSE Report, India*, 1–15.
Metcalf, R. L. (2000). Insect control. In *Ullmann's encyclopedia of industrial chemistry*. Wiley-VCH Verlag GmbH & Co. KGaA.
Motekar, S. C. (2011). Ground water contamination by organochlorine insecticide residues in the Godavari plain of Nanded district. *Recent Research in Science and Technology, 3*(12), 04–06.
Mugambi, J. M., Kanja, L., Maitho, T. E., Skaare, J. U., & Lökken, P. (1989). Organochlorine pesticide residues in domestic fowl (Gallus domesticus) eggs from central Kenya. *Journal of the Science of Food and Agriculture, 48*(2), 165–176.
Muir, D. C., Wagemann, R., Hargrave, B. T., Thomas, D. J., Peakall, D. B., & Norstrom, R. J. (1992). Arctic marine ecosystem contamination. *Science of the Total Environment, 122*(1–2), 75–134.
Muralidharan, S. (1993). Aldrin poisoning of Sarus cranes (Grus antigone) and a few granivorous birds in Keoladeo National Park, Bharatpur, India. *Ecotoxicology, 2*, 196–202.
Mustafa, M. D., Pathak, R., Tripathi, A. K., Ahmed, R. S., Guleria, K., & Banerjee, B. D. (2010). Maternal and cord blood levels of Aldrin and Dieldrin in Delhi population. *Environmental Monitoring and Assessment, 171(1), 633–638.* https://doi.org/10.1007/s10661-010-1307-9
Mutiyar, P. K., & Mittal, A. K. (2013). Status of organochlorine pesticides in Ganga river basin: Anthropogenic or glacial? *Drinking Water Engineering and Science, 6*(2), 69–80.
Mutiyar, P. K., Mittal, A. K., & Pekdeger, A. (2011). Status of organochlorine pesticides in the drinking water well-field located in the Delhi region of the flood plains of river Yamuna. *Drinking Water Engineering and Science, 4*(1), 51–60.
Nair, A., Dureja, P., & Pillai, M. K. K. (1991). Levels of aldrin and dieldrin in environmental samples from Delhi, India. *Science of the Total Environment, 108*(3), 255–259.
Nair, A., Dureja, P., & Pillai, M. K. K. (1992). Aldrin and dieldrin residues in human fat, milk and blood serum collected from Delhi. *Human & Experimental Toxicology, 11*(1), 43–45.
NIP. (2011). *Government of India National Implementation Plan, Stockholm Convention on Persistent Organic Pollutants*. Published by MoEF and UNIDO, Delhi, India. http://www.indiaenvironmentportal.org.in/files/NIP%20final%20draft.pdf
Pang, S., Lin, Z., Li, J., Zhang, Y., Mishra, S., Bhatt, P., & Chen, S. (2022). Microbial degradation of aldrin and dieldrin: Mechanisms and biochemical pathways. *Frontiers in Microbiology, 13*, 713375.
Polanco Rodríguez, Á. G., Inmaculada Riba López, M., Angel DelValls Casillas, T., León, J. A. A., Anjan Kumar Prusty, B., & Álvarez Cervera, F. J. (2017). Levels of persistent organic pollutants in breast milk of Maya women in Yucatan, Mexico. *Environmental Monitoring and Assessment, 189*, 1–13.
Raghunathan, K., Sa Swamy, D., Marathe, A. S., Thawale, P., & Kumari, K. (2022). Degradation pathways of POPs in the environment and various mitigation techniques for the revival of contaminated sites. In *Persistent organic pollutants: Gaps in management and associated challenges* (1st ed., pp. 155–180). CRC Press.
Rissato, S. R., Galhiane, M. S., Ximenes, V. F., De Andrade, R. M., Talamoni, J. L., Libânio, M., et al. (2006). Organochlorine pesticides and polychlorinated biphenyls in soil and water samples in the Northeastern part of São Paulo State, Brazil. *Chemosphere, 65*(11), 1949–1958.
Ritter, L., Solomon, K. R., Forget, J., Stemeroff, M., & O'leary, C. (1995). A review of selected persistent organic pollutants. In *International Programme on Chemical Safety (IPCS). PCS/95.39* (Vol. 65, p. 66). World Health Organization.
Saranya, B., & Kanaga, P. (2019). Waste water treatment by electrochemical oxidation of organic pollutants. *Research Journal of Engineering and Technology, 10*(1), 50–56.

Saxena, M. C., Siddiqui, M. K. J., Bhargava, A. K., Krishna Murti, C. R., & Kutty, D. (1981). Placental transfer of pesticides in humans. *Archives of Toxicology, 48*, 127–134.

Siddarth, M., Datta, S. K., Mustafa, M. D., Ahmed, R. S., Banerjee, B. D., Kalra, O. P., & Tripathi, A. K. (2014). Increased level of organochlorine pesticides in chronic kidney disease patients of unknown etiology: Role of GSTM1/GSTT1 polymorphism. *Chemosphere, 96*, 174–179.

Singh, R. P. (2001). Comparison of organochlorine pesticide levels in soil and groundwater of Agra, India. *Bulletin of Environmental Contamination and Toxicology, 67*, 126–132.

Singh, P. B., & Singh, V. (2008). Pesticide bioaccumulation and plasma sex steroids in fishes during breeding phase from north India. *Environmental Toxicology and Pharmacology, 25*(3), 342–350.

Singh, K. P., Malik, A., Mohan, D., & Sinha, S. (2005). Persistent organochlorine pesticide residues in alluvial groundwater aquifers of Gangetic Plains, India. *Bulletin of Environmental Contamination and Toxicology, 74*, 162–169.

Singh, K. P., Malik, A., & Sinha, S. (2007). Persistent organochlorine pesticide residues in soil and surface water of northern Indo-Gangetic alluvial plains. *Environmental Monitoring and Assessment, 125*, 147–155.

Singh, N. K., Chhillar, N., Banerjee, B. D., Bala, K., Basu, M., & Mustafa, M. (2013). Organochlorine pesticide levels and risk of Alzheimer's disease in north Indian population. *Human & Experimental Toxicology, 32*(1), 24–30.

Sruthi, S. N., Shyleshchandran, M. S., Mathew, S. P., & Ramasamy, E. V. (2017). Contamination from organochlorine pesticides (OCPs) in agricultural soils of Kuttanad agroecosystem in India and related potential health risk. *Environmental Science and Pollution Research, 24*, 969–978.

Stockholm Convention. (2001). Listing of POPs in Stockholm Convention. Retrieved on 7 Sept 2022 from http://chm.pops.int/TheConvention/ThePOPs/ListingofPOPs/tabid/2509/Default.aspx

Sundararajan, S., Karthikeyan, R., Jabakumar, P. P., & Raghumaran, S. (2015). Aldrin contamination in Ennore estuary and coastal waters of Chennai, East Coast of India. *Asian Journal of Chemistry, 27*(4), 1417.

Toxics Link. (2006). Retrieved on 5 Feb 2023 from https://ipen-test.clients.clerestory.com/sites/default/files/documents/4ind_india_country_situation_report-en.pdf

Toxics Link. (2018). Retrieved in 18 Oct 2022 from https://ipen.org/sites/default/files/documents/POPs%20Country%20Situation%20report.pdf

UNEP/POPS/INC.1/INF/10. Intergovernmental Negotiating Committee for an international legally binding instrument for implementing international action on certain persistent organic pollutants, First session Montreal, 29 June–3 July 1998. Retrieved on 4 July 2023 from http://chm.pops.int/Portals/0/docs/from_old_website/documents/meetings/inc1/RITTER-En.html

World Health Organization. (2003). *Aldrin and dieldrin in drinkingwater: Background document for development of WHO guidelines for drinking-water quality*. WHO/SDE/WSH/03.04/73.

Xu, S., Jiang, X., Dong, Y., Sun, C., Feng, J., Wang, L., et al. (2000). Polychlorinated organic compounds in Yangtse River sediments. *Chemosphere, 41*(12), 1897–1903.

Zhang, Z., Hong, H., Zhou, J. L., Yu, G., Chen, W., & Wang, X. (2002). Transport and fate of organochlorine pesticides in the River Wuchuan, Southeast China. *Journal of Environmental Monitoring, 4*(3), 435–441.

Zhang, Z. L., Hong, H. S., Zhou, J. L., Huang, J., & Yu, G. (2003). Fate and assessment of persistent organic pollutants in water and sediment from Minjiang River Estuary, Southeast China. *Chemosphere, 52*(9), 1423–1430.

Zhou, R., Zhu, L., Yang, K., & Chen, Y. (2006). Distribution of organochlorine pesticides in surface water and sediments from Qiantang River, East China. *Journal of Hazardous Materials, 137*(1), 68–75.

Chlordane

Kanchan Kumari and Senerita Swamy

1 Introduction

Since the mid of 1940s until 1980s, chlordane, an organochlorine pesticide, was applied to agricultural crops and used in home foundations to combat termites. Chlordane is also known as 1,2,4,5,6,7,8,8-octachloro-2,3,3a,4,7,7a hexahydro-4,7-methano-1H-indene according to the IUPAC. Chlordane is mostly composed of the isomers cis- and trans-chlordane (Table 1). The primary component of chlordanes was cis-chlordane, indicating that recent applications of chlordane in this field were dubious.

Technically, chlordane is a chlorinated cyclodiene composed up of at least 147 different molecules, including isomers, chlorinated hydrocarbons, and byproducts. The manufacturing process affects the composition. It includes between 43% and 75% of both trans- and cis-chlordane (TC and CC, commonly known as -chlordane) and low amount (10–20%) of heptachlor, cis- and trans-nonachlor, and chlordenes. The uses have been mainly identified in controlling termites along with uses as insecticide of broad-spectrum applied on different agricultural plants. Chlordane have been reported to persist in soil with half-life of approximately 1 year. Chlordane have been declared as possible human carcinogen affecting the body's immune

K. Kumari (✉)
CSIR-National Environmental Engineering Research Institute (NEERI), Kolkata Zonal Center, Kolkata, West Bengal, India
e-mail: k_kumari@neeri.res.in

S. Swamy
CSIR-National Environmental Engineering Research Institute (NEERI), Nagpur, India

K. Kumari, *Pollutants of Global Concern*, Emerging Contaminants and Associated Treatment Technologies, https://doi.org/10.1007/978-3-031-50996-4_2

Table 1 Physico-chemical properties of chlordane

Details	Description (chlordane)
Structure	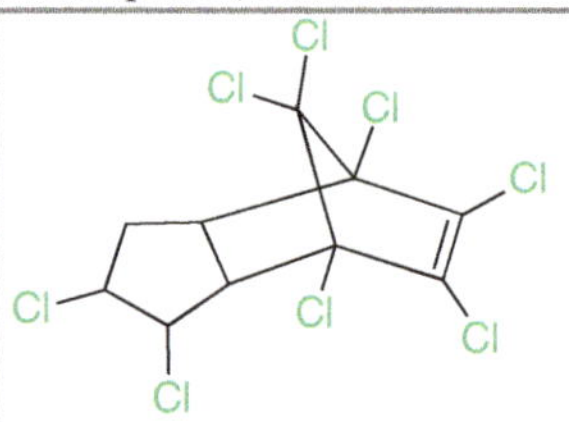
IUPAC name	1,3,4,7,8,9,10,10-Octachlorotricyclo[5.2.1.$0^{2,6}$]dec-8-ene
Molecular formula	$C_{10}H_6Cl_8$
CAS No.	12789-03-6
Molecular weight	409.8 g/mol
Boiling point	175 °C @ 1 mm Hg
Melting point	106.0 °C
Flash point	132 °F
Solubility	In water, 0.056 mg/L @ 25 °C
Density	1.59–1.63 g/cm^3
Vapor pressure	2.05×10^{-05} mmHg
Octanol/water partition coefficient	Log k_{ow} = 6.16
Henry's law constant	2.93×10^{-04} atm-m^3/mole
Stability/shelf life	Dehydrohalogenates in presence of alkali
Decomposition	Near its boiling point, it breaks up to hydrogen chloride chlorine and phosgene
Viscosity	69 poises at 25 °C
Corrosivity	Corrosive to metals, as slowly form hydrogen chloride during storage
Safety and hazards	Irritant, health hazard and environmental hazard
Toxicity	Toxic

system. The main source of exposure is through air and it has been found in Japan and US's indoor air influencing health or residence. Mostly utilized for non-agricultural uses, chlordane is a multipurpose, broad-spectrum contact insecticide (primarily used for safety of structures, however also utilized on ornamental trees, lawn and turf, and drainage ditches). Corn, potatoes, and cattle are also treated with it. It is injected beneath the soil's surface when used to control termites. Chlordane use has becoming increasingly limited in several nations (FAO, 1985; ATSDR, 1989; Worthing, 1991). The Governing Council of the United Nations Environment Programme identified chlordane as a persistent organic pollution in 1997. (UNEP, 1997).

2 Technical Details of Chlordane

Table 2 Uses, sources, and regulation in India

POP	Uses and sources	Annex	Year of listing by SC
Chlordane	Pesticide used as agricultural insecticide and fungicide and also used to treat termites	A	2001

Source: Kumari (2021)

3 Status of Usage Within the Country

Since September 20, 1996, the manufacturing, use, and import of the pesticide chlordane have been prohibited. It is used in agriculture as a fungicide and insecticide. In India, no stockpiles of chlordane were reported. Table 2 represents the status of chemical listed as POPs under the Stockholm Convention.

4 Effects, Toxicity, and Persistence of Chlordane

Effects The detrimental effects reported on avian fauna and fishes are species specific, but chlordane being highly toxic for aquatic animals can lead to mortality in pink shrimps, mallard ducks, and bobwhite quail. Disorders of various organs like kidney, liver, and blood disorders have been linked with this pesticide further damaging cardiovascular, endocrine, and reproductive systems.

Toxicity Declared as possible human carcinogens, evidences are available for disruption of immune system misbalancing endocrine function. Moderate level of acute toxicity has been reported for mammals. Researchers have attributed chlordane exposure to the metabolic syndrome (Lee et al., 2006, 2007; Son et al., 2010), although the underlying molecular pathways are still unclear. According to Wu et al. (2021), exposure to chlordane being an endocrine disruptor impacts growth and development, increases systemic glucose and tri-glycerides levels, and impairs insulin signaling in Drosophila. These effects were likely due to lipid peroxidation and disrupted redox homeostasis under the influence of chlordane. MRL levels suggested by FAO/WHO in milk fat are 0.002 mg/kg and 0.5 mg/kg for poultry fat. Water quality criteria range from 1.5 to 6 μg/L.

Persistence/Fate Chlordane accumulates in biota and degrades slowly in the environment. Water or light do not easily break it down. Chlordane's half-life in soil is thought to be 350 days, while it can be at between 37 and 3500 days. Chlordane has little capacity to disperse in soil, according to studies. Despite having a limited ability to travel through soil, chlordane has been found in groundwater by researchers. Researchers found that the concentration of chlordane in fat was three times higher than the concentration in the meal in a 56-day trial in which rats were fed food containing the chemical. This finding suggests that chlordane does accumulate in the body (US EPA, 2000). Being highly lipophilic, chlordane can be stored in fat tissue for a long time and first builds up in the liver and kidney (Singh et al., 2019).

5 In-house (Countrywide) Scientific Work Done on thc Chemical

In India, there had been complete ban on the production and use of chlordane from 20 September 1996 and no stockpiles have been reported of chlordane in India. According to Dearth and Hites (1991), the cis (α)-chlordane to trans (γ)-chlordane ratio for technical grade chlordane is typically 0.79, and γ-chlordane dissipates more rapidly in the ecosystem than α-chlordane. Hence, γ-chlordane predominates over α-chlordane with respect to level and frequency of identification in groundwater (Singh et al., 2005; Malik et al., 2009). Chlordane application that is recent is indicated by a ratio of α/γ chlordane that is 0.8 (Bidleman et al., 2002; Singh et al., 2007). In surface water from the Unnao, UP, a low α/γ chlordane ratio of 0.04–0.3 was found (Fig. 1), indicating the application of chlordane in the past (Singh et al., 2007). However, an average α/γ chlordane ratio of 1.48 was found in the soils of India's largest cities, showing that chlordane usage is still prevalent (Chakraborty et al., 2015). According to Devi et al. (2015) and Parween et al. (2014), India lacks uniform standards for pesticide residues in soil and sediments. The most widely used guidelines for sediments are consensus-based sediment quality guidelines (CBSQG) from Wisconsin and the National Oceanography and Atmospheric Administration, which describe two categories of concentration levels: threshold effect level (TEL) and probable effect level (PEL) (Mishra et al., 2013; Sarkar et al., 2008). Rare risks for negative effects exist at concentrations below TEL levels. Chlordane in soil has a TEL of 4.50 ng/g. In order to assess the presence of OCPs, including chlordanes, 81 surface soil samples (0–20 cm depth) from urban, suburban, and rural transects in each of the seven most significant Indian cities—New Delhi along with Agra in the north, Kolkata in the east, Mumbai as well as Goa in the west, and Chennai along with Bangalore in the south—were collected and analyzed. About 35% and 25% of the total chlordanes were from trans and cis-chlordanes, respectively. The examined soils had residual chlordane concentrations ranging from 0.01 to 30 ng/g, with the Goa rural site having the highest level. Considering that the reported range is higher than the TEL level, this range suggests a high risk of adverse effects. As evident from the data shown in Fig. 1, there is a rare risk of adverse effects because the mean concentration of chlordane found in

Fig. 1 Residual OCPs concentration in water, sediment, and soil in India

the Yamuna River is lower than the TEL value (Parween et al., 2014). On the contrary, some researchers found higher concentrations of chlordane in soil in significant areas of India when compared to TEL values, which suggests a high risk of adverse effects (Sruthi et al., 2017; Singh & Singh, 2017; Chakraborty et al., 2015). Similar to this, the recommended maximum residual limit (MRL) for water is 0.2 ng/g (WHO, 2021). Chlordane levels were found to be above MRL in significant rivers in India, including the Gomti River and the Ganga River (Sah et al., 2020; Selvaraj et al., 2021). This indicates a potential severe adverse effect on the environment as a whole. Similar levels were identified to be significantly higher than MRL values in the surface water of the Unnao district and the rainwater of Lucknow. Data on the amount of chlordane in various environmental biota are lacking or very scanty (Table 3). As a result, monitoring the bioaccumulation and transit of chlordane in the food chain is still a challenge.

Table 3 Residual chlordane concentration in biota in India (Senthilkumar et al., 2001)

Location	Environmental compartment	∑Chlordane range
South India	Aquatic biota	ND-3.7
	Bats	<0.01–1.4
	Birds (resident)	ND-8.8
	Birds (local migrants)	0.6–1.5
	Birds (short distance migrants)	ND-12
	Birds (long distance migrants)	0.5–5.7

6 International Scientific Work on Chlordane

According to the United Nations Environment Programme (UNEP), the Stockholm Convention's goal is to preserve the environment and human health from persistent organic pollutants (POPs) (UNEP, 2019). Passive air monitoring was done as part of the UNEP/GEF (Global Environment Facility) projects "Implementation of the POPs Global Monitoring Plan (GMP) under the Stockholm Convention" (GMP2) in the Asia Region and "Continuing Regional Support for the POPs Global Monitoring Plan under the Stockholm Convention" in the Africa, Pacific, and Latin America and Caribbean Regions, and as part of capacity building and data generation for these projects, high quality data was reported for these projects. de Boer et al. (2023) reported that Chlordane was the highest POP in Nigeria, Palau, and the Philippines (7, 7, and 15 ng/PUF, respectively). These rather diverse observations suggest at least chlordane have been used more recently in one country than another. However, also within a country, their emissions might have been specific and limited to one or two locations only. In any case the relatively high level of chlordane in air in the Philippines requires further action, because since many years, application of this pesticide is not allowed. Chlordane was most often recorded in Nigeria (NGA), Niue, and Palau in both GMPs. Chlordane had greater levels in GMP2 than in GMP1. The overall regional rise in chlordane levels reveals that in both Africa and the Pacific regions, two countries—Ethiopia and Nigeria in Africa and Palau and the Solomon Islands in the Pacific Islands—appear to be responsible for these regional trends. Chlordane had a larger worldwide mean and median value in GMP2 (1% and 27%) but on a smaller scale (Fiedler et al., 2023). Thus, we may conclude that chlordane's overall worldwide trend is still rising, which is quite alarming. Tables 4, 5, and 6 summarize the reported values of chlordane in sediments, water, and biota samples collected from different geographical locations, respectively.

7 Adverse Effects of Chlordane

Chlordane has a variety of fatal effects on fish and birds, and it can also kill pink shrimp, bobwhite quail, and mallard ducks. It is also carcinogenic and affects the human immune system. In addition to triggering thyroid disorders, airborne

Table 4 Comparison of chlordane levels in sediments from other developing countries

Location	Environmental compartment	∑Chlordane	References
River Ravi and its tributaries (Pakistan)	Sediments (ng/g)	0.65–159.33	Baqar et al. (2018)
River Colorado (Mexico)	Sediments (ng/g)	2.24	Lugo-Ibarra et al. (2011)
River Bosna (Bosnia and Herzogovina)	Sediments (ng/g)	<0.01–0.09	Harman et al. (2013)
Mocu and Xnoha (Mexico)	Sediments (μg/g)	ND-0.2267	Hinojosa-Garro et al. (2016)
Manzala Lake (Egypt)	Sediments (ng/g)	ND-2.35	Barakat et al. (2012)
Qarun Lake (Egypt)	Sediments (ng/g)	0.25–21.1	Barakat et al. (2013)
Daye lake (China)	Sediment	0.0–1.78	Bhutto et al. (2021)
Daye lake (China)	Soil	0.0–0.12	Bhutto et al. (2021)
Shrimp farm Northern Coast of central Java	Sediment	ND-10	Hidayati et al. (2021)
Fish pond Ado-Ekiti, Nigeria	Sediment	0.004	Adeyeye et al. (2021)

Table 5 Comparison of chlordane levels in water from other developing countries

Location	Environmental compartment	∑Chlordane	References
River Ravi and its tributaries (Pakistan)	Surface water (ng/L)	0.60–1.59	Baqar et al. (2018)
Citarum River Estuary, West Java, Indonesia	River water (μg/L)	0.000–108.046	Oginawati et al. (2021)
Fish pond Ado-Ekiti, Nigeria	Surface water (μg/L)	0.001	Adeyeye et al. (2021)
Three Gorge Reservoir, China	Surface water (ng/L)	0.027–0.063	Qiao et al. (2010)

Table 6 Comparison of chlordane levels in biota from other developing countries

Location	Environmental compartment	∑Chlordane	References
Côte d'Ivoire	Human milk	5.7–3.6 ng/g	Pierre et al. (2018)
Northern Tanzania	Human milk	nd– 12.9 ng/g	Müller et al. (2019)
Mexico	Human milk	<1.9 ng/g	Martínez et al. (2022)
Aquafarm Taiwan	Codfish	4.76–5.75 ng/g	Chang (2018)
Qena city, Egypt	Nile tilapia fish African catfish Long fin catfish	22.97–25.35 ppb 8.91–58.94 ppb 81.81–82.73 ppb	Hassan et al. (2020)
Lake Taabo, Cote d'Ivoire	Nile tilapia fish	43.73 μg/L	Roche and Tidou (2009)
Densu, River, Ghana	Fish	0. 5 ± 0.10–10.2 ± 1.20 μg/L	Afful et al. (2010)

chlordane has also been linked to irregular respiratory movements, seizures, and liver lesions in rats and hepatomegaly in monkeys (ATSDR, 1994). The liver and central nervous system have both been shown to be impaired in other animal experiments. Humans are exposed to chlordane through the air, and exposure to these pollutants can result in cancer and a host of other ailments. These build up in human and animal lipids and are resistant to environmental deterioration.

Bronchitis, sinus congestion, and migraines appear to be correlated with chlordane exposure in the air of termite-treated homes in a dose-dependent manner. A dose-related rise in respiratory diseases was observed in a study of 261 adults exposed to different amounts of indoor air (Kharitonov & Barnes, 2001). Cerebrovascular illness occurred at twice the predicted rate in a study conducted on employees in industries related to the production of chlordane (Brown, 1992). Chlordane is extremely persistent in its environment since it is difficult to break down. Due to its hydrophobic characteristics, it adheres to soil particles and flows gently through groundwater. It takes an extended amount of time to breakdown, is highly toxic, and bio-accumulates in mammals and aquatic life. The SC on POPs banned chlordane in 2001 and classified them as part of the "dirty dozen" (Fuentes et al., 2016). Breast cancer, prostate cancer, lymphoma, obesity, testicular cancer, and diabetes are among the possible outcomes.

I. *Carcinogenic potential*: Park et al. (2020) observed that even decades after its manufacturing and usage were outlawed, lung cancer risk was still shown to be correlated with blood concentrations of chlordane. Chlordane is believed to promote tumor growth and cancer. In mice and rats, it results in both normal and cancerous liver tumors. Chlordane is listed by the US EPA as a potential carcinogen (Sruthi et al., 2017) while being categorized as a potential human carcinogen by the International Agency for Research on Cancer (IARC).
II. *Neurological effects*: In a study of residents of an apartment block treated with termite spray, 216 people were shown to have unusual equilibrium, choice reaction time, verbal recall, cognitive function, and lowering of motor speed (Kilburn & Warshaw, 1995). Using a logistic regression model, Kao et al. (2019) discovered a correlation between the socioemotional scores and trans-chlordane levels in breast milk. Correlation that is significantly negative. The effects of chlordane include slight tremor, rigidity in the back legs, convulsive symptoms, and a dose-dependent drop in rectal temperature. Quantification of neurotransmitters in the brain at various time points revealed a consistent drop in cortical and striatal AChE levels as well as somewhat elevated levels of AChE activity. The hypothermic response is linked to a clear and substantial decrease in NE in the brain stem, along with a rise in 5-HIAA, whereas 5-HT remained unaltered (Green et al., 2003). In humans, severe chlordane poisoning manifests as tremors and convulsions. Human serum chlordane concentrations that are deadly are greater than 5 ppm and convulsive concentrations are 3 ppm. Chlordane's toxic action manifests within 2 hours of an acute dose, progressing after 24 hours, and regenerating after 48 hours. Chlordane intoxication does not result in persistent neurotoxic side effects (Ahmed et al., 2021).

III. *Reproductive effects*: Chlordane may imitate sex steroids or alter a person's response to them. Chlordane leads to estrogenic, anti-estrogenic, and anti-androgenic effects (Glynn et al., 2000). A study conducted on groups of rat to understand the chronic effect of low-level exposure of chlordane revealed those hormonal disturbances at a dose of 100 and 500 ng/g (Weiss, 1997). Additionally, cis-chlordane is connected to menstrual features, hormone consumption, and infertility therapy (Chen et al., 2018). Just like estrogen, chlordane might reflect differential and complex biological functions (Cranmer et al., 1984; Cassidy et al., 1994).

8 Remedial Action Taken for Management of Chlordane

Chlordane in workroom air has a threshold limit value-time weighted average (TLV-TWA) of 0.5 mg/m^3, according to the American Conference of Governmental Industrial Hygienists (ACGIH, 1981). The 15-minute short-term exposure cap was set at 2 mg/m^3. The permitted working exposure level established by the Occupational Safety and Health Administration in 1981 is 0.5 mg/m^3. Both organizations stated that dermal exposure to chlordane should be avoided because it is absorbed via the skin. A total of 5 g/m^3 was proposed as an interim recommendation for airborne chlordane in military housing by the Committee on Toxicology (Committee on Toxicology, 1982). Under the Insecticides Act, 1968 (46 of 1968), the Indian government established an expert committee to look into the usage of chlordane and any potential dangers to humans' and animals' health. The registration committee issued an order banning the production, distribution, consumption, of chlordane in India as of September 20, 1996 as a result of its suggestions to the committee to discontinue the use of chlordane, which is likely detrimental to humans and animals. Only a few large industrial units in India have self-contained hazardous waste treatment and disposal facilities; commercial hazardous waste treatment and disposal facilities do not exist. As per prevention of Food adulteration rules, 1955, the prescribed tolerance limit of chlordane in food grains is 0.02 mg/kg, vegetables 0.2 mg/kg, fruits 0.1 kg, whereas in milk/milk products, it is 0.05 mg/kg on a fat basis. The WHO has established worldwide drinking water standards, and the limit value for chlordane is 0.0002 mg L^{-1} (WHO, 2018). Soil response level (Dutch Intervention Levels) is 4 mg/kg. Screening levels defined for surface water quality as per Canadian Water Quality Guidelines are 0.006 μg/L for the protection of aquatic life and 7 μg/L for Agriculture (CPCB, 2015).

9 Conclusion

The persistence of chlordane may contribute to the high prevalence of metabolic illnesses like hyperlipidemia and diabetes. Chlordane is quickly absorbed via the skin and has the potential to enter the body through the digestive system. It can then harm the liver, kidneys, and central nervous system, leading to a number of disorders. Thus, finding ways to lessen the danger posed by chlordane or other POPs is of utmost importance. The overall trend for chlordane is still rising on a global scale. In GMP2, chlordane had a higher global mean and median value (1% and 27%), but on a smaller scale. In the context of India, some researchers have observed significant levels of chlordane, raising concerns about possible adverse impacts on the environment. No evident impact has been noticed in India following the ban on chlordane, indicating a lack of progress in the struggle against its eradication. The level of chlordane contamination is being monitored, but no specific standards have been established. Additionally, there are not many studies being conducted presently on the photocatalytic and microbiological degradation of chlordane. India needs to establish the infrastructure for monitoring chlordane and implement appropriate regulations to completely outlaw its use. A critical step that will aid in the nation's understanding of chlordane is the establishment of appropriate assessment guidelines.

References

ACGIH, (American Conference of Governmental Industrial Hygienists (1981), 1981. In. National Research Council (US) Committee on Toxicology. *An Assessment of the Health Risks of Seven Pesticides Used for Termite Control*. Washington (DC): National Academies Press (US); 1982. CHLORDANE. Available from: https://www.ncbi.nlm.nih.gov/books/NBK217617/

Adeyeye, E. I., Ibigbami, O. A., Adesina, A. J., Azeez, M. A., Olaleye, A. A., Olatoye, R. A., & Gbolagade, Y. A. (2021). Assessment of pesticides residues in water, sediment and fish parts: Case study of fish pond in Ado-Ekiti, Nigeria. *Asian Journal of Microbiology Biotechnology Environmental Science, 23*(1), 42–50.

Afful, S., Anim, A. K., & Serfor-Armah, Y. (2010). Spectrum of organochlorine pesticide residues in fish samples from the Densu Basin. *Research Journal of Environmental and Earth Sciences, 2*(3), 133–138.

Ahmed, H., Sharif, A., Bakht, S., Javed, F., & Hassan, W. (2021). Persistent organic pollutants and neurological disorders: From exposure to preventive interventions. In *Environmental contaminants and neurological disorders* (pp. 231–247). Springer International Publishing.

ATSDR. (1989). *Toxicological profile for chlordane*. Atlanta, GA, us department of health and human services, public health service, agency for toxic substances and disease registry.

ATSDR. (1994). *Toxicological profile for chlordane*. Atlanta, GA, US Department of Health and Human Services, Public Health Service, Agency for Toxic Substances and Disease Registry.

Baqar, M., Sadef, Y., Ahmad, S. R., Mahmood, A., Li, J., & Zhang, G. (2018). Organochlorine pesticides across the tributaries of River Ravi, Pakistan: Human health risk assessment through dermal exposure, ecological risks, source fingerprints and spatio-temporal distribution. *Science of the Total Environment, 618*, 291–305.

Barakat, A. O., Khairy, M., & Aukaily, I. (2013). Persistent organochlorine pesticide and PCB residues in surface sediments of Lake Qarun, a protected area of Egypt. *Chemosphere, 90*(9), 2467–2476.

Barakat, A. O., Mostafa, A., Wade, T. L., Sweet. S. T., & El Sayed, N. B. (2012). Assessment of persistent organochlorine pollutants in sediments from Lake Manzala, Egypt. *Marine Pollution Bulletin, 64*(8), 1713–1720.

Bhutto, S. U. A., Xing, X., Shi, M., Mao, Y., Hu, T., Tian, Q., et al. (2021). Occurrence and distribution of OCPs and PAHs in water, soil and sediment of Daye Lake. *Journal of Geochemical Exploration, 226*, 106769.

Bidleman, T. F., Jantunen, L. M. M., Helm, P. A., Brorström-Lundén, E., & Juntto, S. (2002). Chlordane enantiomers and temporal trends of chlordane isomers in arctic air. *Environmental Science & Technology, 36*, 539–544.

Brown, D. P. (1992). Mortality of workers employed at organochlorine pesticide manufacturing plants—An update. *Scandinavian Journal of Work, Environment & Health, 18*, 155–161.

Cassidy, R. A., Vorhees, C. V., Minnema, D. J., & Hastings, L. (1994). The effects of chlordane exposure during pre-and postnatal periods at environmentally relevant levels on sex steroid-mediated behaviors and functions in the rat. *Toxicology and Applied Pharmacology, 126*(2), 326–337.

Chakraborty, P., Zhang, G., Li, J., Sivakumar, A., & Jones, K. C. (2015). Occurrence and sources of selected organochlorine pesticides in the soil of seven major Indian cities: Assessment of air–soil exchange. *Environmental Pollution, 204*, 74–80. https://doi.org/10.1016/j.envpol.2015.04.006

Chang, G. R. (2018). Persistent organochlorine pesticides in aquatic environments and fishes in Taiwan and their risk assessment. *Environmental Science and Pollution Research, 25*, 7699–7708.

Chen, M. W., Santos, H. M., Que, D. E., Gou, Y. Y., Tayo, L. L., Hsu, Y. C., et al. (2018). Association between organochlorine pesticide levels in breast milk and their effects on female reproduction in a Taiwanese population. *International Journal of Environmental Research and Public Health, 15*(5), 931.

Cranmer, J. M., Cranmer, M. F., & Goad, P. T. (1984). Prenatal chlordane exposure: Effects on plasma corticosterone concentrations over the lifespan of mice. *Environmental Research*, 35(1), 204–210.

CPCB. (2015). Guidance document for assessment and remediation of contaminated sites in India, *National Program for Rehabilitation of Polluted Sites in India. central pollution control board*, https://cpcb.nic.in/uploads/hwmd/MoEFCC guidelines contaminatedsites.pdf

Dearth, M. A., & Hites, R. A. (1991). Chlordane accumulation in people. *Environmental Science & Technology, 25*(7), 1279–1285.

de Boer, J., van Dijk, R., Abad, E., & Abalos, M. (2023). Persistent organic pollutants in air from Asia, Africa, Latin America and the Pacific Islands. *Chemosphere, 324*, 138271. https://doi.org/10.1016/j.chemosphere.2023.138271

Devi, N. L., Yadav, I. C., Raha, P., Shihua, Q., & Dan, Y. (2015). Spatial distribution, source apportionment and ecological risk assessment of residual organochlorine pesticides (OCPs) in the Himalayas. *Environmental Science and Pollution Research, 22*(24), 20154–20166.

FAO. (1985). *Pesticide residues in food — 1984 evaluations. Rome, food and agriculture organization of the United Nations, joint FAO/WHO meeting on pesticide residues* (FAO plant production and protection paper 67).

Fiedler, H., Abad, E., & de Boer, J. (2023). Preliminary trends over ten years of persistent organic pollutants in air – Comparison of two sets of data in the same countries. *Chemosphere, 324*, 138299.

Fuentes, M. S., Colin, V. L., Amoroso, M. J., & Benimeli, C. S. (2016). Selection of an actinobacteria mixed culture for chlordane remediation. Pesticide effects on microbial morphology and bioemulsifier production. *Journal of Basic Microbiology, 56*(2), 127–37. https://doi.org/10.1002/jobm.201500514. Epub 2015 Nov 10. PMID: 26554742.

Glynn, A. W., Glynn, A. W., Michaëlsson, K., Lind, P. M., Wolk, A., Aune, M., et al. (2000). Organochlorines and bone mineral density in Swedish men from the general population. *Osteoporosis International, 11*, 1036–1042.

Green, A. R., Mechan, A. O., Elliott, J. M., O'Shea, E., & Colado, M. I. (2003). The pharmacology and clinical pharmacology of 3, 4-methylenedioxymethamphetamine (MDMA, "ecstasy"). *Pharmacological Reviews, 55*(3), 463–508.

Harman, C., Grung, M., Djedjibegovic, J., Marjanovic, A., Sober, M., Sinanovic, K., et al. (2013). Screening for Stockholm Convention persistent organic pollutants in the Bosna River (Bosnia and Herzogovina). *Environmental Monitoring and Assessment, 185*, 1671–1683.
Hassan, M. A. E., Ahmed, A. M., Marzouk, N. M., & Maky, M. A. (2020). Detection of organochlorine pesticides residues in Nile fish and its risks in Qena city. *SVU-International Journal of Veterinary Sciences, 3*(1), 51–65.
Hidayati, N. V., Asia, L., Khabouchi, I., Torre, F., Widowati, I., Sabdono, A., et al. (2021). Ecological risk assessment of persistent organic pollutants (POPs) in surface sediments from aquaculture system. *Chemosphere, 263*, 128372.
Hinojosa-Garro, D., Burgos Chan, A. M., & Rendón-von Osten, J. (2016). Organochlorine pesticides (OCPs) in sediment and fish of two tropical water bodies under different land use. *Bulletin of Environmental Contamination and Toxicology, 97*, 105–111.
Kao, C. C., Que, D. E., Bongo, S. J., Tayo, L. L., Lin, Y. H., Lin, C. W., et al. (2019). Residue levels of organochlorine pesticides in breast milk and its associations with cord blood thyroid hormones and the offspring's neurodevelopment. *International Journal of Environmental Research and Public Health, 16*(8), 1438.
Kharitonov, S. A., & Barnes, P. J. (2001). Exhaled markers of pulmonary disease. *American Journal of Respiratory and Critical Care Medicine, 163*(7), 1693–1722.
Kilburn, K. H., & Warshaw, R. H. (1995). Hydrogen sulfide and reduced-sulfur gases adversely affect neurophysiological functions. *Toxicology and Industrial Health, 11*(2), 185–197.
Kumarasamy, P., Govindaraj, S., Vignesh, S., Rajendran, R. B., & James, R. A. (2012). Anthropogenic nexus on organochlorine pesticide pollution: A case study with Tamiraparani river basin, South India. *Environmental Monitoring and Assessment, 184*(6), 3861–3873. https://doi.org/10.1007/s10661-011-2229-x
Kumari, K. (Ed.). (2021). *Persistent organic pollutants: Gaps in management and associated challenges* (1st ed.). CRC Press. https://doi.org/10.1201/9781003046806
Lee, D. H., Lee, I. K., Song, K., Steffes, M., Toscano, W., Baker, B. A., & Jacobs, D. R., Jr. (2006). A strong dose-response relation between serum concentrations of persistent organic pollutants and diabetes: Results from the National Health and Examination Survey 1999–2002. *Diabetes Care, 29*(7), 1638–1644.
Lee, D. H., Lee, I. K., Porta, M., et al. (2007). Relationship between serum concentrations of persistent organic pollutants and the prevalence of metabolic syndrome among non-diabetic adults: Results from the National Health and Nutrition Examination Survey 1999–2002. *Diabetologia, 50*, 1841–1851. https://doi.org/10.1007/s00125-007-0755-4
Lugo-Ibarra, K. C., Daesslé, L. W., Macías-Zamora, J. V., & Ramírez-Álvarez, N. (2011). Persistent organic pollutants associated to water fluxes and sedimentary processes in the Colorado River delta, Baja California, México. *Chemosphere, 85*(2), 210–217.
Malik, A., Ojha, P., & Singh, K. P. (2009). Levels and distribution of persistent organochlorine pesticide residues in water and sediments of Gomti River (India)—A tributary of the Ganges River. *Environmental Monitoring and Assessment, 148*, 421–435. https://doi.org/10.1007/s10661-008-0172-2
Martínez, C., Arroyo, A. M., Alemán, D. B., García, A. G., Caba, M., Garcidueñas, A. L. C., et al. (2022). Persistent organic compounds in human milk and evaluation of the effectiveness of the Stockholm convention in Mexico. *Environmental Advances, 8*, 100190.
Mishra, K., Sharma, R. C., & Kumar, S. (2013). Contamination profile of DDT and HCH in surface sediments and their spatial distribution from North-East India. *Ecotoxicology and Environmental Safety, 95*, 113–122. https://doi.org/10.1016/j.ecoenv.2013.05.029
Müller, M. H. B., Polder, A., Brynildsrud, O. B., Grønnestad, R., Karimi, M., Lie, E., et al. (2019). Prenatal exposure to persistent organic pollutants in Northern Tanzania and their distribution between breast milk, maternal blood, placenta and cord blood. *Environmental Research, 170*, 433–442.
National Academy of Sciences (US), & Committee on Toxicology. (1982). *An assessment of the health risks of seven pesticides used for termite control* (p. 0080). National Academy of Sciences.

Oginawati, K., Susetyo, S. H., Rahmawati, S. I., Kurniawan, S. B., & Abdullah, S. R. S. (2021). Distribution of organochlorine pesticide pollution in water, sediment, mollusk, and fish at Saguling Dam, West Java, Indonesia. *Toxicology Research, 38*(2), 149–157. https://doi.org/10.1007/s43188-021-00094-1. PMID: 35419272; PMCID: PMC8960551.

Park, E. Y., Park, E., Kim, J., Oh, J. K., Kim, B., Hong, Y. C., & Lim, M. K. (2020). Impact of environmental exposure to persistent organic pollutants on lung cancer risk. *Environment International, 143*, 105925.

Parween, M., Ramanathan, A., Khillare, P. S., & Raju, N. J. (2014). Persistence, variance and toxic levels of organochlorine pesticides in fluvial sediments and the role of black carbon in their retention. *Environmental Science and Pollution Research, 21*(10), 6525–6546. https://doi.org/10.1007/s11356-014-2531-6

Pierre, M., Jean-Baptiste, A. A., & DjÃ, D. (2018). Organochlorine pesticides contamination in human milk in Abidan (Cte dIvoire). *Journal of Toxicology and Environmental Health Sciences, 10*(2), 10–14.

Qiao, M., An, T., Zeng, X., Zhang, D., Li, G., Sheng, G., et al. (2010). Safety assessment of the source water within the Pearl River Delta on the aspect of organochlorine pesticides contamination. *Journal of Environmental Monitoring, 12*(9), 1666–1677.

Roche, H., & Tidou, A. (2009). First ecotoxicological assessment assay in a hydroelectric reservoir: The Lake Taabo (Côte d'Ivoire). *Bulletin of Environmental Contamination and Toxicology, 82*, 322–326.

Sah, R., Baroth, A., & Hussain, S. A. (2020). First account of spatio-temporal analysis, historical trends, source apportionment and ecological risk assessment of banned organochlorine pesticides along the Ganga River. *Environmental Pollution, 263, Part B*, 114229. https://doi.org/10.1016/j.envpol.2020.114229. ISSN 0269-7491.

Sarkar, S. K., Binelli, A., Riva, C., Parolini, M., Chatterjee, M., Bhattacharya, A. K., Bhattacharya, B. D., & Satpathy, K. K. (2008). Organochlorine pesticide residues in sediment cores of Sunderban Wetland, Northeastern Part of Bay of Bengal, India, and their ecotoxicological significance. *Archives of Environmental Contamination and Toxicology, 55*(3), 358–371. https://doi.org/10.1007/s00244-008-9133-6

Selvaraj, S., Gaonkar, O., Kumar, B., Cincinelli, A., & Chakraborty, P. (2021). Legacy persistent organochlorine pollutants and polycyclic aromatic hydrocarbons in the surface soil from the industrial corridor of South India: Occurrence, sources and risk assessment. *Environmental Geochemistry and Health, 43*. https://doi.org/10.1007/s10653-020-00786-x

Senthilkumar, K., Kannan, K., Subramanian, A., & Tanabe, S. (2001). Accumulation of organochlorine pesticides and polychlorinated biphenyls in sediments, aquatic organisms, birds, bird eggs and bat collected from South India. *Environmental Science and Pollution Research, 8*(1), 35–47. https://doi.org/10.1007/BF02987293

Singh, T., & Singh, D. (2017). Phytoremediation of organochlorine pesticides: Concept, method, and recent developments. *International Journal of Phytoremediation, 19*, 834–843.

Singh, K. P., Malik, A., Mohan, D., & Sinha, S. (2005). Persistent organochlorine pesticide residues in alluvial groundwater aquifers of Gangetic Plains, India. *Bulletin of Environmental Contamination and Toxicology, 74*(1), 162–169. https://doi.org/10.1007/s00128-004-0563-1

Singh, K. P., Malik, A., & Sinha, S. (2007). Persistent organochlorine pesticide residues in soil and surface water of northern Indo-Gangetic alluvial plains. *Environmental Monitoring and Assessment, 125*(1–3), 147–155. https://doi.org/10.1007/s10661-006-9247-0

Singh, K., Nong, A., Feeley, M., & Chan, H. M. (2019). Development of biomonitoring equivalents for chlordane and toxaphene with application to the general Canadian population. *Regulatory Toxicology and Pharmacology, 106*, 262–269. https://doi.org/10.1016/j.yrtph.2019.05.015. ISSN 0273-2300.

Son, H. K., Kim, S. A., Kang, J. H., Chang, Y. S., Park, S. K., Lee, S. K., et al. (2010). Strong associations between low-dose organochlorine pesticides and type 2 diabetes in Korea. *Environment International, 36*(5), 410–414.

Sruthi, S. N., Shyleshchandran, M. S., Mathew, S. P., & Ramasamy, E. V. (2017). Contamination from organochlorine pesticides (OCPs) in agricultural soils of Kuttanad agroecosystem in India and related potential health risk. *Environmental Science and Pollution Research, 24*. https://doi.org/10.1007/s11356-016-7834-3

UNEP. (1997). *World atlas of desertification*. United Nations Environment Programme.

UNEP. (2019). *Guidance on the global monitoring plan for persistent organic pollutants*. UNEP/POPS/COP.9/INF/36. http://chm.pops.int/TheConvention/ConferenceoftheParties/Meetings/COP9/tabid/7521/Default.aspx

US EPA. (2000). *Guidance for assessing chemical contaminant data for use in fish advisories*. Volume 2. Risk Assessment and Fish Consumption Limits. 3rd Ed. EPA 823-B-00-008.

Weiss, B. (1997). Pesticides as a source of developmental disabilities. *Mental Retardation and Developmental Disabilities Research Reviews, 3*(3), 246–256.

WHO. (2018). *Chlordane in drinking-water background document for development of WHO guidelines for drinking-water quality*. WHO/SDE/WSH/03.04/84.

World Health Organization. (2021). *Pesticide residues in food–2019*. Evaluation: Part II, Toxicological. Food & Agriculture Org.

Worthing, CR (Ed.). (1991). *The pesticide manual: A world compendium,* 9th Edition. British Crop Protection Council: Farnham, England, UK. Illus. Isbn 0-948404-42-6.; 0 (0). 1991. Xlvii+ 1141p. (1991): 6.

Wu, Q., Du, X., Feng, X., Cheng, H., Chen, Y., Lu, C., Wu, M., & Tong, H. (2021). Chlordane exposure causes developmental delay and metabolic disorders in Drosophila melanogaster. *Ecotoxicology and Environmental Safety, 225*, 112739. https://doi.org/10.1016/j.ecoenv.2021.112739. ISSN 0147-6513.

Dichlorodiphenyltrichloroethane (DDT)

Kanchan Kumari and Senerita Swamy

1 Introduction

Dichlorodiphenyltrichloroethane, also known as DDT or 1,1,1-trichloro-2,2-bis(p-chlorophenyl) ethane, is a manmade pesticide belonging to the family of organic halogen compounds. It is effective against a variety of insects and acts as a contact poison that reportedly disrupts the nervous system to cause its effects. DDT is prepared in the presence of sulfuric acid by reacting chloral with chlorobenzene. Austrian chemist Othmar Zeidler for the first time created DDT in 1874, and Paul Hermann Müller, a Swiss chemist, first identified its insecticidal properties in 1939 (Rizqi et al., 2023). It was employed to combat typhus and malaria during World War II, and the US Food and Drug Administration (FDA) approved its use as a public insecticide in 1945 (Rizqi et al., 2023). DDT was one of the first and best-known organopesticides. When its insecticidal properties were discovered after 1939, it was placed into extensive industrial production. Around 400,000 tonnes of DDT were used annually in the early 1960s, with 70–80% going to agriculture (Timokhina et al., 2021). In the U.S., DDT use for commercial and agricultural purposes spread after 1945. DDT, being a chlorinated hydrocarbon, gained early acceptance because of its affordable price, potency, durability, and adaptability. DDT was domestically used for 30 years before it was banned, totalling roughly 1,350,000,000 pounds. The use of DDT in the USA drastically decreased after 1959, going from a peak of roughly 80 million pounds in that year to less than 12 million pounds in the early

K. Kumari (✉)
CSIR-National Environmental Engineering Research Institute (NEERI),
Kolkata Zonal Center, Kolkata, West Bengal, India
e-mail: k_kumari@neeri.res.in

S. Swamy
CSIR-National Environmental Engineering Research Institute (NEERI), Nagpur, India

K. Kumari, *Pollutants of Global Concern*, Emerging Contaminants and Associated Treatment Technologies, https://doi.org/10.1007/978-3-031-50996-4_3

1970s (USEPA, 1975). Over 80% of the pesticide used in 1970–1972 was applied to cotton crops; the remaining 20% was primarily used on peanut and soybean crops. Sweden was the first nation to outlaw the use of DDT effective in January 1970 due to environmental concerns (Nygård et al., 2019). Its use has had a significant negative impact on animals' health and caused widespread contamination of water and soil resources. DDT is still used for indoor residual spraying (IRS) for disease vector control, primarily malaria, in a number of nations, including South Africa, despite being outlawed in many others (Tangena et al., 2020). The technical details of DDT has been mentioned in Table 1.

2 Technical Details of DDT

Table 1 Physico-chemical properties of DDT

Details	Description (DDT)
Structure	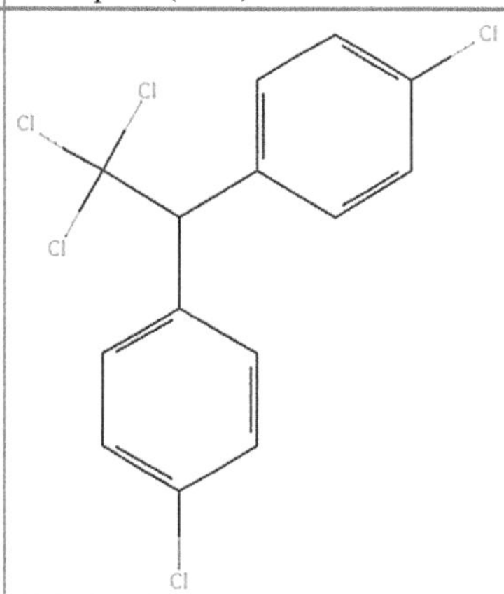
IUPAC name	1-Chloro-4-[2,2,2-trichloro-1-(4-chlorophenyl) ethyl] benzene
Molecular formula	$C_{14}H_9Cl_{15}$
CAS No.	50-29-3
Molecular weight	354.5 g/mol
Color/form	Colorless or white crystal
Odor	Slight aromatic odor
Boiling point	186 °C at 0.05 mm Hg
Melting point	108.5 °C
Flash point	162–171 °F (closed cup)
Solubility	Dissolve in acetone, benzene, ether, carbon tetrachloride, dioxane, kerosene, and pyridine
Density	1.6 g/cm^3
Vapor pressure	1.60e-07 mmHg
Octanol/water partition coefficient	log Kow = 6.91

(continued)

Table 1 (continued)

Details	Description (DDT)
Octanol-air partition coefficient	9.82
Henry's law constant	8.32×10^{-6} atm-m^3/mole
Stability/shelf life	Stable to oxidation as well damaging to iron, dehydrochlorinated above the melting point to transform into non-insecticidal DDE, a chemical reaction that catalyzed by iron(III) or aluminum chlorides in presence of ultraviolet light and alkaline solutions
Decomposition	Decompose at 110 °C; dehydrochlorinates in alkali or organic bases when in organic solvents
Corrosivity	Corrosive to iron metal
Safety and hazards	Acute toxic, hazardous to the aquatic environment, carcinogenicity
Toxicity	Toxic

3 Status of Works and Usage in India

In order to limit the use of DDT and other persistent organic pollutants used globally, the EPA has been taking part in discussions on a global scale since 1996. The Stockholm Convention allows for a restricted exception for the use of DDT (Table 2) in the control of mosquitoes that spread the bacterium that causes malaria, a disease that still claims millions of lives annually throughout the globe (Kumari et al., 2021). In African nations where malaria is still a serious health concern, the World Health Organization (WHO) said in September 2006 that it supported the indoor use of DDT, stating that the pesticide's advantages exceed its hazards to human health and the environment. The WHO's stance is in line with the Stockholm Convention on POPs, which forbids the use of DDT for all purposes other than preventing malaria. The global ban of DDT by the end of year 2020 was first proposed in the sixth meeting of COP (Conference of the Parties) held for the Stockholm Convention on Persistent Organic Pollutants (POPs) in Geneva, India, being the greatest producer of DDT, however opposed the initiative (Lallas, 2001). DDT is a pesticide that was solely produced in India at that time. In 2007, China stopped manufacturing the chemical. The sole DDT manufacturer in India is Hindustan Insecticide Limited (HIL), a public sector company. In the FY 2020–2021, the Company was working to supply DDT 75% WP to Zimbabwe (128 MT) and Zambia (113 MT) (Ministry of Chemicals and Fertilizers by PIB Delhi 21 July 2020).

India is currently the world's largest user of DDT (Van den Berg, 2009), with roughly 2786 tonnes produced for disease control (UNEP, 2016). DDT has a distribution pattern that is comparable to that of other OCPs. The percentage chemical composition of DDT with its metabolites has been represented in the Fig. 1.

DDT along with DDE and DDD can be transformed via biodegradation and photodegradation in environmental matrices such as water. This is because shorter wavelength radiation does not penetrate deep into water; photolysis happens predominantly on the surface of water and depends on water transparency. DDT and

Table 2 Uses, sources, and regulation as per Stockholm Convention

Identified POP	Uses and sources	Annex	Year of listing by SC	Regulation in India
DDT	Previously used as agricultural pest control, now majorly used to eliminated disease causing insect	B	2001	Used to eradicate vectors spreading malaria

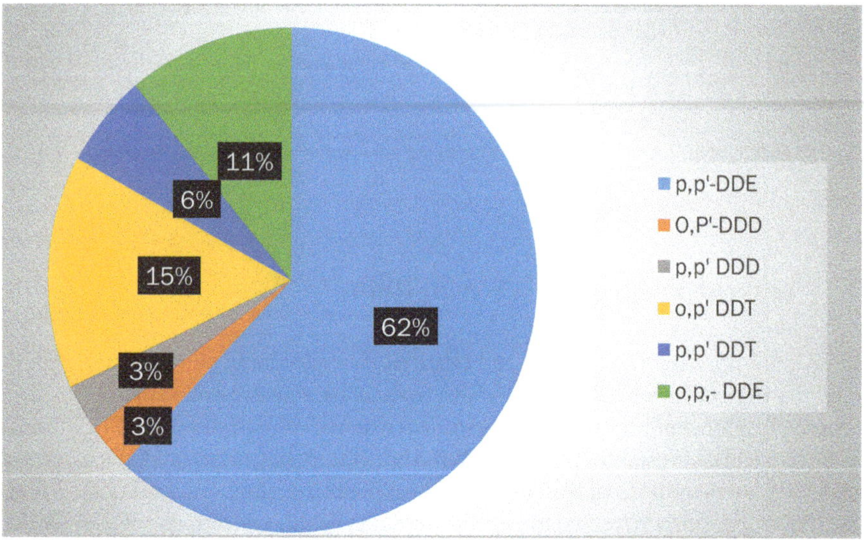

Fig. 1 Percentage chemical composition of DDT with its metabolite

DDD photolysis is very slow in water bodies, with expected half-lives of more than 150 years (EPA, 1979). DDE's photolysis will change based on photosynthetic rate and illumination, which leads to various half-lives. DDE's half-life over the United States is around one day in the summer and six days in the winter. When DDE is exposed to sunshine, photoisomerization also occurs. When compared to DDE, photolysis of DDE photoisomers is slower by at least an order of magnitude. Studies using DDT at lower wavelengths indicate that the Cl 2C-Cl link dissociates as a result of the first reaction. The indirect photolysis of DDT is the subject of some information, but data is not available with respect to indirect photolysis of DDD or DDE (Coulston, 1985; EPA, 1979; Zepp et al., 1977). The photo-induced 1,2 addition of DDT to methyl oleate, a model lipid, suggests that DDT additions to unsaturated fatty acids in plant waxes and cutins may be widespread (Schwack, 1988). DDT hydrolyzes by a mechanism that is catalyzed by a base, giving it an 81-day half-life at pH 9. The end product of hydrolysis is DDE. DDE and DDD hydrolysis are not important destiny step (EPA, 1979). According to reports, one minor method of transformation for DDT is biodegradation in water (ORIMBO, 2016). DDE and DDD biodegrade more slowly in the aquatic environment than DDT does (EPA, 1979).

About half of the DDT present in atmosphere is bound to particulate matter, while the other half is present as vapor (Bidleman, 1988). DDT interacts with photochemically generated hydroxyl radicals in the vapor phase (Meylan & Howard, 1993). Higher vapor pressures than DDT are shared by DDE and DDD, and less of these substances will bind to particle matter than DDT. DDE and DDD's vapor-phase half-lives are thought to be 17 h and 30 h, respectively. The atmosphere may also undergo direct photolysis. The majority of DDT residue losses from soils have been attributed to four mechanisms: volatilization, harvesting (e.g., plants after absorbing the residue), runoff of water, and chemical change (Fishbein, 1973). Three mechanisms are mainly transportation operations, while the fourth, chemical transformation, can be brought on by both biotic and abiotic mechanisms. On the surfaces of soil or after sediment adsorption, photooxidation of DDT and DDE is known to occur (Lichtenstein & Schulz, 1959; Baker & Applegate, 1970; Miller & Zepp, 1979). In a 90-day trial, exposure to sunshine accelerated the conversion of DDT to DDE in soil, with the sample exposed to light retaining 65% of its original DDT content compared to the unexposed dark control's 91% (Racke et al., 1997). In soils, DDE and DDD are both created as a result of the biodegradation of DDT. Both metabolites may continue to change, but how much and how quickly depends on the soil's characteristics and maybe the microbial communities that exist there. By Aislabie et al. (1997) and Zook and Feng (1999), the breakdown mechanisms of DDT in aerobic and anaerobic circumstances have been reviewed. The process of co-metabolism in which the microorganisms obtain nutrients for energy and growth from the sources other than the substance of concern is involved in the biodegradation of DDT and its metabolites. Under aerobic and anaerobic circumstances, DDE, the main DDT metabolite detected, frequently resists biodegradation (Strompl & Thiele, 1997). DDT has been proven to degrade to DDE and DDD in laboratory trials using marine sediments. In China and Sudan, the half-life of DDT varied from 22 to 365 days. One exception observed in Brazil, where the half-life was >672 days in a highly acidic environment (pH 4.5). The study found that DDT disappeared far more quickly under tropical circumstances than it did under temperate conditions. Under tropical circumstances, volatilization, biochemical degradation, and occasionally adsorption were the main modes of dissipation. There have been reports of comparable half-lives in temperate climates ranging from 837 to 6087 days (Stewart & Chisholm, 1971; Racke et al., 1997). According to one researcher, DDT had a mean lifespan of roughly 5.3 years in temperate U.S. soils (Racke et al., 1997).

4 In-House (Countrywide) Scientific Work Done on the Chemical

In the Indian population, DDT is one of the POPs that is most commonly observed. It is apparent that the emphasis on organochlorine pesticides is connected to the extensive historical manufacturing and consumption done in India during the pre-SC epoch. Figure 2, Table 3, and Fig. 3 summarize the total concentration of DDT in water, soil, and biota samples collected from various locations of India respectively.

In most investigations including the Indian population from various geographic locations and socioeconomic backgrounds, high amounts of DDT were found in human milk and blood. Among the different DDT metabolites, p,p′ DDE was the

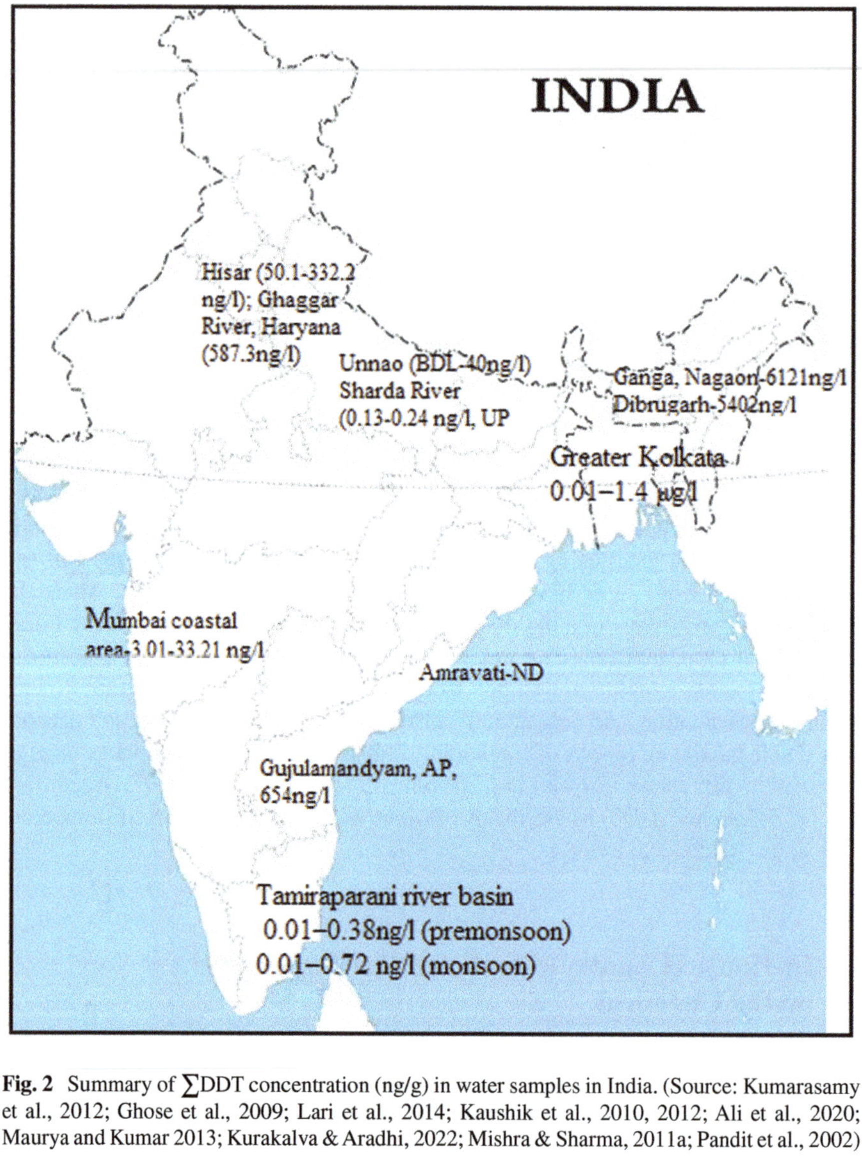

Fig. 2 Summary of ∑DDT concentration (ng/g) in water samples in India. (Source: Kumarasamy et al., 2012; Ghose et al., 2009; Lari et al., 2014; Kaushik et al., 2010, 2012; Ali et al., 2020; Maurya and Kumar 2013; Kurakalva & Aradhi, 2022; Mishra & Sharma, 2011a; Pandit et al., 2002)

Table 3 Summary of ∑DDT concentration (ng/g) in soil samples in India

Site	Sample type	Observed range	Reference
Tapti river, Gujrat	Sediment	0.52–0.72 ng/g	Hashmi et al. (2020)
River Yamuna Delhi	Sediment	18.3–49 ng/g	Pandey et al. (2011)
Palakkad district	Soil	BDL – 55 ng/g	Gopalan & Chenicherry (2018)
Thamirabarani river system of southern peninsular India	Soil	ND – 1194 ng/g	Arisekar et al. (2019)
Cardamom Hill Reserve, Kerala	Soil	ND – 43.62 ng/g	Joseph et al. (2020)
Yamuna flood plains	Sediment	0.41–18 μg/kg	Kumar et al. (2021)
Korba	Soil	2.1–315 μg/kg	Kumar et al. (2014)
Gwalior, Madhya Pradesh	Soil	1.30–27.41 μg/kg	Kumar (2018)
South West Coast	Agricultural soil	ND – 148 ng/g	Khuman et al. (2020)
	Coastal soil	ND – 141 ng/g	
	Backwaters soil	ND – 9 ng/g	
Hooghly River Estuary	Sediment	0.33–18.58 ng/g	Mitra et al. (2019)
Vellar River, Tamil Nadu	Soil	1314–1976 pg/m^3	Chakraborty et al. (2019)

most abundant due to its fat solubility and hence progressive bioaccumulation in living beings, as evidenced by previous investigations conducted in Punjab, where p,p′ DDE, p,p′ DDT, and p,p′ DDD residues were detected in human breast milk samples (Bhanti & Taneja, 2007). DDT contributed the highest in Guwahati (21%) and Ludhiana (27%) respectively, with a similar trend in Udaipur, where the percentage fraction of DDT is 27%. When compared to affluent countries, India has a higher level of DDT contamination in numerous environmental matrices due to the country's ongoing usage of the chemical to combat vector-borne diseases, which is an unavoidable challenge here (Mansouri et al., 2017).

5 International Data Versus the Indian Scenario

DDT levels were extensively investigated in environmental matrix (Table 4) and found to be much greater than those of other OCPs. The Brazilian beaches have the greatest DDT concentrations (mean 7.22 ng g $^{-1}$ dw). High DDT concentrations were also found in sediments from the Peruvian coast, with a mean overall value of 5.23 ng g^{-1} dw. The observed concentrations, nevertheless, matched sediments

Fig. 3 Residual ∑DDT concentration in biota in India

Table 4 Residual ∑DDT concentration in various matrices worldwide

Site	Sample type	Observed range	Reference
Luvuvhu River System	Surface water	BDL – 0.007	Barnhoorn et al. (2009)
	Domestic chicken	11.5–36.5 mg/kg	
	Terrestrial bird	0.01–0.54 mg/kg	
Cananeia Estuary	Dolphin	93.95 µg/g	Yogui (2003)
Callao Coast, Peru	*Semimytilus algosus*	5.14 ng/g	Torres Jimenez et al. (2006)
Okanagan Valley Fruit Orchards, Canada	Earthworm	118.7 ± 18.144 µg/g	Kesic et al. (2021)
	American robin	9.65 ± 2.94 µg/g	

measured in 1998, a few years after DDT was outlawed. The average DDT concentration in Argentinian sediments was just 0.7 ng g^{-1} dw (Petrus et al., 2014; Rizzi et al., 2017; Sánche et al., 1998; Tombes et al., 2018). The spatio-temporal pattern of DDT dispersion in seawater is substantially distinct from the pattern of application in the adjacent continental regions. Although the maximum burden of DDT was not reached until 1990, it is still accumulating in the Equatorial Ocean, sections of the Arctic Ocean, and the Gulf Stream (including the North Atlantic Drift) area. In the Arctic Ocean, volatilization is decreased due to low mean sea surface temperatures, and DDT sources (deposition and in flow) predominate. Globally, the amount of DDT in the ocean below 100 m is gradually escalating. Data available on global DDT production revealed a 32% decrease throughout the reporting period. Similarly, global consumption of DDT for malaria and leishmaniosis management fell by 30% between 2001 and 2014. By far the greatest producer and consumer of DDT has been India. DDT is employed in some countries in reaction to the development of susceptibility in malaria vectors to pyrethroid and carbamate insecticides. In accordance with the SC, some other nations have discontinued the use of DDT. The majority of nations apparently have measures in place that prohibit or restrict the manufacturing, import, export, and use of DDT, indicating that progress has been achieved in adopting or revising national legislative restrictions on DDT.

6 Adverse Effects of DDT

In the past, millions of illnesses and fatalities caused by diseases transmitted by insects may have been avoided with the use of DDT. Humans are subjected to DDT and DDE through indoor residual spraying can lead to significant exposure thus, DDT may provide a concern to human health. However, very few researches have examined the consequences of DDT/DDE exposure at levels seen in populations exposed by indoor residual spraying, and studies have seldom ever quantified body burdens of both DDE and DDT. Additionally, there is little data on DDT/DDE exposure during times when consequences like cancer are most likely to occur. The SC on POPs, which stresses that DDT should be utilize cautiously, when it is necessary, or when there is no adverse effect, safe, inexpensive alternatives and easily available, is supported by the most recent research on DDT exposed in to human populations and its potential adverse effects. According to the SC, each nation that presently uses DDT is required to submit a management and implementation strategy to restrict the use of DDT to the control of disease vectors and to lessen dependency on the usage of DDT. Ultimately, countries should be helped so that they can count on other suitable malaria control approaches, tactics, and strategies.

I. *Carcinogenic Potential* – Over a hundred cohort and case-control studies from different nations have looked into the correlations between cancer and DDT exposure. In China, case-control studies that were nested and population-based indicated that, after adjusting for possible confounders, there were significant dose-relate relationships between blood DDT levels and liver cancer. In a historical cohort analysis of eight males who received DDT during an Italian campaign to combat malaria, there was no increased incidence of liver cancer. Numerous cohort and case-control studies on non-Hodgkin lymphoma in North America and Europe have indicated favorable connections, whereas other investigations have found no correlation. There was no conclusive relation between breast cancer and DDT or DDE assessed in adult blood or adipose tissue samples, despite the evaluation of more than 40 research published since 1993; nonetheless, the potential significance of early-life DDT exposure is still up for debate. There is very weak evidence for DDT's carcinogenicity in humans from studies on non-Hodgkin lymphoma, liver, and testicular malignancies. The carcinogenic nature of DDT and its degradation products DDE and 1-chloro-4- [2,2-dichloro-1-(4-chlorophenyl) ethyl] benzene in experimental animals was sufficiently demonstrated in several experiments in mice, rats, and hamsters (mostly oral administration) (DDD).

II. *Neurobehavioral Effects* – High doses of DDT are acutely absorbed, causing hypersensitivity to external stimuli, abnormal susceptibility to fear, and motor unrest as well as more frequent spontaneous movements (light, touch, sound). This is followed by the emergence of fine tremors, which then turn coarse and eventually lead to tonic-clonic convulsions. After exposure, symptoms typically start to appear a few hours later, and 24–72 h later, death, usually from respiratory failure, may occur (Joy, 1982). Similar signs and symptoms of poisoning can be seen in most animal species, with dogs showing the most pronounced convulsions. Voltage-gated sodium channels (Nav) remain open as a result of the insecticidal effects of DDT, which causes the nervous system to be persistently depolarized and hyperactive. DDT is only mildly hazardous (rat oral LD50 = 113 mg/kg) (Gaines, 1969), while oral treatment of 35 mg DDT per day to humans for up to 18 months had no overt toxic effects or neurological effects (Cueto Jr. et al., 1956; Morgan & Roan, 1971). However, there have been worries expressed regarding a number of possible negative health consequences due to the possibility of chronic DDT exposure (Eskenazi et al., 2009; Rogan & Chen, 2005). Sadly, the possible neurotoxicity of DDT has not been well investigated in many human research. Despite the fact that neither research had access to measurements of DDT or DDE, two studies (Misra et al., 1984) indicated that employees who sprayed DDT exhibited cognitive impairment. According to a short investigation, DDT was more commonly detected in the brains of Alzheimer's disease (AD; $n = 7$) patients than in those of controls ($n = 14$) (Fleming et al., 1994). Most recently, a research that used information from the NHANES population found an association between blood DDE levels and reduced cognitive function in older US citizens, indicating that non-occupational exposures to DDT can also result in cognitive impairments (Kim et al., 2015). According to a recent analysis by Richardson and colleagues (Richardson et al., 2014), AD patients have blood levels of DDE that are around four times higher than normal, and this is linked to a greater probability of receiving an AD diagnosis (odds ratio [OR] = 4.18). Furthermore, when compared to those without an APOE 4 allele, those with high DDE levels and an APOE 4 allele had higher cognitive impairment (Gaffney et al., 2005).

III. *Reproductive Effects* – A research by García-Salazar et al. (2004) offered some proof that male employees' reproductive health is impacted by workplace exposure to DDT. The relevance of reported effects at lower levels, however, remains unclear because the estimated measures of doses are much higher than the exposure of the general inhabitants where DDT has been sprayed to treat malaria. The dose-response connections were inconsistent. Rogan and Chen (2005) suggested that DDT exposure could make PTB lose the advantage of reducing malaria infant mortality at the doses required for malaria control. In 100% infertile women, Bastos et al. (2013) found higher concentrations of a DDT metabolite (p, p′-DDE). There has also been a connection between pesticide exposure and the reduced fecundability ratio (Hanke & Jurewicz, 2004). The risk of congenital malformations such as birthmarks in the form of hemangiomas, orofacial clefts, nervous system impairment, and musculoskeletal defects may increase if one or both parents work in agriculture (Kumar et al., 2019). Only weak correlations between DDT/DDE levels and menstrual cycle characteristics were found in two recent studies (Cooper et al., 2005; Chen et al., 2005). However, one of the studies from the INUENDO population found that Polish women had a three times higher probability of having prolong menstrual cycles than women from other European nations (Toft et al., 2008). DDT and DDE concentrations were linked to a notably early age of menopause, according to results of a different study that used information from the Hispanic Health and Nutrition Examination Survey (HHANES) (Akkina et al., 2004). Among 289 Californian women born between 1960 and 1963, the only research to look at in exposure to DDT in the uterus and time to conception was carried out (CHDS). Researchers discovered that each 10 g/L rise in mothers pregnancy-related blood levels of p,p′-DDT and p,p′-DDE was linked to, respectively, a 32% decrease and a 16% elevation in their daughters' chances of becoming pregnant each cycle (Cohn et al., 2003). When maternal serum samples' p,p′-DDT to p,p′-DDE ratio increased, it was shown that daughters' times to conception were longer. According to the research, the underlying risk factor was likely recent or direct pesticide exposure as opposed to long-term exposure to p,p′-DDE in the food web (Eskenazi et al., 2009).

IV. *Other Effects* – DDT exposure at low environmental doses may have unknown effects on human health. Human symptoms after exposure to high doses may include nausea, tremors or shakiness, and seizures. The liver and reproduction were affected, according to laboratory animal studies. DDT and DDE body burdens may be related to the prevalence of diabetes, according to some Swedish and American studies. Despite the fact that thyroid hormones are crucial for healthy brain growth (Dunn, 1993), research suggests that DDE and likely DDT may lower levels of triiodothyronine (T3) and thyroxine (T4) in the blood of pregnant women, newborns, and young children (Abdelouahab et al., 2008; Alvarez-Pedrerol et al., 2008). Additionally, congenital hypothyroid cases had higher DDT levels than controls (Nagayama et al., 2007). Several markers, including interleukin-4 and interleukin-13, plasma levels of type 1 (interferon-) reaction to mitogen in breast feeding mothers, white blood cell counts (WBC), and various lymphocyte phenotypes, as well as immunoglobulin (Ig) A, G, and E levels, have shown that DDT and especially DDE have the ability to modulate the immunity response in humans (Daniel et al., 2002; Brooks et al., 2007; Cooper et al., 2004; Karmaus et al., 2005). Aplastic anemia in DDT exposure (Ahamed et al., 2006), asthma in DDE exposure (Sunyer et al., 2005), otitis media (with DDE) (Dallaire et al., 2004), and farmer's lung in DDE exposure are just a few examples of conditions associated with immune system-related conditions. Table 5 highlights the effects, toxicity, and persistence related notable consequences.

7 Management and Remedial Measures

Since 1996, the EPA has taken part in discussions on a global scale to limit the utilization of DDT and other POPs. Countries collaborated and negotiated SC to implement international bans or limitations on POPs, a category that included DDT, under the aegis of the United Nations Environment Programme. The application of DDT to control vector borne disease malaria, a global threat that still claims millions of lives, is allowed under the Convention with some restrictions. It is expected that projects being carried out or proposed by the WHO, UNEP, and UNDP will increase the body of evidence supporting DDT's superiority over alternatives (WHO, 2007). The findings will be important in making decisions about future vector management strategies for malaria prevention. The World Health Organization (WHO) recommends DDT as one of the effective IRS chemicals to reduce the threat of malaria mosquitoes, and Southern African nations like South Africa, Zimbabwe, Zambia, Namibia, and Mozambique as well as India frequently use it.

Table 5 Effects, toxicity, and persistence of DDT

DDT	
Effects	DDT exposure at low ambient concentrations may have unclear consequences on human health. Human symptoms after exposure to large levels may include nausea, tremors or shakiness, and seizures. The liver and reproductive system were affected, according to laboratory animal research. DDT is thought to be a potential human carcinogen (Beard 2006)
Toxicity	DDT's hepatotoxicity and carcinogenicity have been confirmed in animals, including liver weight gain, hepatic hypertrophy, induction of microsomal enzyme comparable to phenobarbital, and hepatocellular tumor induction. DDT's reproductive toxicity and endocrine effects may differ between animal species/strains based on susceptibility and dosage levels (ATSDR, 2002)
Persistence/fate	DDT has a long environmental persistence. DDT has a soil half-life ranging from 2 to 15 years. DDT has a half-life of roughly 150 years in aquatic environments (Harris & McLellan, 1994). DDT is progressively dehydrochlorinated to DDE and then reduced to DDD by liver microsomal enzymes (Burr 2014)

8 Conclusion

The Stockholm Convention on POPs, which restricted the use of DDT for all purposes other than preventing malaria, is consistent with the WHO's stance. The WHO has recommended 12 pesticides for indoor residual spray programs, including DDT. Whether or not to use DDT is a decision left to each nation. The EPA interacts with other agencies and nations on how DDT programs are established and maintained with the intention that DDT be utilize only within the standards set by the programs known as IVM (Integrated Vector Management). Major problems in periodic reporting by SC Parties, as well as flaws in reporting to the DDT Register, limit the SC's ability to achieve its DDT objectives. Since the SC's implementation, global DDT production and consumption have decreased slightly. India, being the largest producer and consumer of DDT, is now unable to entirely outlaw DDT because of the urgent need to eradicate vector-borne illness. However, India has taken the major move to eliminate dependency on the insecticide DDT by promoting the Long-Lasting Insecticidal Net (LLIN) and Neem-based and bacteria-based biopesticides. For the time being, India should concentrate on developing more environmentally friendly alternatives. This is advocated for the future, and more remedial efforts to remove DDT residue from heavily polluted sites should be made. This would eventually prevent DDT from bioaccumulating in different environmental compartments, assisting India in accomplishing sustainable goals.

References

Abdelouahab, N., Mergler, D., Takser, L., Vanier, C., St-Jean, M., Baldwin, M., et al. (2008). Gender differences in the effects of organochlorines, mercury, and lead on thyroid hormone levels in lakeside communities of Quebec (Canada). *Environmental Research, 107*, 107.

Ahamed, M., Anand, M., Kumar, A., & Siddiqui, M. K. (2006). Childhood aplastic anaemia in Lucknow, India: Incidence, organo-chlorines in the blood and review of case reports following exposure to pesticides. *Clinical Biochemistry, 39*(7), 762–766.

Aislabie, J. M., Richards, N. K., & Boul, H. L. (1997). Microbial degradation of DDT and its residues – A review. *New Zealand Journal of Agricultural Research, 40*, 269–282.

Akkina, J., Reif, J., Keefe, T., & Bachand, A. (2004). Age at natural menopause and exposure to organochlorine pesticides in Hispanic women. *Journal of Toxicology and Environmental Health. Part A, 67*(18), 1407–1422.

Ali, S. S., Al-Tohamy, R., Xie, R., El-Sheekh, M. M., & Sun, J. (2020). Construction of a new lipase- and xylanase-producing oleaginous yeast consortium capable of reactive azo dye degradation and detoxification. *Bioresource Technology, 313*, 123631. https://doi.org/10.1016/j.biortech.2020.123631

Alvarez-Pedrerol, M., Ribas-Fito, N., Torrent, M., Carrizo, D., Grimalt, J. O., & Sunyer, J. (2008). Effects of PCBs, *p,p′*-DDT, *p,p-′* DDE, HCB and beta-HCH on thyroid function in pre-school children. *Occupational and Environmental Medicine, 65*(7), 452–457.

ATSDR. (2002). *Toxicological profile for DDT, DDE, and DDD*. Agency for Toxic Substances and Disease Registry.

Arisekar, U., Shakila, R. J., Jeyasekaran, G., Shalini, R., Kumar, P., Malani, A. H., & Rani, V. (2019). Accumulation of organochlorine and pyrethroid pesticide residues in fish, water, and sediments in the Thamirabarani river system of southern peninsular India. *Environmental Nanotechnology, Monitoring & Management, 11*, 100194.

Baker, R. D., & Applegate, H. G. (1970). Effect of temperature and ultraviolet radiation on the persistence of methyl parathion and DDT in soils. *Agronomy Journal, 62*, 509–512.

Barnhoorn, I. E., Bornman, M. S., Jansen van Rensburg, C., & Bouwman, H. (2009). DDT residues in water, sediment, domestic and indigenous biota from a currently DDT-sprayed area. *Chemosphere, 77*(9), 1236–1241. https://doi.org/10.1016/j.chemosphere.2009.08.045

Bastos, A. M. X., Souza Mdo, C. B. D., Almeida Filho, G. L. D., Krauss, T. M., Pavesi, T., Silva, L. E. D., et al. (2013). Organochlorine compound levels in fertile and infertile women from Rio de Janeiro, Brazil. *Arquivos Brasileiros de Endocrinologia e Metabologia, 57*, 346–353.

Beard, J. (2006). DDT and human health. *Science of the Total Environment, 355*, 78–89.

Bidleman, T. F. (1988). Atmospheric processes. Wet and dry deposition of organic compounds are controlled by their vapor-particle partitioning. *Environmental Science & Technology, 22*, 361–367.

Bhanti M, Taneja A (2007) Contamination of vegetables of different seasons with organophosphorous pesticides and related health risk assessment in Northern India. Chemosphere 69:63–68.

Brooks, K., Hasan, H., Samineni, S., Gangur, V., & Karmaus, W. (2007). Placental *p,p′*-dichlorod iphenyldichloroethylene and cord blood immune markers. *Pediatric Allergy and Immunology, 18*(7), 621–624.

Burr, S. A. (2014). *DDT (Dichlorodiphenyltrichloroethane) encyclopedia of toxicology* (3rd ed., pp. 1141–1143). https://doi.org/10.1016/B978-0-12-386454-3.00122-6

Chakraborty, P., Zhang, G., Li, J., Sampathkumar, P., Balasubramanian, T., Kathiresan, K., ... & Jones, K. C. (2019). Seasonal variation of atmospheric organochlorine pesticides and polybrominated diphenyl ethers in Parangipettai, Tamil Nadu, India: Implication for atmospheric transport. *Science of the Total Environment, 649*, 1653–1660.

Chen, A., Zhang, J., Zhou, L., Gao, E. S., Chen, L., Rogan, W. J., et al. (2005). DDT serum concentration and menstruation among young Chinese women. *Environmental Research, 99*(3), 397–402.

Cohn, B. A., Cirillo, P. M., Wolff, M. S., Schwingl, P. J., Cohen, R. D., Sholtz, R. I., et al. (2003). DDT and DDE exposure in mothers and time to pregnancy in daughters. *Lancet, 361*(9376), 2205–2206. https://www.ncbi.nlm.nih.gov/pubmed/12842376.

Cooper, G. S., Martin, S. A., Longnecker, M. P., Sandler, D. P., & Germolec, D. R. (2004). Associations between plasma DDE levels and immunologic measures in African-American farmers in North Carolina. *Environmental Health Perspectives, 112*, 1080–1084.

Cooper, G. S., Klebanoff, M. A., Promislow, J., Brock, J. W., & Longnecker, M. P. (2005). Polychlorinated biphenyls and menstrual cycle characteristics. *Epidemiology, 16*(2), 191–200.

Coulston, F. (1985). Reconsideration of the dilemma of DDT for the establishment of an acceptable daily intake. *Regulatory Toxicology and Pharmacology, 5*(4), 332–383. https://doi.org/10.1016/0273-2300(85)90002-9

Cueto, C., Jr., Durham, W. F., & Hayes, W. J., Jr. (1956). The effect of known repeated oral doses of chlorophenothane (DDT) in man. *Journal of the American Medical Association, 162*(9), 890–897.

Dallaire, F., Dewailly, É., Muckle, G., Vezina, C., Jacobson, S. W., Jacobson, J. L., et al. (2004). Acute infections and environmental exposure to organochlorines in Inuit infants from Nunavik. *Environmental Health Perspectives, 112*, 1359–1365.

Daniel, V., Huber, W., Bauer, K., Suesal, C., Conradt, C., & Opelz, G. (2002). Associations of dichlorodiphenyltrichloroethane (DDT) 4.4 and dichlorodiphenyldichloroethylene (DDE) 4.4 blood levels with plasma IL-4. *Archives of Environmental Health, 57*(6), 541–547.

Dunn, J. T. (1993). Iodine supplementation and the prevention of cretinism. *Annals of the New York Academy of Sciences, 678*, 158–168.

EPA. (1979). *Water-related environmental fate of 129 priority pollutants: Volume II: Halogenated aliphatic hydrocarbons, halogenated ethers, monocyclic aromatics, phthalate esters, polycyclic aromatic hydrocarbons, nitrosamines, and miscellaneous compounds.*

U.S. Environmental Protection Agency. EPA440479026b. https://nepis.epa.gov/Exe/ZyPURL.cgi?Dockey=2000K6JL.txt. 10 Aug 2021

Eskenazi, B., Chevrier, J., Rosas, L. G., Anderson, H. A., Bornman, M. S., Bouwman, H., Chen, A., Cohn, B. A., de Jager, C., Henshel, D. S., Leipzig, F., Leipzig, J. S., Lorenz, E. C., Snedeker, S. M., & Stapleton, D. (2009). The Pine River statement: human health consequences of DDT use. *Environmental Health Perspectives, 117*(9), 1359–1367. https://doi.org/10.1289/ehp.11748

Fishbein, L. (1973). Mutagens and potential mutagens in the biosphere. I. DDT and its metabolites, polychlorinated biphenyls, chlorodioxins, polycyclic aromatic hydrocarbons, haloethers. *Science of the Total Environment, 4*, 305–340.

Fleming, L., Mann, J. B., Bean, J., Briggle, T., & Sanchez-Ramos, J. R. (1994). Parkinson's disease and brain levels of organochlorine pesticides. *Annals of Neurology, 36*(1), 100–103.

Gaffney, S. H., Curriero, F. C., Strickland, P. T., Glass, G. E., Helzlsouer, K. J., & Breysse, P. N. (2005). Influence of geographic location in modeling blood pesticide levels in a community surrounding a U.S. Environmental Protection Agency superfund site. *Environmental Health Perspectives, 113*, 1712–1716.

Gaines, T. (1969). Acute toxicity of pesticides. *Toxicology and Applied Pharmacology, 14*, 515–534.

García-Salazar, F., Gallardo-Díaz, E., Cerón-Mireles, P., Loomis, D., & Borja-Aburto, V. H. (2004). Reproductive effects of occupational DDT exposure among male malaria control workers. *Environmental Health Perspectives, 112*(5), 542–547. https://doi.org/10.1289/ehp.112-1241918

Ghose, N. C., Saha, D., & Gupta, A. S. (2009). Synthetic detergents (surfactants) and organochlorine pesticide signatures in surface water and groundwater of greater Kolkata, India. *Journal of Water Resource and Protection, 1*, 290–298.

Gopalan, N. K., & Chenicherry, S. (2018). Fate and distribution of organochlorine insecticides (OCIs) in Palakkad soil, India. *Sustainable Environment Research, 28*, 179–185. https://doi.org/10.1016/j.serj.2018.01.007

Hanke, W., & Jurewicz, J. (2004). The risk of adverse reproductive and developmental disorders due to occupational pesticide exposure: An overview of current epidemiological evidence. *International Journal of Occupational Medicine and Environmental Health, 17*, 223–243.

Harris, O., & McLellan, W. (1994). *Toxicological profile for 4, 4'-DDT, 4, 4'-DDE, 4, 4'-DDD* (update). Atlanta, GA: Agency for Toxic Substances and Disease Registry.

Hashmi, T. A., Qureshi, R., Tipre, D., & Menon, S. (2020). Investigation of pesticide residues in water, sediments, and fish samples from Tapi River, India as a case study and its forensic significance. *Environmental Forensics, 21*(1), 1–10.

Joseph, L., Paulose, S. V., Cyril, N., Santhosh, S. K., Varghese, A., Nelson, A. B., ... & Kasu, S. (2020). Organochlorine pesticides in the soils of Cardamom Hill Reserve (CHR), Kerala, India: Geo spatial distribution, ecological and human health risk assessment. *Environmental Chemistry and Ecotoxicology, 2*, 1–11.

Joy, R. M. (1982). Chlorinated hydrocarbon insecticides. In D. J. Ecobichon & R. M. Joy (Eds.), *Pesticides and neurological diseases* (pp. 91–150). CRC Press.

Karmaus, W., Brooks, K. R., Nebe, T., Witten, J., Obi-Osius, N., & Kruse, H. (2005). Immune function biomarkers in children exposed to lead and organochlorine compounds: A cross-sectional study. *Environmental Health, 4*, 1–5. https://doi.org/10.1186/1476-069X-4-5

Kaushik, A., Sharma, H. R., Jain, S., Dawra, J., & Kaushik, C. P. (2010). Pesticide pollution of river Ghaggar in Haryana, India. *Environmental Monitoring and Assessment, 160*, 61–69.

Kaushik, C. P., Sharma, H. R., & Kaushik, A. (2012). Organochlorine pesticide residues in drinking water in the rural areas of Haryana, India. *Environmental Monitoring and Assessment, 184*, 103–112. https://doi.org/10.1007/s10661-011-1950-9

Kesic, R., Elliott, J. E., Fremlin, K. M., Gauthier, L., Drouillard, K. G., & Bishop, C. A. (2021). Continuing persistence and biomagnification of DDT and metabolites in northern temperate fruit orchard avian food chains. *Environmental Toxicology and Chemistry, 40*(12), 3379–3391. https://doi.org/10.1002/etc.5220

Khuman, S. N., Vinod, P. G., Bharat, G., Kumar, Y. M., & Chakraborty, P. (2020). Spatial distribution and compositional profiles of organochlorine pesticides in the surface soil from the agricultural, coastal and backwater transects along the south-west coast of India. *Chemosphere, 254*, 126699.

Kim, S. A., Lee, Y. M., Lee, H. W., Jacobs Jr, D. R., & Lee, D. H. (2015). Greater cognitive decline with aging among elders with high serum concentrations of organochlorine pesticides. *PloS One, 10*(6), e0130623.

Kumar, S. (2018). Organochlorines in urban soils from Central India: probabilistic health hazard and risk implications to the human population. *Environmental geochemistry and health, 40*, 2465–2480.

Kumar, S., Sharma, A., & Kshetrimayum, C. (2019). Environmental & occupational exposure & female reproductive dysfunction. *The Indian Journal of Medical Research, 150*(6), 532–545. https://doi.org/10.4103/ijmr.IJMR_1652_17

Kumar, B., Verma, V. K., Mishra, M., Gaur, R., Kumar, S., & Sharma, C. S. (2014). DDT and HCH (organochlorine pesticides) in residential soils and health assessment for human populations in Korba, India. *Human and Ecological Risk Assessment, 20*, 1538–1549.

Kumar, B., Verma, V. K., Mishra, M., Piyush, Kakkar, V., Tiwari, A., ... & Gargava, P. (2021). Assessment of persistent organic pollutants in soil and sediments from an urbanized flood plain area. *Environmental Geochemistry and Health, 43*, 3375–3392.

Kumarasamy, P., Govindaraj, S., Vignesh, S., Rajendran, R. B., & James, R. A. (2012). Anthropogenic nexus on organochlorine pesticide pollution: A case study with Tamiraparani River Basin, South India. *Environmental Monitoring and Assessment, 184*(6), 3861–3873. https://doi.org/10.1007/s10661-011-2229-x

Kumari, K., Swamy, S., & Singh, A. (2021). *Global monitoring plan on persistent organic pollutants (POPs). Persistent organic pollutants: Gaps in management and associated challenges* (1st ed., pp. 227–252). CRC Press.

Kurakalva, R. M., & Aradhi, K. K. (2022). Occurrence and distribution of HCHs and DDTs in surface water and groundwater from the Gajulamandyam region along the Swarnamukhi River Basin, Andhra Pradesh, India. *International Journal of Environmental Analytical Chemistry, 102*(18), 6955–6969. https://doi.org/10.1080/03067319.2020.1818735

Lallas. (2001). The Stockholm convention on persistent organic pollutants. *American Journal of International Law, 95*(3), 692. https://doi.org/10.2307/2668517

Lari, S. Z., Khan, N. A., Gandhi, K. N., Meshram, T. S., & Thacker, N. P. (2014). Comparison of pesticide residues in surface water and ground water of agriculture intensive areas. *Journal of Environmental Health Science and Engineering, 12*(1), 11. https://doi.org/10.1186/2052-336X-12-11

Lichtenstein, E., & Schulz, K. (1959). Persistence of some chlorinated hydrocarbon insecticides as influenced by soil types, rate of application and temperature. *Journal of Economic Entomology, 52*(1), 124–131.

Mansouri, A., Cregut, M., Abbes, C., Durand, M. J., Landoulsi, A., & Thouand, G. (2017). The environmental issues of DDT pollution and bioremediation: a multidisciplinary review. *Applied biochemistry and biotechnology, 181*, 309–339.

Maurya, A. K., & Kumar, A. (2013). Organochlorine pesticides in the surface waters from Sharda River Region, Uttar Pradesh, India. *Advances in Space Research & Earth Exploration, 1*(1), 1–7.

Meylan, W. M., & Howard, P. H. (1993). Computer estimation of the atmospheric gas-phase reaction rate of organic compounds with hydroxyl radicals and ozone. *Chemosphere, 26*(2), 2293–2299.

Miller, G. C., & Zepp, R. G. (1979). Photoreactivity of aquatic pollutants sorbed on suspended sediments. *Environmental Science & Technology, 13*(7), 860–863. https://doi.org/10.1021/es60155a019

Mishra, K., & Sharma, R. C. (2011b). Assessment of organochlorine pesticides in human milk and risk exposure to infants from North-East India. *Science of the Total Environment, 409*(23), 4939–4949. https://doi.org/10.1016/j.scitotenv.2011.07.038

Misra, U. K., Nag, D., & Murti, C. R. (1984). A study of cognitive functions in DDT sprayers. *Industrial Health, 22*(3), 199–206.
Mitra, S., Corsolini, S., Pozo, K., Audy, O., Sarkar, S. K., & Biswas, J. K. (2019). Characterization, source identification and risk associated with polyaromatic and chlorinated organic contaminants (PAHs, PCBs, PCBzs and OCPs) in the surface sediments of Hooghly estuary, India. *Chemosphere, 221*, 154–165.
Morgan, D., & Roan, C. (1971). Absorption, storage and metabolic conversation of ingested DDT and DDT metabolites in man. *Archives of Environmental Health, 22*, 301–308. (as cited in ATSDR, 2002 and WHO, 1979).
Nagayama, J., Kohno, H., Kunisue, T., Kataoka, K., Shimomura, H., Tanabe, S., et al. (2007). Concentrations of organochlorine pollutants in mothers who gave birth to neonates with congenital hypothyroidism. *Chemosphere, 68*(5), 972–976.
Nygård, T., Sandercock, B. K., Reinsborg, T., & Einvik, K. (2019). Population recovery of peregrine falcons in Central Norway in the 4 decades since the DDT-ban. *Ecotoxicology, 28*, 1160–1168. https://doi.org/10.1007/s10646-019-02111-4
ORIMBO, C. (2016). *Dissipation rate of chlorpyrifos within mosquito larvae habitats inn Maseno and its activity against the larvae relative to dichlorodiphenyltrichloroethane* (Doctoral dissertation, Maseno University).
Pandey, P., Khillare, P. S., & Krishan K. (2011). Assessment of organochlorine pesticide residues in the surface sediments of River Yamuna in Delhi, India. *Journal of Environmental Protection, 2*(5), 511.
Pandit, G. G., Sahu, S. K., & Sadasivan, S. (2002). Distribution of HCH and DDT in the coastal marine environment of Mumbai, India. *Journal of Environmental Monitoring, 4*(3), 431–434.
Petrus, G., Bernhard, H., Renan, L., Renato, D. P., Machado T. J. P., Olaf, M., et al. (2014). Partition of organochlorine concentrations among suspended solids, sediments and brown mussel Perna perna, in tropical bays. *Chemosphere, 114*, 9–5.
Racke, K., Skidmore, M., Hamilton, D., et al. (1997). Pesticide fate in tropical soils. *Pure and Applied Chemistry, 69*(6), 1349–1371.
Richardson, J. R., Roy, A., Shalat, S. L., von Stein, R. T., Hossain, M. M., Buckley, B., Gearing, M., Levey, A. I., & German, D. C. (2014). Elevated serum pesticide levels and risk for Alzheimer disease. *JAMA Neurology, 71*, 284–290. https://doi.org/10.1001/jamaneurol.2013.603
Rizqi, H. D., Purnomo, A. S., & Ulfi, A. (2023). The effect of bacteria addition on DDT biodegradation by BROWN-ROT fungus Gloeophyllum trabeum. *Heliyon, 9*(7), e18216. https://doi.org/10.1016/j.heliyon.2023
Rizzi, J., Pérez-Albaladejo, E., Fernandes, D., Contreras, J., Froehner, S., & Porte, C. (2017). Characterization of quality of sediments from Paranaguá Bay (Brazil) by combined in vitro bioassays and chemical analyses. *Environmental Toxicology and Chemistry, 36*(7), 1811–1819.
Rogan, W. J., & Chen, A. (2005). Health risks and benefits of bis(4-chlorophenyl)-1,1,1-trichloroethane (DDT). *Lancet, 366*(9487), 763–773.
Sánchez, G. R., Moreyra, R. I., & Jacinto Tayco, M. E. (1998). Estado de la contaminación marina en el litoral peruano en 1994 y 1995. *Informe Instituto del Mar del Perú. 136*, 7–22 (in Spanish).
Schwack, W. (1988). Photoinduced addition of pesticides to biomolecules.ill Model the reaction of the captain. *Chemosphere 21*, 905–912
Stewart, D. K. R., & Chisholm, D. (1971). Long-term persistence of BHC, DDT and chlordane in a sandy loam soil. *Canadian Journal of Soil Science, 51*, 379–383.
Strompl, C., & Thiele, J. H. (1997). Comparative fate of 1,1-diphenylethylene (DPE), 1,1-dichloro-2,2-bis(4-Chlorophenyl)-ethylene (DDE), and pentachlorophenol (PCP) under alternating aerobic and anaerobic conditions. *Archives of Environmental Contamination and Toxicology, 33*(4), 350–356. https://doi.org/10.1007/s002449900264
Sunyer, J., Torrent, M., Munoz-Ortiz, L., Ribas-Fito, N., Carrizo, D., Grimalt, J., et al. (2005). Prenatal dichlorodiphenyldichloroethylene (DDE) and asthma in children. *Environmental Health Perspectives, 113*, 1787–1790.

Tangena, J. A., Hendriks, C. M. J., Devine, M., Tammaro, M., Trett, A. E., Williams, I., DePina, A. J., Sisay, A., Herizo, R., Kafy, H. T., Chizema, E., Were, A., Rozier, J., Coleman, M., & Moyes, C. L. (2020). Indoor residual spraying for malaria control in sub-Saharan Africa 1997 to 2017: An adjusted retrospective analysis. *Malaria Journal, 19*(1), 150. https://doi.org/10.1186/s12936-020-03216-6

Timokhina, E. P., Yaglov, V. V., & Nazimova, S. V. (2021). Dichlorodiphenyltrichloroethane and the adrenal gland: From toxicity to endocrine disruption. *Toxics, 9*(10), 243. https://doi.org/10.3390/toxics9100243

Toft, G., Axmon, A., Lindh, C. H., Giwercman, A., & Bonde, J. P. (2008). Menstrual cycle characteristics in European and Inuit women exposed to persistent organochlorine pollutants. *Human Reproduction, 23*(1), 193–200.

Tombesi, N., Pozo, K., Arias, A., Alvarez, M., Pribylova, P., Audy, O., et al. (2018). Records of organochlorine pesticides in soils and sediments on the southwest of Buenos Aires Province, Argentina. *Environmental Earth Sciences, 77*, 403.

Torres Jimenez, M., Campoy Folgoso, C., Cañabate Reche, F., Rivas Velasco, A., Cerrillo Garcia, I., Mariscal Arcas, M., & Olea-Serrano, F. (2006). Organochlorine pesticides in serum and adipose tissue of pregnant women in Southern Spain giving birth by cesarean section. *Science of the Total Environment, 372*(1), 32–38. https://doi.org/10.1016/j.scitotenv.2006.07.009

UNEP. (2016). *Report of the sixth expert group meeting on DDT*. UNEP/POPS/DDT-EG.6, Stockholm Convention on Persistent Organic Pollutants. November 8, 2016. Archived from the original on March 5, 2018. Retrieved March 4, 2018.

United States Environmental Protection Agency (USEPA). (1975). *DDT regulatory history: A brief survey (to 1975)*. USEPA. http://www.epa.gov/history/topics/ddt/02.htm. 20 July 2005.

Van den Berg, H. (2009). Global status of DDT and its alternatives for use in vector control to prevent disease. *Environmental Health Perspectives, 117*, 1656–1663. 3.4 degradation and environmental fate.

WHO (World Health Organization). (2007). *The use of DDT in Malaria vector control. WHO position statement*. WHO.

Yogui, G. T., de Oliveira Santos, M. C., & Montone, R. C. (2003). Chlorinated pesticides and polychlorinated biphenyls in marine tucuxi dolphins (Sotalia fluviatilis) from the Cananéia estuary, Southeastern Brazil. *Science of the Total Environment, 312*(1–3), 67–78. https://doi.org/10.1016/S0048-9697(03)00198-0

Zepp, R. G., Wolfe, N. L., Azarraga, L. V., et al. (1977). Photochemical transformation of the DDT and methoxychlor degradation products, DDE and DMDE, by sunlight. *Archives of Environmental Contamination and Toxicology, 6*(2–3), 305–314. https://doi.org/10.1007/BF02097771

Zook, C., & Feng, J. (1999). *1,1,1-trichloro-2,2-bis-(4′-chlorophenyl)ethane (DDT) pathway map*. University of Minnesota. http://www.labmed.umn.edu/umbdd/ddt/ddt_map.html. 23 June 1999

Dieldrin

Kanchan Kumari and Shaheen Akhtar

1 Introduction

Dieldrin is a common cyclodiene organochlorine pesticide that is composed of 1,2,3,4,10,10-hexachloro-6,7-epoxy-1,4,4a,5,6,7,8,8aoctahydro-exo-1,4-endo-5,8-dimethanonaphthalene. It's also a byproduct of the related pesticide aldrin. Dieldrin has a water solubility of 0.1 to 0.25 mg/L at 208 °C and a log K_{ow} of 5.48. Dieldrin has a density of 1.7 g/cm^3 at 208 °C and a half-life of approximately 1.5 years in soil (Connell et al., 2002). Because of its particularly long half-life, it is listed under persistent organic pollutants (POPs) and was designated by USEPA in 1987 as B2 carcinogen. Therefore, this insecticide has long been prohibited in a number of nations. Breakdown of pesticides is functionally dependent on certain parameters. Aldrin if present in soil is reported to oxidize at slow pace to dieldrin by around 75% per year in temperate conditions, with more rapid oxidation happening in tropical climates; hence, dieldrin is detected more frequently in environmental matrix than aldrin. According to the studies, the concentration of dieldrin pollution varies with soil depth, with the top soil having the highest concentration of the chemical residues (Mandocdoc & Primo David, 2008).

Dieldrin has been used to manage insects living in agricultural soils, in addition to controlling termites and textile pests. Because the insecticide aldrin quickly changes to dieldrin, dieldrin concentrations in the environment are greater than dieldrin usage alone would imply (Kumari et al.,, 2022). Dieldrin is hazardous to fish and other aquatic species, notably frog embryos, which can acquire spinal abnormalities following low-level exposure. Dieldrin remnants have been discovered in

K. Kumari (✉) · S. Akhtar
CSIR-National Environmental Engineering Research Institute (NEERI),
Kolkata Zonal Center, Kolkata, West Bengal, India
e-mail: k_kumari@neeri.res.in

K. Kumari, *Pollutants of Global Concern*, Emerging Contaminants and Associated Treatment Technologies, https://doi.org/10.1007/978-3-031-50996-4_4

the environment, the air, water, soil, fish, birds, and animals, including people. Food is the most common source of exposure for the general population. In a US examination of pasteurized milk, for example, dieldrin was detected as second most prevalent pesticide identified (Stockholm Convention, 2001).

2 Technical Details of Dieldrin

Physicochemical properties of dieldrin are shown in Table 1.

Table 1 Physico-chemical properties of dieldrin

Details	Description (dieldrin)
Structure	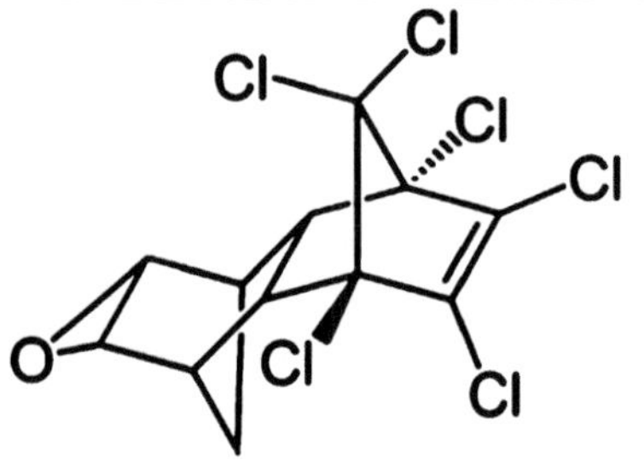
IUPAC name	(1*R*,2*S*,3*S*,6*R*,7*R*,8*S*,9*S*,11*R*)-3,4,5,6,13,13-Hexachloro-10-oxapentacyclo[6.3.1.$1^{3,6}$.$0^{2,7}$.$0^{9,11}$]tridec-4-ene
Molecular formula	$C_{12}H_8Cl_6O$
CAS No.	60-57-1
Molecular weight	380.9 g/mol
Color/form	Colorless crystals
Odor	Slight, aromatic odor
Boiling point	330.0 °C
Melting point	175.5 °C
Solubility	In water, 0.195 mg/l at 25 °C
Density	1.7 g/cm^3
Vapor pressure	4.20e-06 mm Hg
Octanol/water partition coefficient	5.40
Octanol-air partition coefficient	8.515
Henry's law constant	8.18e-06 atm-m^3/mole
Stability/shelf life	Resistant to moisture and alkalis
Decomposition	When heated to decomposition, emits toxic fumes of hydrogen chloride
Corrosivity	Corrosive to metals
Safety and hazards	Acute toxic, health hazard and environmental hazard

3 Status of Works and Usage in India

Dieldrin, a common soil pest control chemical listed in the Annex A in the Stockholm Convention (2001), has been banned from import, manufacture, and use in India since 17 July 2001. No reports on dieldrin manufacturing in India have been published after its ban on the production and use in 2001 by the Government of India. However, records suggest that India has imported and exported very little dieldrin in the last decade.

The states of Maharashtra, Rajasthan, and Gujarat reported a total of 33,111 kg of technical grade dieldrin and 20,744 liters of dieldrin 18% EC. Figure 1 displays the dieldrin distribution in three prominent states of India. However, there is no current information on stockpile's disposal of dieldrin. In contrast, around 0.04 thousand kgs of dieldrin was imported into India from Germany and the United States in 2014–2015, and there was no import of dieldrin after that year (NIP, 2011).

POPs pollution of the environment is a major source of worry in India. Even after the prohibition was implemented, air, water, and soil testing continue to reveal significant contamination. Dieldrin has been found in various soil sediment samples, indicating the likelihood of agricultural pilferage, runoff from the soil, and eventually detection in water and soil sediments (Fig. 2).

Dieldrin was also a pesticide that was regularly found in foods. Chourasiya et al. (2015) had studied the organochlorine pesticide exposure health risk evaluation by dietary consumption of vegetables cultivated in Delhi's peri-urban areas. They had reported the occurrence of residual dieldrin in brinjal (3.5 ng/g), okra (4.5 ng/g),

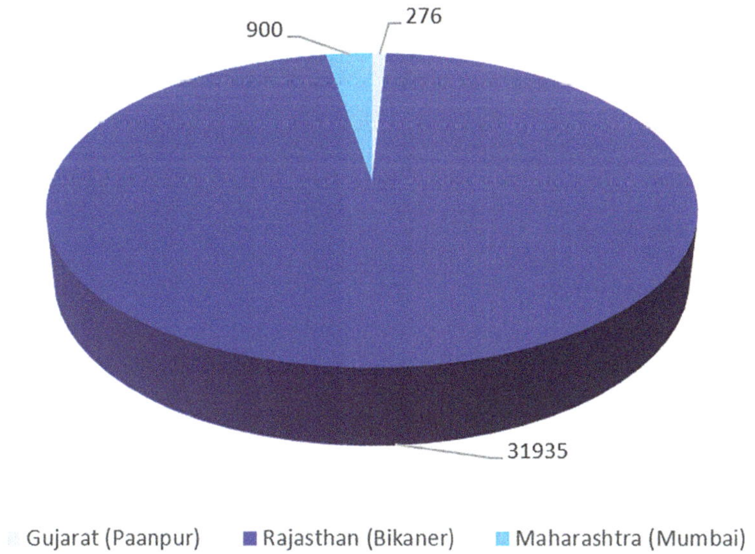

Fig. 1 Stockpile of dieldrin (technical) (kg) in three different states of India (NIP 2011)

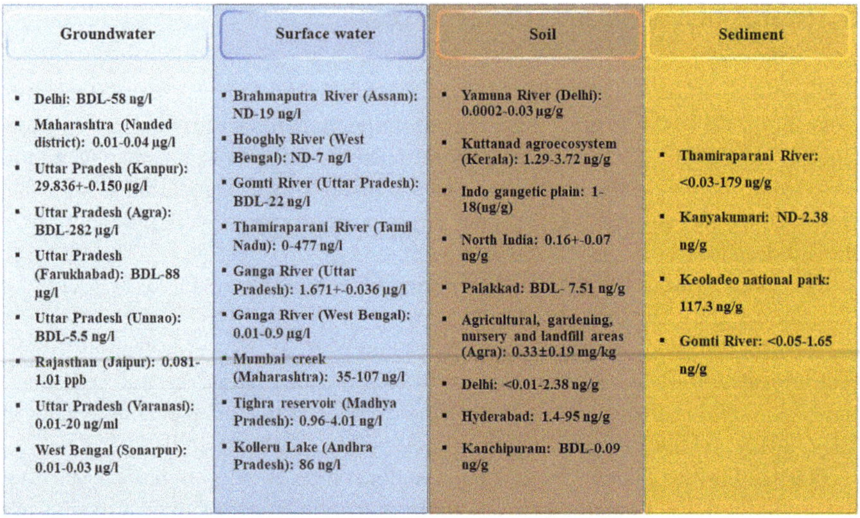

Fig. 2 Residues of dieldrin reported in different environmental compartments of Indian Peninsula. (Source: Singare, 2016; Mamta et al., 2015; Sreenivasa Rao and Ramamohana Rao, 2000; Mutiyar et al., 2011; Motekar, 2011; Sankararamakrishnan et al., 2005; Mohapatra et al., 1995; Sharma and Khan, 2009; Singh et al., 2005; Singh et al., 2007; Ghose et al., 2009; Chakraborty et al., 2016; Malik et al., 2009; Arisekar et al., 2021; Nair et al., 1991; Sruthi et al., 2017; Kumar et al., 2011; Gopalan and Chenicherry, 2018; Singh, 2001; Kata et al., 2014; Selvaraj et al., 2021; Kumaraswamy et al., 2012; Jeyakumar et al., 2014; Bhadouria et al., 2012) *BDL: below detection limit

Table 2 Residues of dieldrin found in fish samples collected from various sites in India

Study site	Dieldrin concentration
Delhi	0.0001–0.2 µg/g
Kerala	0.07–0.16 ng/g
Karnataka	BDL – 4.4 ng/g
Punjab	3 ng/g
Tamil Nadu	BDL – 10 ng/g

BDL - below detection limit
Source: (Nair et al., 1991; Sankar et al., 2006; Dhananjayan and Muralidharan, 2010; Kaur et al., 2008; Samidurai et al., 2019)

radish (2.8 ng/g), and cauliflower (50 ng/g). Various investigations have also established the occurrence of dieldrin in aquatic and terrestrial animals. Recent results from various research show substantial levels of dieldrin in both inland and offshore fish samples (Table 2) as well as in avian fauna. Dieldrin residues in different body parts of birds retrieved dead from three bird sanctuaries between 2008 and 2010, namely, Vedanthangal (Chengalpattu district), Koonthankulam (Tirunelveli district), and Vettangudi bird sanctuary (Sivagangai district) of Tamil Nadu were surveyed (Jayakumar et al., 2020). The species of birds that were reported to contain

Table 3 Residues of dieldrin reported in blood/serum in humans

Study site	Gender	Medical condition	Dieldrin concentration
Delhi	Women	Pre-term delivery (maternal)	1.49 ppb
Delhi	Women	Pre-term delivery (cord)	0.663 ppb
Delhi	Women	Healthy	0.002 ppb
New Delhi	Women	Maternal blood	2.25 ng/g
New Delhi	Women	Placental	5.388 ng/g
North India	Both	Parkinson's disease	2.1–28.63 ppb
North India	Both	Alzheimer's disease	9.63±6.47 ppb

Source: (Mustafa et al., 2012; Nair et al., 1992; Tyagi et al., 2015; Chhillar et al., 2013; Singh et al., 2013)

residues of dieldrin included Indian Pond Heron (3.4–63.4 ng/g), Asian Openbill (BDL-10.2 ng/g), Little Egret (1.4–23.0 ng/g), Black-headed Ibis (5.8–16.0 ng/g), Painted Stork (BDL-17.1 ng/g), Intermediate Egret (BDL-19.0 ng/g), Cattle Egret (13.8–19.0 ng/g), Indian Cormorant (3.1–50.2 ng/g), Eurasian Spoonbill (0.4–26.6 ng/g), Glossy Ibis (3.0–169.1 ng/g), Grey Heron (2.3–20.7 ng/g), Little Cormorant (15.4–59.4 ng/g), Spot-billed Pelican (BDL-2.1 ng/g), and Black-crowned Night Heron (3.6–82.9 ng/g).

Because organochlorine pesticides (OCPs) are prevalent in the environment, human exposure to them is unavoidable. Due to their xenoestrogenic nature, OCPs such as dieldrin may function as endocrine disruptors, disrupting normal balance of estrogen and progesterone levels during pregnancy resulting in unfavorable outcomes such as recurrent miscarriages, preterm birth (PTB), low birth weight, and so on. These OCPs bioaccumulate in adipose tissues and hence impact women more because of their larger percentage of body fat, thereby amplifying the pesticidal effects. OCPs in placental tissue and maternal blood provide prenatal exposure risks to fetuses due to extended bioaccumulation and low removal, which may have a negative impact on health (Table 3). A study was done to investigate the burden of OC pesticides and their impact on thyroid function in women. In 1997–1998, 123 women from Jaipur were examined. One hundred women had normal thyroid hormone levels, whereas 23 had low T4 and high TSH levels (high TSH levels suggest hypothyroidism). Dieldrin levels were found to be considerably elevated in hypothyroid patients. It was thus proposed that chlorinated hydrocarbons are goitrogenic substances that alter hormone action. Pesticide levels were greater in women of lower age groups. This might be owing to increased exposure of the younger generation as well as pesticide excretion by older women during menstruation and delivery (Rathore et al., 2002).

4 Global Scenario

Historically, aldrin and dieldrin were derived from cyclopentadiene, a synthetic rubber waste product, produced by the Diels-Alder reaction. Julius Hyman discovered aldrin and dieldrin, but a lawsuit led to the company's acquisition. Shell Chemical produced aldrin and dieldrin in 1948, with Shell being the sole manufacturer until AMVAC began in 1968. According to statistics from domestic usage in the United States, output was around 90% aldrin and 10% dieldrin. Aldrin/dieldrin was the second most commonly used agricultural pesticide in the USA, used on around 90 crops. It was advised to use it on crops, mostly wheat, maize, hay, barley, rye, and oats, as well as orchards and vegetables. It was employed on cotton, tobacco, and citrus crops in the South. In 1964, more than half of the dieldrin produced was utilized for pest control rather than agriculture. This included soil treatment for termite control and mothproofing during the production of wool carpets and apparel. Pest management applications also included harvest and fire ant control, as well as aerial spraying for spruce budworm and gipsy moths (Jorgenson, 2001).

Table 4 represents residual level of dieldrin in various water bodies of the world. Dieldrin has been found in a variety of environmental compartments including soil, sediments, water, atmosphere, and organisms. It is majorly due to its persistent chemical nature and the usage of formulations containing dieldrin or counterfeit goods containing it. Several studies have revealed that dieldrin are hazardous to biological species and threat to the integrity of ecosystem owing to stable structure and fat-soluble nature. Their tendency is to bioaccumulate, bioconcentrate, and biomagnify as they go up the food chain, causing vertebrate and invertebrate toxicity in non-target biota including humans (Table 5). Researches have shown concentrations of dieldrin in various biota namely blue mussels (0.21 ng/g) and crabs (0.10 ng/g) collected from Baltic Sea. Moreover, fishes collected from the same sampling location have shown dieldrin residual concentration ranging from 0.83 ng/g to 6.6 ng/g while eels have shown a range of 0.84 ng/g to 1.4 ng/g of dieldrin residue (Falandysz et al., 2001). Other than that, Ogbeide et al. (2015) have reported 0.08 to 0.53 μg/g in fishes in Owan River, Nigeria, while studies conducted on different aquatic species from Montery Harbour and Point Sur, California (USA), have reported 0.38 ng/g (Fat innkeeper worms), 0.96 ng/g (Mussel), 0.12 ng/g (Red urchin), 0.04 ng/g (Red abalone), 0.12 ng/g (Kelp crab), and 0.09 ng/g (Turban snail) of dieldrin (Kannan et al., 2004).

Table 4 Residues of dieldrin found in various water bodies of the world

Study site	Location	Concentration
Red River	Vietnam	BDL – 18.6 ng/l
River Wuchuan	Southeast China	1.78–21.14 ng/l
Buffalo River	South Africa	10 ± 0.003 ng/l
Minjiang River Estuary	Southeast China	5.39–74.03 ng/l
Qiantang River	East China	<0.15–42.06 ng/l
Northeastern part of Sao Paulo	Brazil	0.005–0.005 ng/l
Gao-ping River	Taiwan	<0.12–1.42 ng/l
River Nile (Rosetta Branch)	Egypt	>0.01 ng/l
Rufiji River	Tanzania	7.8 ng/l
Owan River	Edo state, Nigeria	0.14 ± 0.05 μg/l

Source: (Hung and Thiemann, 2002; Zhang et al., 2002; Awofolu and Fatoki, 2003; Zhang et al., 2003; Zhou et al., 2006; Rissato et al., 2006; Doong et al., 2008; Ahmed et al., 2008; Mwevura et al., 2021; Ogbeide et al., 2015)

Table 5 Residues of dieldrin found in human blood/serum

Country	Gender	Medical condition	Concentration (ppb)
Mexico	Women (Maya Community)	Exposed to contaminated water	121
Costa rica	Both	Parkinson's disease	7.58
Sudan	Both	Healthy	82
Spain	Women	Breast cancer	12.2 ± 30.0
Tanzania	Women	Healthy	0.50 ± 0.07
Tanzania	Men	Subfertility	0.55 ± 0.09
Germany	Women	Healthy	0.02 ± 0.01
Germany	Men	Healthy	0.08 ± 0.01
USA	Both	Healthy	0.72 ± 0.02

Source: (Polanco Rodríguez et al., 2017; Steenland et al., 2014; Elbashir et al., 2015; Boada et al., 2012; Weiss et al., 2006; Gaffney et al., 2005)

5 Adverse Effects of Dieldrin

Dieldrin, while extremely toxic, is also highly species-specific. As with other organochlorine contaminants, most animal studies showed that liver damage is the predominant consequence. In animal populations, this pesticide produces convulsions, hypersensitivity, tremors, neural degeneration, temporary hypothermia, and anorexia (ATSDR, 1993). They cause negative enzymatic and hormonal alterations in fish, resulting in decreased reproductive capabilities (Environmental Health Criteria, 1989). In humans acutely exposed, aldrin and dieldrin showed similar adverse effects. Among these were dizziness, headache, irritability, nausea, muscle twitching, loss of appetite, convulsions, loss of consciousness, and possible death (ATSDR, 1993). These conditions, though not permanent, may persist for few weeks after the exposure. The lethal dose of dieldrin for an adult male has been estimated to be 40 to 50 times as toxic as DDT (Ritter et al., 1995).

There are plenty of researches, which indicated that chronic exposure of mice to dieldrin develops liver tumors either benign or malignant (Kolaja et al., 1996). Simultaneously, testing on rats found no substantial effects on hepatic tissues. A significant rise in hepatic and biliary tract malignancies was reported among a sample of employees exposed at workplaces to dieldrin, aldrin, and endrin, albeit the study was constrained by a lack of information with respect to quantitative exposure (Ritter et al., 1995). Dieldrin is also classified as a potential human carcinogen by the EPA. Adults have developed neurological symptoms following acute and intermediate-length exposure, while two infants experienced severe convulsions after ingesting unknown amounts of dieldrin. Dizziness, headaches, anorexia, nausea and vomiting, muscular twitching, myoclonic jerks, and, in most critical cases, aberrant EEG recordings have been documented in workers involved in the production or application of dieldrin. Furthermore, two investigations of people exposed to dieldrin at work revealed unusually high rates of psychiatric disease. One study found aberrant electromyograms (EMGs), indicating peripheral neuropathy (ATSDR, 1993). Studies have also linked dieldrin levels in the brain with Parkinson's disease (Fleming et al., 1994).

Dieldrin has been shown in animal studies to have estrogenic characteristics; however, fetotoxicity has only been detected at doses where maternal toxicity is evident (Soto et al., 1995). Large, single doses of dieldrin given to mice and hamsters during mid-gestation period resulted in physical abnormalities in the fetus, including cleft lip, foot webbing, and open-eye. There was also an increase in mortality (ATSDR, 1993). In rats exposed to high doses, extensive kidney impairment has been recorded. Chronic dieldrin exposure causes hepatic effects similar to halogenated hydrocarbon pesticides, such as liver enlargement, increased enzymatic activity at microsomal level, increased cellular smooth endoplasmic reticulum, and vacuolization (ATSDR, 1993).

6 Management and Remedial Measures

According to the order of Ministry of Agriculture published on 17 July 2001 (S.O. 682 (E) sub-sec (2) of Sec 27 of the Insecticides Act, 1968), an order was issued by the Indian government banning dieldrin to manufacture, use, export, and import on the same day. However, marketing and restricted usage (locust control) of dieldrin were authorized for 2 years from the date of the prohibition, or until the date of expiration, whichever came first (Toxics link, 2018).

Owing to its health hazards that critically consist of liver and biliary cancer, efforts have been made to find adequate remediation techniques. Among those, microbial degradation has found prevalence where *Burkholderia* sp., *Cupriavidus* sp., *Pseudonocardia* sp., *Mucor racemosus*, etc. have been documented to have effectively degraded dieldrin via deoxygenation (Morillo & Villaverde, 2017; Ritter et al., 1995). Additionally certain microbes were found to be capable of converting dieldrin into metabolic chemicals. Researchers have found that dieldrin may be mineralized to CO2. In general, fungi degrade dieldrin quicker than bacteria, whose degradation process is well documented. Many variables influence the efficacy of decomposing microorganisms placed into polluted locations, including pH, carbon and nitrogen supplies, enzymes, hormones, and light. Some enzymes have also been discovered to be involved in the breakdown of dieldrin (Purnomo, 2017).

7 Conclusion

As discussed in the present chapter, dieldrin is a toxic organochlorine pesticide that was initially used for insect control in agriculture and public health programs. Its persistence in the environment, particularly in aquatic ecosystems, has led to its accumulation in the food chain and affected various biotic species. Dieldrin is considered a potential human toxin by several health agencies and has been linked to nervous system disorders, developmental effects, and reproductive problems. Many countries banned or restricted dieldrin's use by the 1970s and 1980s, and the

Stockholm Convention on Persistent Organic Pollutants classified it as one of the "dirty dozen" chemicals targeted for global elimination. Despite its ban, dieldrin remains a risk due to its persistence in the environment, with residues found in soil, sediments, and wildlife.

Dieldrin concentrations may still be found in several locations of India. The allowable value of dieldrin in India is quite low, according to established guidelines. As a result, no further review of this OCP is required. Additionally, a critical point could be observed that even though India has outlawed few pesticides usage such as Endosulfan, DDT, aldrin, and dieldrin, among others, surveys have reported that these banned pesticides are still making its entry by illicit means in the market. It was also discovered that additional pesticides, such as DDT and Endosulfan, are being sold in the guise of prohibited pesticides, which needs to be explored further. Furthermore, it appears that there is a paucity of information among official organizations about the unlawful export and import of POPs (Toxics link, 2018). Also, the scattered available data of its usage and availability trend is a firm cause of improper monitoring and examining, an area that needs proper focus specially to keep a record of its residue accumulation and further its ecotoxic consequences.

References

Ahmed, A., Barbary, E., Yehia, M. M., Mohamed, M., & Bouraie, E. (2008). Evaluation of organochlorine pesticides (OCPs) in surface water and bed sediment samples from the river Nile at Rosetta Branch, Egypt. *Journal of Applied Sciences Research, 4*(12), 1985–1993.

Arisekar, U., Shakila, R. J., Shalini, R., & Jeyasekaran, G. (2021). Pesticides contamination in the Thamirabarani, a perennial river in peninsular India: The first report on ecotoxicological and human health risk assessment. *Chemosphere, 267*, 129251.

ATSDR (Agency for Toxic Substances and Disease Registry. (1993). *Cancer policy framework*. U.S. Department of Health and Human Services, Public Health Service.

Awofolu, R. O., & Fatoki, O. S. (2003). Persistent organochlorine pesticide residues in freshwater systems and sediments from the Eastern Cape, South Africa. *Water SA, 29*(3), 323–330.

Bhadouria, B. S., Mathur, V. B., & Kaul, R. (2012). Monitoring of organochlorine pesticides in and around Keoladeo National Park, Bharatpur, Rajasthan, India. *Environmental Monitoring and Assessment, 184*, 5295–5300.

Boada, L. D., Zumbado, M., Henríquez-Hernández, L. A., Almeida-González, M., Álvarez-León, E. E., Serra-Majem, L., & Luzardo, O. P. (2012). Complex organochlorine pesticide mixtures as determinant factor for breast cancer risk: A population-based case–control study in the Canary Islands (Spain). *Environmental Health, 11*, 1–9.

Chakraborty, P., Khuman, S. N., Selvaraj, S., Sampath, S., Devi, N. L., Bang, J. J., & Katsoyiannis, A. (2016). Polychlorinated biphenyls and organochlorine pesticides in river Brahmaputra from the outer Himalayan Range and River Hooghly emptying into the Bay of Bengal: Occurrence, sources and ecotoxicological risk assessment. *Environmental Pollution, 219*, 998–1006.

Chhillar, N., Singh, N. K., Banerjee, B. D., Bala, K., Mustafa, M., Sharma, D., & Chhillar, M. (2013). Organochlorine pesticide levels and risk of Parkinson's disease in north Indian population. *International Scholarly Research Notices, 2013*.

Chourasiya, S., Khillare, P. S., & Jyethi, D. S. (2015). Health risk assessment of organochlorine pesticide exposure through dietary intake of vegetables grown in the periurban sites of Delhi, India. *Environmental Science and Pollution Research, 22*, 5793–5806.

Connell, D. W., Miller, G., & Anderson, S. (2002). Chlorohydrocarbon pesticides in the Australian marine environment after banning in the period from the 1970s to 1980s. *Marine Pollution Bulletin, 45*(1–12), 78–83.

Dhananjayan, V., & Muralidharan, S. (2010). Organochlorine pesticide residues in inland wetland fishes of Karnataka, India and their implications on human dietary intake. *Bulletin of Environmental Contamination and Toxicology, 85*, 619–623.

Doong, R. A., Lee, S. H., Lee, C. C., Sun, Y. C., & Wu, S. C. (2008). Characterization and composition of heavy metals and persistent organic pollutants in water and estuarine sediments from Gao-Ping River, Taiwan. *Marine Pollution Bulletin, 57*(6–12), 846–857.

Elbashir, A. B., Abdelbagi, A. O., Hammad, A. M., Elzorgani, G. A., & Laing, M. D. (2015). Levels of organochlorine pesticides in the blood of people living in areas of intensive pesticide use in Sudan. *Environmental Monitoring and Assessment, 187*, 1–10.

Environmental Health Criteria. (1989). *International programme on chemical safety, Environmental Health Criteria 91, Aldrin and Dieldrin*. World Health Organization, Geneva. Retrieved on 02.03.2023 from https://apps.who.int/iris/bitstream/handle/10665/37536/9241542918 eng.pdf ?sequence=1&isAllowed=y

Falandysz, J., Strandberg, L., Puzyn, T., Gucia, M., & Rappe, C. (2001). Chlorinated cyclodiene pesticide residues in blue mussel, crab, and fish in the Gulf of Gdańsk, Baltic Sea. *Environmental Science & Technology, 35*(21), 4163–4169.

Fleming, L., Mann, J. B., Bean, J., Briggle, T., & Sanchez-Ramos, J. R. (1994). Parkinson's disease and brain levels of organochlorine pesticides. *Annals of Neurology: Official Journal of the American Neurological Association and the Child Neurology Society, 36*(1), 100–103.

Gaffney, S. H., Curriero, F. C., Strickland, P. T., Glass, G. E., Helzlsouer, K. J., & Breysse, P. N. (2005). Influence of geographic location in modeling blood pesticide levels in a community surrounding a US Environmental Protection Agency Superfund site. *Environmental Health Perspectives, 113*(12), 1712–1716.

Ghose, N. C., Saha, D., & Gupta, A. (2009). Synthetic detergents (surfactants) and organochlorine pesticide signatures in surface water and groundwater of Greater Kolkata, India. *Journal of Water Resource and Protection, 2009*(4), 227–298. 290.

Gopalan, N. K., & Chenicherry, S. (2018). Fate and distribution of organochlorine insecticides (OCIs) in Palakkad soil, India. *Sustainable Environment Research, 28*(4), 179–185.

Hung, D. Q., & Thiemann, W. (2002). Contamination by selected chlorinated pesticides in surface waters in Hanoi, Vietnam. *Chemosphere, 47*(4), 357–367.

Jayakumar, S., Muralidharan, S., & Dhananjayan, V. (2020). Organochlorine pesticide residues among colonial nesting birds in Tamil Nadu, India: A maiden assessment from their breeding grounds. *Archives of Environmental Contamination and Toxicology, 78*, 555–567.

Jeyakumar, T., Kalaiarasi, I., Rajavel, A., Anbu, M., & Kumar, R. (2014). Levels of organochlorine pesticide residues in water and sediment from selected agricultural sectors of Kanyakumari District, Tamil Nadu, India. *International Journal of Environmental Research, 8*(2), 493–500.

Jorgenson, J. L. (2001). Aldrin and dieldrin: A review of research on their production, environmental deposition and fate, bioaccumulation, toxicology, and epidemiology in the United States. *Environmental Health Perspectives, 109*(suppl 1), 113–139.

Kannan, K., Kajiwara, N., Watanabe, M., Nakata, H., Thomas, N. J., Stephenson, M., et al. (2004). Profiles of polychlorinated biphenyl congeners, organochlorine pesticides, and butyltins in southern sea otters and their prey. *Environmental Toxicology and Chemistry: An International Journal, 23*(1), 49–56.

Kata, M., Srinivasa Rao, S., & Rama Mohan, K. (2014). Spatial distribution, ecological risk evaluation and potential sources of organochlorine pesticides from soils in India. *Environment and Earth Science, 74*, 4031–4038.

Kaur, M., Sharma, J. K., Gill, J. P., Aulakh, R. S., Bedi, J. S., & Joia, B. S. (2008). Determination of organochlorine pesticide residues in freshwater fish species in Punjab, India. *Bulletin of Environmental Contamination and Toxicology, 80*, 154–157.

Kolaja, K. L., Stevenson, D. E., Johnson, J. T., Walborg, E. F., Jr., & Klaunig, J. E. (1996). Subchronic effects of dieldrin and phenobarbital on hepatic DNA synthesis in mice and rats. *Fundamental and Applied Toxicology, 29*(2), 219–228.

Kumar, B., Kumar, S., Gaur, R., Goel, G., Mishra, M., Singh, S. K., et al. (2011). Persistent organochlorine pesticides and polychlorinated biphenyls in intensive agricultural soils from North India. *Soil and Water Research, 6*(4), 190–197.

Kumarasamy, P., Govindaraj, S., Vignesh, S., Rajendran, R. B., & James, R. A. (2012). Anthropogenic nexus on organochlorine pesticide pollution: A case study with Tamiraparani river basin, South India. *Environmental Monitoring and Assessment, 184*, 3861–3873.

Kumari, K., Swamy, S., & Singh, A. (2022). *Global monitoring plan on persistent organic pollutants (POPs). Persistent organic pollutants: Gaps in management and associated challenges* (1st ed., pp. 227–252). CRC Press.

Malik, A., Ojha, P., & Singh, K. P. (2009). Levels and distribution of persistent organochlorine pesticide residues in water and sediments of Gomti River (India) – A tributary of the Ganges River. *Environmental Monitoring and Assessment, 148*, 421–435.

Mamta, Rao, R. J., & Wani, K. A. (2015). Monitoring of organochlorine and organophosphorus pesticide residues in water during different seasons of Tighra reservoir Gwalior, Madhya Pradesh, India. *Environmental Monitoring and Assessment, 187*(11), 684.

Mandocdoc, M., & Primo David, C. (2008). Dieldrin contamination of the groundwater in a former US military base (Clark Air Base, Philippines). *CLEAN - Soil, Air, Water, 36*(10–11), 870–874.

Mohapatra, S. P., Kumar, M., Gajbhiye, V. T., & Agnihotri, N. P. (1995). Ground water contamination by organochlorine insecticide residues in a rural area in the Indo-Gangetic plain. *Environmental Monitoring and Assessment, 35*, 155–164.

Morillo, E., & Villaverde, J. (2017). Advanced technologies for the remediation of pesticide-contaminated soils. *Science of the Total Environment, 586*, 576–597.

Motekar, S. C. (2011). Ground water contamination by organochlorine insecticide residues in the Godavari plain of Nanded district. *Recent Research in Science and Technology, 3*(12).

Mustafa, M. D., Banerjee, B. D., Ahmed, R. S., Tripathi, A. K., & Guleria, K. (2012). Gene–environment interaction in preterm delivery with special reference to organochlorine pesticides. *MHR: Basic Science of Reproductive Medicine, 19*(1), 35–42.

Mutiyar, P. K., Mittal, A. K., & Pekdeger, A. (2011). Status of organochlorine pesticides in the drinking water well-field located in the Delhi region of the flood plains of river Yamuna. *Drinking Water Engineering and Science, 4*(1), 51–60.

Mwevura, H., Kylin, H., Vogt, T., & Bouwman, H. (2021). Dynamics of organochlorine and organophosphate pesticide residues in soil, water, and sediment from the Rufiji River Delta, Tanzania. *Regional Studies in Marine Science, 41*, 101607.

Nair, A., Dureja, P., & Pillai, M. K. K. (1991). Levels of aldrin and dieldrin in environmental samples from Delhi, India. *Science of the Total Environment, 108*(3), 255–259.

Nair, A., Dureja, P., & Pillai, M. K. K. (1992). Aldrin and dieldrin residues in human fat, milk and blood serum collected from Delhi. *Human & Experimental Toxicology, 11*(1), 43–45.

NIP. (2011). *Government of India national implementation plan, Stockholm convention on persistent organic pollutants.* MoEF&CC and UNIDO. http://www.indiaenvironmentportal.org.in/files/NIP%20final%20draft.pdf

Ogbeide, O., Tongo, I., & Ezemonye, L. (2015). Risk assessment of agricultural pesticides in water, sediment, and fish from Owan River, Edo State, Nigeria. *Environmental Monitoring and Assessment, 187*, 1–16.

Polanco Rodríguez, Á. G., Inmaculada Riba López, M., Angel DelValls Casillas, T., León, J. A. A., Anjan Kumar Prusty, B., & Álvarez Cervera, F. J. (2017). Levels of persistent organic pollutants in breast milk of Maya women in Yucatan, Mexico. *Environmental Monitoring and Assessment, 189*, 1–13.

Purnomo, A. S. (2017). Microbe-assisted degradation of aldrin and dieldrin. *Microbe-Induced Degradation of Pesticides*, 1–22.

Rathore, M., Bhatnagar, P., Mathur, D., & Saxena, G. N. (2002). Burden of organochlorine pesticides in blood and its effect on thyroid hormones in women. *Science of the Total Environment, 295*(1–3), 207–215.
Rissato, S. R., Galhiane, M. S., Ximenes, V. F., De Andrade, R. M., Talamoni, J. L., Libânio, M., et al. (2006). Organochlorine pesticides and polychlorinated biphenyls in soil and water samples in the Northeastern part of São Paulo State, Brazil. *Chemosphere, 65*(11), 1949–1958.
Ritter, L., Solomon, K. R., Forget, J., Stemeroff, M., & O'leary, C. (1995). *A review of selected persistent organic pollutants. International Programme on Chemical Safety (IPCS). PCS/95.39* (Vol. 65, p. 66). World Health Organization.
Samidurai, J., Subramanian, M., & Venugopal, D. (2019). Levels of organochlorine pesticide residues in fresh water fishes of three bird sanctuaries in Tamil Nadu, India. *Environmental Science and Pollution Research, 26*, 1983–1993.
Sankar, T. V., Zynudheen, A. A., Anandan, R., & Nair, P. V. (2006). Distribution of organochlorine pesticides and heavy metal residues in fish and shellfish from Calicut region, Kerala, India. *Chemosphere, 65*(4), 583–590.
Sankararamakrishnan, N., Sharma, A. K., & Sanghi, R. (2005). Organochlorine and organophosphorous pesticide residues in ground water and surface waters of Kanpur, Uttar Pradesh, India. *Environment International, 31*(1), 113–120.
Selvaraj, S., Gaonkar, O., Kumar, B., Cincinelli, A., & Chakraborty, P. (2021). Legacy persistent organochlorine pollutants and polycyclic aromatic hydrocarbons in the surface soil from the industrial corridor of South India: Occurrence, sources and risk assessment. *Environmental Geochemistry and Health, 43*, 2105–2120.
Sharma, A., & Khan, T. I. (2009). Organochlorine pesticides in irrigation water of Jaipur City (India). *Journal of Environmental Science & Engineering, 51*(3), 175–178.
Singare, P. U. (2016). Distribution and risk assessment of suspected endocrine-disrupting pesticides in creek water of Mumbai, India. *Marine Pollution Bulletin, 102*(1), 72–83.
Singh, R. P. (2001). Comparison of organochlorine pesticide levels in soil and groundwater of Agra, India. *Bulletin of Environmental Contamination and Toxicology, 67*, 126–132.
Singh, K. P., Malik, A., Mohan, D., & Sinha, S. (2005). Persistent organochlorine pesticide residues in alluvial groundwater aquifers of Gangetic Plains, India. *Bulletin of Environmental Contamination and Toxicology, 74*, 162–169.
Singh, K. P., Malik, A., & Sinha, S. (2007). Persistent organochlorine pesticide residues in soil and surface water of northern Indo-Gangetic alluvial plains. *Environmental Monitoring and Assessment, 125*, 147–155.
Singh, N. K., Chhillar, N., Banerjee, B. D., Bala, K., Basu, M., & Mustafa, M. (2013). Organochlorine pesticide levels and risk of Alzheimer's disease in north Indian population. *Human & Experimental Toxicology, 32*(1), 24–30.
Soto, A. M., Sonnenschein, C., Chung, K. L., Fernandez, M. F., Olea, N., & Serrano, F. O. (1995). The E-SCREEN assay as a tool to identify estrogens: An update on estrogenic environmental pollutants. *Environmental Health Perspectives, 103*(suppl 7), 113–122.
Sreenivasa Rao, A., & Ramamohana Rao, P. (2000). Kolleru lake water pollution by pesticides. *Indian Journal of Environmental Health, 42*(4), 169–175.
Sruthi, S. N., Shyleshchandran, M. S., Mathew, S. P., & Ramasamy, E. V. (2017). Contamination from organochlorine pesticides (OCPs) in agricultural soils of Kuttanad agroecosystem in India and related potential health risk. *Environmental Science and Pollution Research, 24*, 969–978.
Steenland, K., Mora, A. M., Barr, D. B., Juncos, J., Roman, N., & Wesseling, C. (2014). Organochlorine chemicals and neurodegeneration among elderly subjects in Costa Rica. *Environmental Research, 134*, 205–209.
Stockholm Convention. (2001). *Listing of POPs in Stockholm Convention*. Retrieved on 07.09.2022 from http://chm.pops.int/TheConvention/ThePOPs/ListingofPOPs/tabid/2509/Default.aspx
Toxics Link. (2018). Retrieved in 18.10.2022 from https://ipen.org/sites/default/files/documents/POPs%20Country%20Situation%20report.pdf

Tyagi, V., Garg, N., Mustafa, M. D., Banerjee, B. D., & Guleria, K. (2015). Organochlorine pesticide levels in maternal blood and placental tissue with reference to preterm birth: A recent trend in North Indian population. *Environmental Monitoring and Assessment, 187*, 1–9.
Weiss, J. M., Bauer, O., Blüthgen, A., Ludwig, A. K., Vollersen, E., Kaisi, M., et al. (2006). Distribution of persistent organochlorine contaminants in infertile patients from Tanzania and Germany. *Journal of Assisted Reproduction and Genetics, 23*, 393–399.
Zhang, Z., Hong, H., Zhou, J. L., Yu, G., Chen, W., & Wang, X. (2002). Transport and fate of organochlorine pesticides in the River Wuchuan, Southeast China. *Journal of Environmental Monitoring, 4*(3), 435–441.
Zhang, Z. L., Hong, H. S., Zhou, J. L., Huang, J., & Yu, G. (2003). Fate and assessment of persistent organic pollutants in water and sediment from Minjiang River Estuary, Southeast China. *Chemosphere, 52*(9), 1423–1430.
Zhou, R., Zhu, L., Yang, K., & Chen, Y. (2006). Distribution of organochlorine pesticides in surface water and sediments from Qiantang River, East China. *Journal of Hazardous Materials, 137*(1), 68–75.

Endrin

Kanchan Kumari and Shaheen Akhtar

1 Introduction

Shell and Velsicol Chemical Corporation developed endrin, an organochlorine molecule with the chemical formula $C_{12}H_8C_{16}O$, in 1950. It was mostly employed as an insecticide, although it was also utilized as a rodenticide and piscicide. Although commercial samples are frequently off-white, it is a colorless, odorless substance. Endrex, an emulsifiable solution, was used to create endrın (Van Esch & van Heemstra-Lequin, 1992). Because of its reputation as a persistent organic pollution, the chemical is now prohibited in several nations.

Endrin appears in the environment as either endrin aldehyde or endrin ketone and is mostly found in the bottom sediments of water bodies (ATSDR, 2022; Hansen, 2003). This pesticide was sprayed on the leaves of cotton and grain crops. It was also used to keep rodents like mice and voles in check. Endrin can be metabolized by animals; therefore, it does not accumulate in their fatty tissue as much as structurally comparable compounds do. However, it has a half-life of 12 years in soil. Furthermore, endrin is toxic to fishes. Sheepshead minnows hatched early and perished on the ninth day after being exposed to high amounts of endrin in the water. The overall human population's major route of exposure is through food, albeit current dietary consumption estimates are below the acceptable levels set by international health authorities (Stockholm Convention, 2001).

Endrin exposure can occur by inhalation, ingestion of compounds containing the chemical, or skin contact (ATSDR, 2022). Once within the body, it can be deposited in fat and serve as a neurotoxic on the central nervous system, causing convulsions,

K. Kumari (✉) · S. Akhtar
CSIR-National Environmental Engineering Research Institute (NEERI), Kolkata Zonal Center, Kolkata, West Bengal, India
e-mail: k_kumari@neeri.res.in

K. Kumari, *Pollutants of Global Concern*, Emerging Contaminants and Associated Treatment Technologies, https://doi.org/10.1007/978-3-031-50996-4_5

seizures, and even death (Revzin, 1970). Despite the fact that endrin is not currently classed as a mutagen (Van Esch & van Heemstra-Lequin, 1992) or a human carcinogen, it is nevertheless a hazardous substance with negative consequences. Because of these harmful consequences, the makers discontinued all usage of endrin in the United States by 1991. Concerns have been raised about food imports since certain nations may still be using endrin as a pesticide (ATSDR, 2022).

2 Technical Details of Endrin

Endrin is a white crystalline substance which has a mild odour (Table 1). When kept at room temperature, technical endrin is stable. It is stable in alkaline agents, emulsifiers, wetting agents, and solvents-containing formulations. Concentrated mineral acids, acid catalysts, acid oxidizing agents, and active metals breakdown it. When heated over 200 °C, endrin transforms into delta-ketoendrin, a less poisonous and less insecticidally active chemical (IPCS, 1992).

The condensation of hexachlorocyclopentadiene with vinyl chloride is the first step in the production of endrin. After that, the product is dehydrochlorinated. Isodrin is generated as a result of the reaction with cyclopentadiene. The final stage in the synthesis of endrin is the creation of epoxides by adding peracetic acid or perbenzoic acid to isodrin. Endrin is a dieldrin stereoisomer with similar characteristics, albeit endrin degrades more easily (CDC, 2013).

3 Status of Works and Usage in India

Endrin, which was primarily used to control insects and rodents, has been prohibited from manufacturing, usage, and import since May 15, 1990, and listed in Stockholm Convention under Annex A in 2001 (Kumari, 2021) under dirty dozen POPs chemicals. Endrin production and usage have been totally halted in India, and no stockpiles of this POPs pesticide have been detected. For the period of 2014–2018, India has not exported or imported endrin (Toxics Link, 2018). Even though stockpiles of endrin in India is not very significant, the popularity of this organochlorine had once made its usage quite extensive. Thus, residues of endrin have been reported by various studies done in different parts of the nation. Table 2 and Fig. 1 represent some of the glimpses of its documented concentrations.

Table 1 Physico-chemical properties of endrin

Details	Description (endrin)
Structure	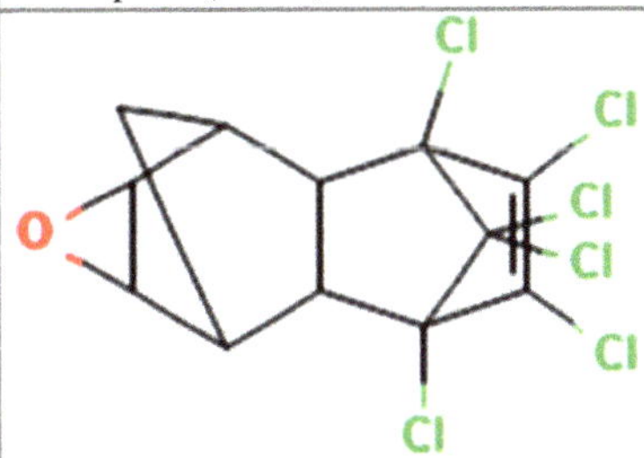
IUPAC name	3,4,5,6,13,13-Hexachloro-10-oxapentacyclo[6.3.1.1^{3,6}.0^{2,7}.0^{9,11}] tridec-4-ene
Molecular formula	$C_{12}H_8Cl_6O$
CAS No.	128-10-9
Molecular weight	380.9 g/mol
Color/form	Colorless crystals
Odor	Mild, chemical odor
Boiling point	245 °C
Melting point	175.5 °C
Flash point	80 °F
Solubility	In water, 0.195 mg/l at 25 °C
Density	1.7 g/cm^3
Vapor pressure	2e-07 mm Hg at 77 °F
Octanol/water partition coefficient	5.40
Stability/shelf life	Resistant to moisture and alkalis
Decomposition	When heated to decomposition, emits toxic fumes of hydrogen chloride
Corrosivity	Slightly corrosive to metals
Safety and hazards	Acute toxic, health hazard and environmental hazard

4 International Data Versus the Indian Scenario

Endrin was a popular pesticide, rodenticide, and avicide. Unlike aldrin and dieldrin, endrin was not commonly employed as a termiticide. Endrin can last for years depending on soil conditions (IPCS, 1992). Keto-endrin is a significant photodegradation product. Endrin has been found in soils and at low levels in sediment and surface waters, owing mostly to previous agricultural application or runoff from contaminated soils (ATSDR, 1993; IPCS, 1992) (Table 3).

Table 2 Endrin concentration reported in various environmental samples in India

Environmental matrix	Study site	Endrin concentration	Reference
Sea water	Mumbai creek	50–97 ng/l	Singare (2016)
Surface water	Tamiraparani river basin	<0.02–61 ng/l	Kumarasamy et al. (2012)
Surface water	Kanyakumari	ND-1.5 ng/l	Jeyakumar et al. (2014)
Surface water	Tighra reservoir	0.57–1.34 ng/l	Mamta et al. (2015)
Surface water	Unnao	BDL-10 ng/l	Singh et al. (2007)
Bore well water	Delhi	0.018–0.058 μg/l	Mutiyar et al. (2011)
Bore well water	Delhi	0.029–0.054 μg/l	Mutiyar et al. (2011)
Surface water	Gomti River	BDL-4.3 ng/l	Malik et al. (2009)
Surface water	Thamiraparani River	140–301 ng/l	Arisekar et al. (2021)
Soil	Kuttanad agroecosystem	6.02–6.95 ng/g (aldehyde)	Sruthi et al. (2017)
Soil	Kuttanad agroecosystem	0.75–7.51 ng/g (ketone)	Sruthi et al. (2017)
Soil	CHR, Kerala	<LOQ-36 ng/g	Joseph et al. (2020)
Soil	Hyderabad	2–47 ng/g	Kata et al. (2015)
Soil	IGP North India	BDL-0.34 ng/g	Singh et al. (2007)
Soil	South west coast	12–113 ng/g	Khuman et al. (2020)
Sediment	Thamiraparani River	<0.02 ng/g	Kumarasamy et al. (2012)
Sediment	Kanyakumari	ND-3.10 ng/g	Jeyakumar et al. (2014)
Sediment	Keoladeo national park	2966 ng/g	Bhadouria et al. (2012)
Sediment	Thamiraparani River	0.40–2 ng/g	Arisekar et al. (2021)
Sediment	Gomti River	<0.05–11.96 ng/g	Malik et al. (2009)
Sediment	Cochin estuarine sytem	BDL-103.1 ng/g	Akhil and Sujatha (2014)

5 Adverse Effects of Endrin

Endrin has a very high acute toxicity in animals. Endrin caused liver problems in mice, rats, and dogs, including widespread degeneration and cell vacuolization (ATSDR, 1993). It causes enlarged kidneys and livers in dogs and rats. Endrin-treated hamster and mouse fetuses showed developmental problems. Endrin was reported to enhance reactive oxygen species in liver and brain tissue in rat studies (Bagchi et al., 1995).

Acute endrin poisoning mostly affects the neurological system. Twitching muscles, mental disorientation, and convulsions have occurred during occupational exposure (ATSDR, 1993). Although no human evidence exist for this impact, endrin may lead to incorrect bone development. Increased progeny mortality in the second and third generations of rats dosed with endrin may indicate that the pesticide is targeting the reproductive system. Endrin causes oxidative stress in fetal and placental tissues, according to research on pregnant mice (Hassoun & Stohs, 1996).

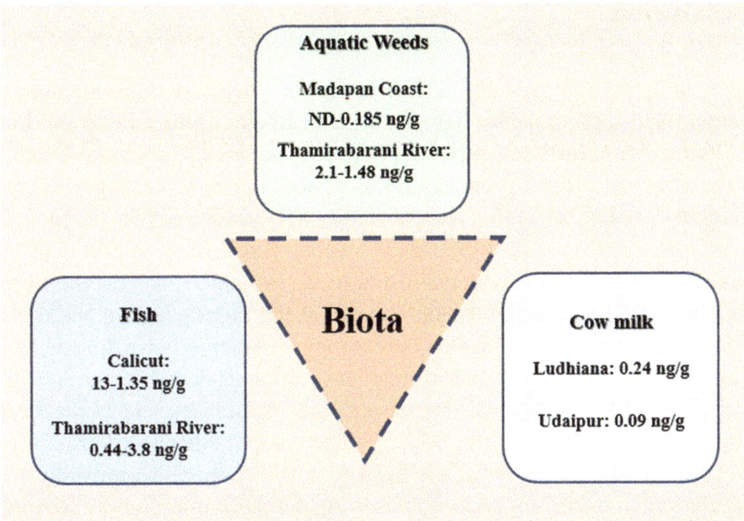

Fig. 1 Concentration of endrin residues in biota reported in India. (Source: Sankar et al., 2006; Arisekar et al., 2021; Minimol et al., 2017; Gill et al., 2020)

Table 3 Endrin concentration reported in various environmental samples

Sample site	Location	Environmental matrix	Endrin concentration	Reference
Northeastern part of Sao Paulo	Brazil	Sediment	<0.05–0.25 ng/g	Rissato et al. (2006)
River Wuchuan	Southeast China	Sediment	0.02–0.13 ng/g	Zhang et al. (2002)
Minjiang River Estuary	Southeast China	Sediment	0.27–2.43 ng/g	Zhang et al. (2003)
Qiantang River	East China	Sediment	<0.16–52.37 ng/g	Zhou et al. (2006)
River Nile (Rosctta Branch)	Egypt	Sediment	>0.01 ng/g	Ahmed et al. (2008)
Horticultural fields	Switzerland	Soil	ND-130 ng/g	Hilber et al. (2008)
Coast	South Korea	Sediment	ND-0.41 ng/g	Hong et al. (2006)
Mountain area	Czech Republic	Soil	ND-1.20 ng/g	Shegunova et al. (2007)
Owan River	Edo State, Nigeria	Sediment	0.66 ± 0.28 μg/g	Ogbeide et al. (2015)

6 Conclusion

Despite its efficiency in pest management, endrin's negative consequences on the environment and human health have resulted in its prohibition or severe limits in many countries. Endrin concentrations may still be found in several locations of India. The allowable value of endrin in India is quite low, according to established guidelines. As a result, no further review of this OCP is required (Toxics Link, 2018).

The regulatory efforts taken to restrict or prohibit the use of endrin highlight the significance of weighing the benefits of chemical compounds against their potential hazards. This serves as a reminder of the need of thoroughly testing and evaluating pesticides and other chemical agents before they are released into the environment. It also emphasizes the significance of continued monitoring and research in order to better understand the long-term impacts of such substances and influence future regulatory choices. The phase-out of endrin and comparable hazardous compounds indicates the rising realization of the need to embrace more ecologically friendly and safer alternatives in the context of sustainable agriculture and environmental protection. This move is consistent with the global endeavor to encourage practices that protect both ecosystems and human well-being, promoting a more sustainable cohabitation of human activities and the natural world.

References

Agency for Toxic Substances and Disease Registry [ATSDR]. (1993). *Cancer policy framework.* U.S. Department of Health and Human Services, Public Health Service.

Agency for Toxic Substances and Disease Registry [ATSDR]. (2022). *Toxicological profile for Aldrin and Dieldrin.* U.S. Department of Health and Human Services, Public Health Service, Agency for Toxic Substances and Disease Registry.

Ahmed, A., Barbary, E., Yehia, M. M., Mohamed, M., & Bouraie, E. (2008). Evaluation of organochlorine pesticides (OCPs) in surface water and bed sediment samples from the river Nile at Rosetta Branch, Egypt. *Journal of Applied Sciences Research, 4*(12), 1985–1993.

Akhil, P. S., & Sujatha, C. H. (2014). Spatial budgetary evaluation of organochlorine contaminants in the sediments of Cochin Estuary, India. *Marine Pollution Bulletin, 78*(1–2), 246–251.

Arisekar, U., Shakila, R. J., Shalini, R., & Jeyasekaran, G. (2021). Pesticides contamination in the Thamirabarani, a perennial river in peninsular India: The first report on ecotoxicological and human health risk assessment. *Chemosphere, 267*, 129251.

Bagchi, M., Ghosh, S., Bagchi, D., Hassoun, E., & Stohs, S. J. (1995). Protective effects of lazaroid U74389F (16-desmethyl tirilazad) on endrin-induced lipid peroxidation and DNA damage in brain and liver and regional distribution of catalase activity in rat brain. *Free Radical Biology and Medicine, 19*(6), 867–872.

Bhadouria, B. S., Mathur, V. B., & Kaul, R. (2012). Monitoring of organochlorine pesticides in and around Keoladeo national park, Bharatpur, Rajasthan, India. *Environmental Monitoring and Assessment, 184*, 5295–5300.

Centre for Disease Control and Prevention [CDC]. (2013). *Biomonitoring summary: Organochlorine pesticides overview* (Endrin Cas. No. 72-20-8). Retrieved on December 13, 2022 from https://www.cdc.gov/biomonitoring/Endrin_BiomonitoringSummary.html

Gill, J. P. S., Bedi, J. S., Singh, R., Fairoze, M. N., Hazarika, R. A., Gaurav, A., et al. (2020). Pesticide residues in peri-urban bovine milk from India and risk assessment: A multicenter study. *Scientific Reports, 10*(1), 8054.

Hansen, D. J. (2003). *Procedures for the derivation of equilibrium partitioning sediment benchmarks (ESBs) for the protection of benthic organisms [microform]*. U.S. Environmental Protection Agency, Office of Research and Development, National Health and Environmental Effects Research Laboratory, Atlantic Ecology Division/Mid-Continent Ecology Division/ Western Ecology Division.

Hassoun, E. A., & Stohs, S. J. (1996). TCDD, endrin and lindane induced oxidative stress in fetal and placental tissues of C57BL/6J and DBA/2J mice. *Comparative Biochemistry and Physiology Part C: Pharmacology, Toxicology and Endocrinology, 115*(1), 11–18.

Hilber, I., Mäder, P., Schulin, R., & Wyss, G. S. (2008). Survey of organochlorine pesticides in horticultural soils and there grown Cucurbitaceae. *Chemosphere, 73*(6), 954–961.

Hong, S. H., Yim, U. H., Shim, W. J., Li, D. H., & Oh, J. R. (2006). Nationwide monitoring of polychlorinated biphenyls and organochlorine pesticides in sediments from coastal environment of Korea. *Chemosphere, 64*(9), 1479–1488.

International Programme on Chemical Safety [IPCS]. (1992). *Health and safety guide no. 60. Endrin: Health and safety guide*. Retrieved on January 6, 2023 from https://www.inchem.org/documents/hsg/hsg/hsg060.htm

Jeyakumar, T., Kalaiarasi, I., Rajavel, A., Anbu, M., & Kumar, R. (2014). Levels of organochlorine pesticide residues in water and sediment from selected agricultural sectors of Kanyakumari District, Tamil Nadu, India. *International Journal of Environmental Research, 8*(2), 493–500.

Joseph, L., Paulose, S. V., Cyril, N., Santhosh, S. K., Varghese, A., Nelson, A. B., et al. (2020). Organochlorine pesticides in the soils of Cardamom Hill Reserve (CHR), Kerala, India: Geo spatial distribution, ecological and human health risk assessment. *Environmental Chemistry and Ecotoxicology, 2*, 1–11.

Kata, M., Srinivasa Rao, S., & Rama Mohan, K. (2015). Spatial distribution, ecological risk evaluation and potential sources of organochlorine pesticides from soils in India. *Environmental Earth Sciences, 74*, 4031–4038.

Khuman, S. N., Vinod, P. G., Bharat, G., Kumar, Y. M., & Chakraborty, P. (2020). Spatial distribution and compositional profiles of organochlorine pesticides in the surface soil from the agricultural, coastal and backwater transects along the south-west coast of India. *Chemosphere, 254*, 126699.

Kumarasamy, P., Govindaraj, S., Vignesh, S., Rajendran, R. B., & James, R. A. (2012). Anthropogenic nexus on organochlorine pesticide pollution: A case study with Tamiraparani river basin, South India. *Environmental Monitoring and Assessment, 184*, 3861–3873.

Kumari, K. (Ed.). (2021). *Persistent organic pollutants: Gaps in management and associated challenges* (1st ed.). CRC Press. https://doi.org/10.1201/9781003046806

Malik, A., Ojha, P., & Singh, K. P. (2009). Levels and distribution of persistent organochlorine pesticide residues in water and sediments of Gomti River (India)—A tributary of the Ganges River. *Environmental Monitoring and Assessment, 148*, 421–435.

Mamta, Rao, R. J., & Wani, K. A. (2015). Monitoring of organochlorine and organophosphorus pesticide residues in water during different seasons of Tighra reservoir Gwalior, Madhya Pradesh, India. *Environmental Monitoring and Assessment, 187*(11), 684.

Minimol, V. A., Anas, K. K., Anandan, R., & Mathew, S. (2017). Monitoring organochlorine pesticide residues in seaweeds from Mandapam coast, India. *FishTech Reporter, 3*, 7–9.

Mutiyar, P. K., Mittal, A. K., & Pekdeger, A. (2011). Status of organochlorine pesticides in the drinking water well-field located in the Delhi region of the flood plains of river Yamuna. *Drinking Water Engineering and Science, 4*(1), 51–60.

Ogbeide, O., Tongo, I., & Ezemonye, L. (2015). Risk assessment of agricultural pesticides in water, sediment, and fish from Owan River, Edo State, Nigeria. *Environmental Monitoring and Assessment, 187*, 1–16.

Revzin, A. M. (1970). *Some acute and chronic effects of endrin on the brain* (Vol. 70, No. 11). Department of Transportation, Federal Aviation Administration, Office of Aviation Medicine.

Rissato, S. R., Galhiane, M. S., Ximenes, V. F., De Andrade, R. M., Talamoni, J. L., Libânio, M., et al. (2006). Organochlorine pesticides and polychlorinated biphenyls in soil and water samples in the Northeastern part of São Paulo State, Brazil. *Chemosphere, 65*(11), 1949–1958.

Sankar, T. V., Zynudheen, A. A., Anandan, R., & Nair, P. V. (2006). Distribution of organochlorine pesticides and heavy metal residues in fish and shellfish from Calicut region, Kerala, India. *Chemosphere, 65*(4), 583–590.

Shegunova, P., Klánová, J., & Holoubek, I. (2007). Residues of organochlorinated pesticides in soils from the Czech Republic. *Environmental Pollution, 146*(1), 257–261.

Singare, P. U. (2016). Distribution and risk assessment of suspected endocrine-disrupting pesticides in creek water of Mumbai, India. *Marine Pollution Bulletin, 102*(1), 72–83.

Singh, K. P., Malik, A., & Sinha, S. (2007). Persistent organochlorine pesticide residues in soil and surface water of northern Indo-Gangetic alluvial plains. *Environmental Monitoring and Assessment, 125*, 147–155.

Sruthi, S. N., Shyleshchandran, M. S., Mathew, S. P., & Ramasamy, E. V. (2017). Contamination from organochlorine pesticides (OCPs) in agricultural soils of Kuttanad agroecosystem in India and related potential health risk. *Environmental Science and Pollution Research, 24*, 969–978.

Stockholm Convention. (2001). *Listing of POPs in Stockholm Convention*. Retrieved on September 7, 2022 from http://chm.pops.int/TheConvention/ThePOPs/ListingofPOPs/tabid/2509/Default.aspx

Toxics Link. (2018). *Persistent organic pollutants (POPs) in India: Country situation report*. Retrieved in October 18, 2022 from https://ipen.org/sites/default/files/documents/POPs%20Country%20Situation%20report.pdf

Van Esch, G. T., & van Heemstra-Lequin, E. A. H. (1992). *Environmental health criteria 130: Endrin*. World Health Organization.

Zhang, Z., Hong, H., Zhou, J. L., Yu, G., Chen, W., & Wang, X. (2002). Transport and fate of organochlorine pesticides in the River Wuchuan, Southeast China. *Journal of Environmental Monitoring, 4*(3), 435–441.

Zhang, Z. L., Hong, H. S., Zhou, J. L., Huang, J., & Yu, G. (2003). Fate and assessment of persistent organic pollutants in water and sediment from Minjiang River Estuary, Southeast China. *Chemosphere, 52*(9), 1423–1430.

Zhou, R., Zhu, L., Yang, K., & Chen, Y. (2006). Distribution of organochlorine pesticides in surface water and sediments from Qiantang River, East China. *Journal of Hazardous Materials, 137*(1), 68–75.

Heptachlor

Kanchan Kumari and Deepak Marathe

1 Introduction

Persistent organic compounds (POPs) are the group of chemicals that persist in the environment for longer period of time, toxic to human health and environment, bioaccumulative in nature, and have long range transport, i.e., those that can migrate longer regions from which they have been initially applied (She et al., 2007). The persistency of these chemicals due to high resistance towards chemical as well as biological degradability, and they are highly bioaccumulative and carcinogenic in nature (Toxic) due to which they have been listed in the Stockholm Convention. Stockholm Convention listed number of chemicals in the category of POPs some of them are heptachlor, chlordane, endrin, aldrin, dieldrin, toxaphene, DDT, mirex, and hexachlorbenzene which are banned for restriction on use as well as production (Grimalt et al., 2001; Arbeli, 2009). There are number of studies that represent the occurrence of these chemicals in air, water, and soil/sediments (Lupi et al., 2016). For example, traces of the organic pollutants have been reported in the water bodies of Mexico (Mainly in coastal regions) which imparting the major detrimental impact on human health and environment (García et al., 2012; Waliszewski et al., 2013; Arellano et al., 2017; Leyva-Morales et al., 2017). In addition, reports are available for the banned pesticides such as DDT, endosulfan, and lindane for persistency and its occurrence for longer period of time. The major reason for the occurrence of these organochlorine pesticides in coastal regions is due to high applicability/

K. Kumari (✉)
CSIR-National Environmental Engineering Research Institute (NEERI),
Kolkata Zonal Center, Kolkata, West Bengal, India
e-mail: k_kumari@neeri.res.in

D. Marathe
CSIR-National Environmental Engineering Research Institute (NEERI), Nagpur, India

K. Kumari, *Pollutants of Global Concern*, Emerging Contaminants and Associated Treatment Technologies, https://doi.org/10.1007/978-3-031-50996-4_6

extensive use of these pesticides in agriculture and they are being deposited to coastal regions due to agricultural runoff through air and water as a medium (Albert et al., 2005, 2014; Ramírez-Elías et al., 2016). In addition to address the high biocuumulative properties of these chemicals, number of steps have been taken place. Consequently, these chemicals are accumulating in soil, water, and sediments where they are highly accumulated in benthic invertebrates, which is the major component of human health as it is the major source of aquatic marine fishes (Leyva-Cardoso et al., 2003; Aguilar et al., 2017). Based on the literature review, it was found that traces of these POPs are abundant in sediments than water as it acts as a sink for accumulation of all types of pollutants (Aguilar et al., 2017; Villanueva et al., 1995; Carvalho et al., 2002). Similarly, heptachlor is one of the known pesticides which was introduced and adopted for use as an insecticide in the year 1952 and its was first isolated from chlordane in the year 1946. There are various brands of heptachlor pesticides, which includes mostly 20% of chlordane and 72% of heptachlor and 28% of other related compounds or products (IARC, 1991a). The heptachlor is highly persistent and have high bioaccumulative properties that can disperse in the environment in longer regions through air. In addition, heptachlor is also produced in animals as heptachlor epoxide and the toxicity of heptachlor epoxide is similar to that of heptachlor, which is majorly found in animal fats. The synthesis of the heptachlor is similar to that of syntheses of chlordane in which hexachlorocyclopentadiene is reacted with cyclopentadiene to from chlordane. Heptachlor is prepared by the free-radical chlorination of chlordane (IARC, 1991a). It is mostly used as insecticide for termination of soil pests and termites (WHO, 1984b). In addition, it has been applied as fighting agent against crop pest, grasshoppers, and cotton bugs (Smith, 1991). It was also being identified as anti-malaria (IARC, 1991a). According to Keith (1997), heptachlor epoxide is the most stable substance present in the environment. It has a strong affinity for binding soil and is not biodegradable in the natural world. The great attraction of these heptachlor epoxides to accumulate in meat, fish, and poultry also contributes to their high likelihood to enter the food chain (USEPA, 1986). The subterranean layer and cultivable layer of the soil profile both contain the organochlorine pesticide (Gupta et al., 2012). In addition, these pollutants, which linger in the air, water, soil, and sediments for a long time, have contaminated many areas of India (Agrawal et al., 2010). In the year 1952, the United States began using heptachlor as an insecticide. Heptachlor was applied to maize crops in the USA in the year 1970 at a rate of 58%, followed by pest control at 27%, seed treatment at 13%, and other uses at 2% for suppressing fire ants and other insects (Environmental Protection Agency, 1976). In 1970, Europe used heptachlor at a rate of 60%, followed by South America (15%), Africa (5%), Asia (15%), Canada (5%), and the United States (5%) (WHO, 1984b). In 1986, it was noted that the average amount of heptachlor used to control termites in the United States was less than 340 tons annually (Environmental Protection Agency, 1987). Since the Agricultural Chemicals Regulation Law's authorization for its commercial manufacture and sale as an agricultural pesticide was removed in 1975, heptachlor has been prohibited. In 1986, "Chlordane," which also contained heptachlor, were designated as "Class 1 Specified Chemical Substances" under the Chemical Substances Control Law. As a result, the production and use of heptachlor were practically prohibited. As a result, Stockholm put heptachlor and its compounds in category appendix A in 2001, and India has prohibited its manufacturing, import, and use since September 20, 1996. Thirty-three tonnes of heptachlor were used annually on average in Korea from 1962 to 1979 (Lee, 1982). The usage of heptachlor has been severely limited in many countries due to the adverse effects it has on both human health and the environment. According to statistics and research, heptachlor and its derivatives were mostly detected in Indian agricultural waste, poultry, crops, soil, sediments, and water. Many nations, including Turkey, Cyprus, Ecuador, the EU, Portugal, Singapore, and Sweden have outlawed the use of heptachlor. In other nations, such as Philippines, Denmark, Canada, Finland, Israel, Japan, New Zealand, Austria, Czechoslovakia, Argentina, USSR, and the USA, it is strictly banned.

2 Technical Details of Heptachlor

Heptachlor is organochlorine pesticide with IUPAC number 1,5,7,8,9,10,10-heptachlorotricyclo [5.2.1.02,6] deca-3,8-diene and molecular weight $C_{10}H_5C_{17}$. Heptachlor occurs in nature as white crystals or tan waxy solid with camphor like odor. The octanol/water partition coefficient (Log k_{OW}) of heptachlor is 6.10. The heptachlor is acute toxic and causes various human health hazards and environmental hazards. Although it decomposes at slow rate, on decomposition, it releases some of the toxic gases which include hydrogen, carbon, chlorine, and carbon monoxide, respectively. Some of the chemical properties of heptachlor are represented in Table 1 which represents the toxicity and non-biodegradability and bioaccumulative properties of heptachlor.

3 Status of Works and Usage in India

In India, heptachlor and its derivatives were once extensively employed as insecticides and pesticides to combat pests, insects, termites, and malaria-carrying mosquitoes (Pozo et al., 2012). However, recognizing their persistence, capacity for long-range transport, high toxicity, bioaccumulative properties, and carcinogenic potential in the environment, India enacted legislation in 1996 to restrict their production and usage. Despite the ban in 1996, traces and residues of this pesticide still endure in soil, water, sediments, plants, and wildlife.

Heptachlor made its debut in the United States in 1952, initially serving as a widely used insecticide. It also found application in structural uses and was employed to combat malaria. Heptachlor is categorized as a non-systemic internal and contact insecticide (WHO, 1984b). In the mid-1970s, the distribution of heptachlor usage in agriculture included maize (58%), pest control operators (27%), seed treatment (13%), and miscellaneous applications like fire ant control (2%).

Table 1 Chemical properties of heptachlor (NCBI, 2023)

Details	Description (heptachlor)
Structure	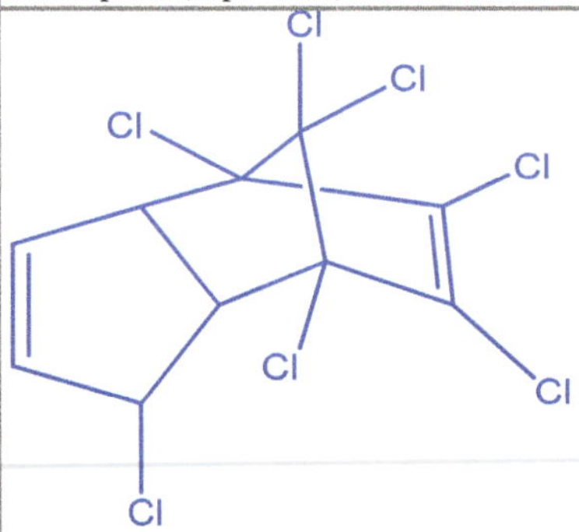
Molecular formula	$C_{10}H_5Cl_7$
CAS No.	76–44-8
IUPAC name	1,5,7,8,9,10,10-Heptachlorotricyclo[5.2.1.$0^{2,6}$]deca-3,8-diene
Molecular weight	373.32 g/mol
Odor	Camphor-like odor
Color	White crystals or tan waxy solid with characteristic odor
Boiling point	310.0 °C
Melting point	95.5 °C
Stability/shelf life	Stable to daylight, air, moisture, and moderate heat (up to 160 °C)
Vapor pressure kPa	3.99×10^{-5} at 25 °C
Solubility	In water, 0.00018 g/l at 25 °C; 0.00010 g/L at 15 °C
Octanol/water partition coefficient	Log k_{OW} = 6.10 and 6.13
Density	1.6 g/cm^3
Octanol-air partition coefficient	Log k_{OA} = 7.64
Henry's law constant	2.94e-04 atm-m^3/mole; 29.75 Pa m^3/mol
Viscosity	–
Toxicity	Toxic to human and environment
Safety and hazards	Acute toxic, health hazard and environmental hazard
Decomposition	Decomposition products: Toxic gases and vapors which include hydrogen chloride and carbon monoxide
Corrosivity	Corrosive to metals

Additionally, it was sprayed on crops such as pineapples and citrus (Environmental Protection Agency, 1976).

Globally, heptachlor utilization was distributed as follows: Europe 60%, Asia 15%, Africa 5%, USA 5%, America 15%, Canada 5%, and South America (WHO, 1984b). Consequently, heptachlor has faced bans in numerous countries (WHO, 1984b). In 1986, approximately less than 340 tons of heptachlor were used for termite control in the USA (Environmental Protection Agency, 1987a). In September 1996, heptachlor was banned for use in India (Toxics link, 2018). The Indian government had taken measures to phase out and prohibit the production, import, sale,

Table 2 Uses, sources, and regulation in India (Kumari, 2021)

Identified POP	Uses and sources	Annex	Year of listing by SC	Regulation in India
Heptachlor	Previously used for agricultural purposes, seed treatment, wood protection, and insect control	A	2001	The use, production, import, and sale of heptachlor and its formulations were prohibited with effect from 20 September 1996 under the Insecticides Act, 1968. This decision was taken due to its high toxicity and persistence in the environment, which poses significant risks to human health and the ecosystem

and use of heptachlor and other persistent organic pollutants (POPs) under the Stockholm Convention on Persistent Organic Pollutants. The Stockholm Convention is an international treaty aimed at protecting human health and the environment from POPs, which include substances like heptachlor, DDT, and PCBs (polychlorinated biphenyls). Various uses, sources, and regulation of heptachlor in India are represented in Table 2. In addition, effects, toxicity, and persistence of heptachlor are represented in Table 3. India is a signatory to this convention and has committed to implementing its provisions to reduce and eliminate the use of these hazardous chemicals.

In India, there is banning on its production and usage since from 20 September 1996; in addition, the stockpile is not reported in India (IPEN POPs, 2018).

4 International Data Versus Indian Scenario

Based upon the data of National Occupational Exposure Survey, there are approximately 1000 workers in the USA who were exposed to heptachlor (Heikkilä & Kauppinen, 1992). Similarly, around 200 workers were affected as a result of heptachlor exposure, which was utilized in specific glues during plywood manufacture in tropical countries (Mussalo-Rauhamaa, 1991). Based upon the facts, pesticide application to the subterranean termites in the USA accounts for exposure of 17 μg/m^3 of chlordane (range: 0.6–116 μg/m^3) and 33 μg/m^3 of heptachlor (range: 2.0–176 μg/m^3). In addition, the exposure of these pesticides has been monitored by measuring the concentration in sterile gauze pads for dermal and inhalation. The results showed that, the concentration of heptachlor in dermal exposure and inhalation process was found to be 1.8 μg/kg bw per h (140 μg/h) and 0.08 μg/kg bw per h (6 μg/h) respectively (Kamble et al., 1992).

Kauppinen et al. (1986), reported the heptachlor concentration in air samples of Finnish plywood mills. Significant concentration of heptachlor was detected in assembling (<10–140 μg/m^3), hot pressing (1–50 μg/m^3), patching of veneers (2–10 μg/m^3), and glue preparation (620 μg/m^3) respectively. In addition, blood samples of these workers was reported 0.30 ng/g of heptachlor and 19.2 ng/g of heptachlor epoxide (Mussalo-Rauhamaa, 1991).

Table 3 Effects, toxicity, and persistence of heptachlor

Heptachlor	Data of India
Effects	The health effects of heptachlor exposure on humans in the Indian scenario could be similar to those observed in other regions where heptachlor has been used. These effects may include acute poisoning, neurotoxicity, reproductive and developmental effects, liver toxicity, and carcinogenic potential (Arisekar et al., 2021). Impedes reproductive potentials in humans
Toxicity	Toxic to wildlife at low concentrations. In birds, it induces behavioral changes, reduced reproductive success, and mortality. It is listed as a possible human carcinogen. The main exposure is through food. MRLs by FAO/WHO are 0.006 mg/kg milk fat and 0.2 mg/kg meat or poultry fat
Persistence/ fate	Heptachlor is a persistent organic pollutant (POP), which means it resists degradation in the environment and can persist for long periods. The persistence and fate of heptachlor depend on several factors, including environmental conditions, chemical properties, and the presence of organisms capable of breaking it down. Here's an overview of the persistence and fate of heptachlor:
	Environmental persistence: Heptachlor is relatively stable in the environment, and its half-life can vary depending on factors like temperature, moisture, and soil type. In soil, heptachlor can have a half-life ranging from months to several years, meaning it can take a long time to degrade naturally
	Bioaccumulation: Heptachlor has the ability to bioaccumulate in living organisms, particularly in the fatty tissues. This phenomenon occurs as animals consume food containing heptachlor residues, and the compound accumulates in their bodies. As predators consume smaller animals, the concentration of heptachlor can increase, leading to biomagnification up the food chain
	Biotransformation: Heptachlor can undergo biotransformation, where it is converted into its more toxic and persistent metabolite, heptachlor epoxide. This transformation occurs in both the environment and living organisms and contributes to its persistence
	Environmental transport: Heptachlor can be transported over long distances by air and water currents. It has been detected in regions far away from its original application sites, demonstrating its potential for global distribution through atmospheric transport
	Degradation: Heptachlor can undergo various degradation processes, such as photodegradation and microbial degradation. Photodegradation occurs when heptachlor is exposed to sunlight, leading to the breakdown of the compound. Microbial degradation involves the action of certain microorganisms that can metabolize and break down heptachlor. However, these processes are relatively slow, and heptachlor's persistence remains a concern
	Environmental concerns: Due to its persistence and ability to bioaccumulate, heptachlor poses significant environmental concerns. It can impact ecosystems, wildlife, and human health even in areas where its use has been discontinued for years
	The persistence of heptachlor and its potential to cause harm to the environment and living organisms led to its inclusion in the Stockholm Convention on persistent organic pollutants. The convention aims to regulate and eliminate the production and use of such chemicals, including heptachlor, to protect human health and the environment (Krithiga et al., 2022)
	Metabolized in soils, plants, and animals to heptachlor epoxide, which is more stable and carcinogenic. It bio concentrates and has a soil half-life of 0.75–2 years in temperate climates

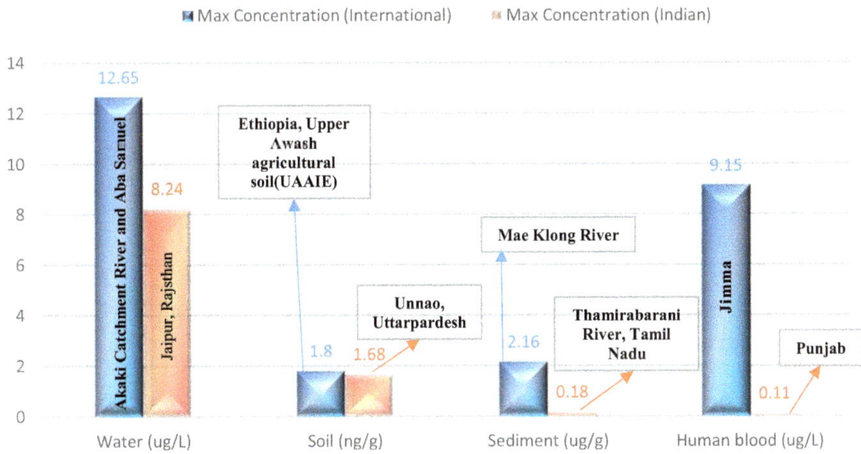

Fig. 1 International vs Indian data on heptachlor concentration amongst different environmental sectors. (Source: Mesfin, 2019; Srimurali et al., 2015; Kassegne et al., 2020; Bakore et al., 2004; Westbom et al., 2008; Singh et al., 2007; Pokethitiyook & Poolpak, 2012; Arisekar et al., 2019; Mekonen et al., 2021; Mathur et al., 2005)

Various environmental monitoring studies/survey were conducted for monitoring the traces of Heptachlor in air, water, sediments, and aquatic animals (fish). In the year 1982, heptachlor was monitored in air. Similarly, the persistency of heptachlor was monitored in environmental matrix in the year 1986 showed that heptachlor was detected 0.2–3.7 ng/g (dry weight) in sediments and 0.001–0.010 g/g (wet weight). In addition, the frequency of heptachlor in sediment and fish was found to be 14 out of 87 sediment samples and 9 out of 110 fish samples respectively. The heptachlor concentration was determined in urban and agricultural passive air samples of Tamil Nadu was ND-41400 pg/m^3 in the year 2009–2010 (Srimurali et al., 2015). High concentration of heptachlor was reported in the vegetables harvested at Jaipur Rajasthan. The concentration of heptachlor and heptachlor epoxide was reported to be 1.5 to16 mg/kg in tomatoes, cabbage, spinach, cauliflower, and okra (Bakore et al., 2002). In addition one study also reported the concentration of heptachlor in buffalo milk samples (John et al., 2001). Moreover, heptachlor concentration was also reported in the human blood of breast cancer patients of Jaipur city of India (Mathur et al., 2002). International vs Indian data on heptachlor concentration amongst different environmental sectors is given in Fig. 1.

5 Adverse Effect of Heptachlor

Heptachlor is a highly toxic pesticide and persistent organic pollutant (POP) that can have a range of adverse effects on human health, wildlife, and the environment (Qing Li et al., 2006). Some of the major adverse effects of heptachlor include:

5.1 *Human Health Effects*

5.1.1 Acute Poisoning

In cases of acute exposure to high levels of heptachlor, individuals can experience symptoms such as nausea, vomiting, abdominal pain, tremors or shakiness, convulsions, and seizures. Severe poisoning can be life threatening.

Carcinogenic Potential: Heptachlor has been classified as a possible human carcinogen by the International Agency for Research on Cancer (IARC). This means that there is limited evidence suggesting its potential to cause cancer in humans, particularly liver cancer and non-Hodgkin's lymphoma.

5.1.2 Neurotoxicity

Chronic exposure to heptachlor can lead to neurotoxic effects on the nervous system, resulting in neurobehavioral changes, cognitive deficits, and neurological disorders.

Reproductive Effects: Heptachlor exposure may disrupt the reproductive system, leading to hormonal imbalances and reproductive disorders.

5.2 *Wildlife and Environmental Effects (Wurl & Obbard, 2005)*

5.2.1 Bioaccumulation

Heptachlor can bioaccumulate in the fatty tissues of animals, leading to higher concentrations as it moves up the food chain. Predators at the top of the food web can experience significant toxic effects due to bio magnification.

5.2.2 Impact on Wildlife Populations

Heptachlor's toxicity can have devastating effects on wildlife populations, particularly birds and aquatic organisms. It can lead to population declines and disruptions in ecosystems.

5.3 *Environmental Persistence*

Heptachlor is resistant to degradation, making it persist in the environment for long periods and posing long-term risks to various organisms. Based upon the published studies, toxicity and its fate in human and animal due to heptachlor are reported in Tables 4 and 5.

Table 4 Toxicological effects of heptachlor exposure (laboratory studies)

	Toxicological effects	Particular	Concentration/interventions	Reference
1	Acute toxicity	Rats (Oral) (LD_{50})	100 to 220 mg/kg	Smith (1991) and Agency for Toxic Substances and Disease Registry (1989)
		Mice (Oral) (LD_{50})	30 to 68 mg/kg	
		Guinea pigs (Oral) (LD_{50})	116 mg/kg	
		Hamsters (Oral) (LD_{50})	100 mg/kg	
		Chicken (Oral) (LD_{50})	62 mg/kg	
		Rats (dermal exposure) (LD_{50})	119 to 320 mg/kg	Kidd and James (1991)
		Rabbit (dermal exposure) (LD_{50})	2000 mg/kg	
2	Chronic toxicity	Rats (1.5 mg/kg/day over 2 years) (Oral)	Mortality of rats increased	Agency for Toxic Substances and Disease Registry (1989)
		Dogs (0.1 mg/kg/day for 60 days) (Oral)	No effect	Smith (1991) and Agency for Toxic Substances and Disease Registry (1989)
		Rats (0.35 mg/kg/day in rats over a 50-week period) (Oral)	Liver damage	Agency for Toxic Substances and Disease Registry (1989)
3	Reproductive effects	Rats (0.25 mg/kg/day heptachlor for 60 days)	Decreased postnatal survival was reported in the progeny of rats	Agency for Toxic Substances and Disease Registry (1989)
		Rats (6.9 mg/kg/day for 3 days)	Decrease in fertility in rats and reduced survival by 1/3rd of young one	Agency for Toxic Substances and Disease Registry (1989)
4	Teratogenic effects	Rats, rabbits, chickens, and beagle dogs (20 mg/kg/day on days 7 to 17)	No teratogenic effects were observed	WHO (1984a)
		Rats (5 mg/kg/day over 3 days)	Developmental abnormalities	Smith (1991) and Agency for Toxic Substances and Disease Registry (1989)

(continued)

Table 4 (continued)

	Toxicological effects	Particular	Concentration/interventions	Reference
5	Mutagenic effects	–	Laboratory tests indicate that neither heptachlor nor heptachlor epoxide are mutageni	Agency for Toxic Substances and Disease Registry, (1989) and WHO (1984a)
6	Carcinogenic effects	Rats (1.2 mg/kg/day)	Incidence of liver carcinomas	Smith (1991) and Agency for Toxic Substances and Disease Registry (1989)

Table 5 Fate in human and animal due to heptachlor exposure

Fate	Response due to heptachlor exposure	Reference
Fate in humans and animals	Heptachlor is readily taken up through the skin, lungs, and gastrointestinal tract	Agency for Toxic Substances and Disease Registry (1989)
	Once absorbed, it is systemically distributed and moves into body fat	Smith (1991) and Agency for Toxic Substances and Disease Registry (1989)
	In mammals, heptachlor is readily converted to its most persistent and toxic metabolite, heptachlor epoxide, in the liver	Smith (1991) and Agency for Toxic Substances and Disease Registry (1989)
	Heptachlor epoxide is stored mainly in fatty tissue, but also in liver, kidney, and muscle tissues	Agency for Toxic Substances and Disease Registry (1989)
	Heptachlor is able to cross the placenta and has been found in human milk. Intense activity or starvation may mobilize the pesticide as body fat is burned, resulting in the reappearance of toxic symptoms long after uptake from the environment	WHO (1984a)
	Heptachlor is generally not detectable in the human population, but heptachlor epoxide has been found in human fat, blood, organs, and milk. In localities where heptachlor was used regularly, it has been found at higher concentrations in human milk than in dairy milk	Solomon and Weiss (2002)
	Rats retained 77% of heptachlor that they inhaled during a 30-min period	Kidd and James (1991)

5.4 *Global Distribution*

Heptachlor is capable of long-range atmospheric transport, leading to its distribution in regions far from its original application sites (WHO, 2003). This global distribution can affect areas where heptachlor has not been used directly.

Because of its dangerous properties and the possibility of detrimental impacts on both human health and the environment, heptachlor has been prohibited or closely controlled in numerous nations, including India. These actions align with international agreements such as the Stockholm Convention on Persistent Organic Pollutants. Proper waste management, monitoring, and the use of safer alternatives are essential to mitigate the adverse effects of heptachlor and prevent further environmental contamination.

6 Management and Remedial Measures

Management and remedial strategies for heptachlor are centered on averting exposure, diminishing environmental pollution, and confronting extant contamination. These tactics encompass a blend of regulatory initiatives, public education, and environmental restoration endeavors. Below are essential approaches for the management and remediation of heptachlor:

6.1 *Ban and Regulation*

As a first step, heptachlor should be banned or strictly regulated to prevent its production, import, sale, and use. This can be achieved through national regulations and adherence to international agreements like the Stockholm Convention on Persistent Organic Pollutants.

One of the most notable international agreements that led to the ban of heptachlor is the Stockholm Convention on Persistent Organic Pollutants. The Stockholm Convention, adopted in 2001 and in force since 2004, is a global treaty aimed at eliminating or restricting the production, use, and release of POPs. Heptachlor is included in Annex A of the convention, which lists specific POPs subject to elimination. Annex A chemicals, including heptachlor, are to be phased out entirely with limited exemptions allowed. Several countries, including India, have enacted domestic regulations to ban heptachlor and other POPs. In India, heptachlor was banned under the Insecticides Act, 1968. The Indian government took this step to protect human health and the environment from the harmful effects of heptachlor exposure. The banning of heptachlor reflects the growing awareness of the dangers posed by persistent organic pollutants and the commitment of the international community to take collective action to safeguard the planet and its inhabitants. It also underscores the importance of finding safer alternatives and adopting sustainable pest management practices to reduce the reliance on hazardous chemicals like heptachlor.

The Insecticides Act, 1968 (46 of 1968) authorized the establishment of an expert panel by the Indian government to look into the usage of heptachlor and any potential dangers to the health of people and animals. The committee suggested limiting the use of the possibly dangerous chemical heptachlor to the registration committee. On September 20, 1996, the registration committee issued an order fully outlawing the manufacturing, use, import, and export of heptachlor in India (IPEN POPs, 2018).

6.2 *Safer Alternatives*

Promoting the utilization of safer and environmentally sustainable alternatives to heptachlor in pest control is of utmost importance. Encouraging the adoption of integrated pest management (IPM) practices presents a viable path to diminishing our dependence on chemical pesticides. In this regard, several alternatives to heptachlor not only effectively combat pests but also present fewer hazards to both human health and the ecosystem. By embracing these alternatives, we can safeguard the environment while ensuring efficient pest management. Some of the commonly used safer alternatives for agricultural sector are presented in Fig. 2.

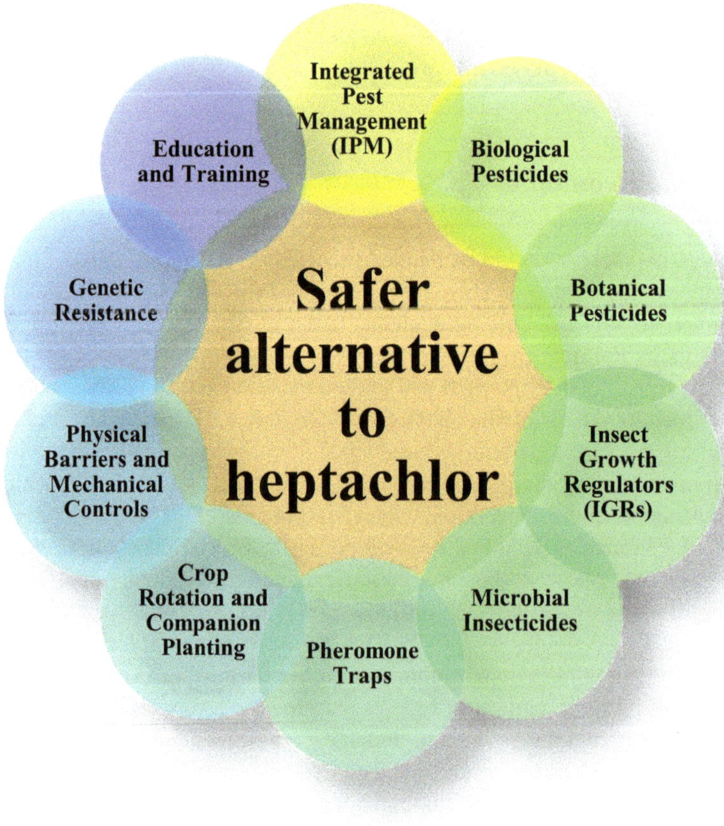

Fig. 2 Safer alternative to heptachlor in terms of agriculture sector

Regularly monitoring environmental media serves as a crucial tool in pinpointing pollution hotspots and assessing the effectiveness of control measures. It is equally vital to educate the public about the hazards associated with heptachlor and potential sources of exposure. By raising awareness, unintentional exposure can be mitigated, particularly amongst farmers, pesticide workers, and the general population, who must be well-informed about proper pesticide handling and disposal practices.

Implementing appropriate waste management protocols for products and waste materials containing heptachlor is paramount. Disposal must strictly adhere to all applicable regulations to prevent any discharge into the environment. In areas previously impacted by heptachlor, environmental cleanup measures may be necessary. Techniques such as soil excavation, bioremediation, and phytoremediation can effectively reduce heptachlor levels, aiding in the restoration of polluted sites and contributing to ecological recovery.

International cooperation is indispensable, as heptachlor and similar persistent organic pollutants can traverse great distances. Enhancing global management of heptachlor entails sharing information, best practices, and collaborative initiatives. Valuable insights from ongoing research into heptachlor's fate in the environment and its implications for human health can guide decision-makers. Continuous monitoring is essential to assess the impact of management efforts and to detect any resurgence of heptachlor contamination.

In order to effectively implement these management and remediation measures, it is imperative for governments, regulatory bodies, environmental agencies, industry stakeholders, and communities to collaborate closely. A unified strategy, backed by sustained commitment, is imperative to shield human health and the environment from the detrimental effects of heptachlor. Figure 3 represents the other safer alternatives of heptachlor.

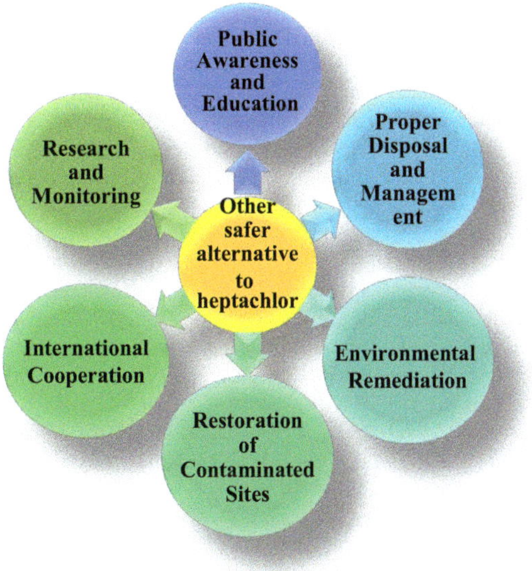

Fig. 3 Other safer alternative for heptachlor management

7 Conclusion

Heptachlor stands as a persistent organic pollutant (POP) that once found extensive use as an insecticide for pest control in India and numerous other nations. Nevertheless, owing to its detrimental impact on both the environment and human health, India and many other countries have imposed a ban on its usage. This ban in India aligns with global initiatives under the Stockholm Convention on Persistent Organic Pollutants, which aims to phase out and limit the application of hazardous chemicals posing substantial risks to the environment and human well-being. India banned and restricted the usage of heptachlor in September, 1996. Adhering to these regulations and adopting safer alternatives remains paramount in preserving the environment, human health, and the overall welfare of the population.

References

Agency for Toxic Substances and Disease Registry. (1989). *Toxicological profile for heptachlor epoxide.*

Agrawal, A., Pandey, R. S., & Sharma, B. (2010). Water pollution with special reference to pesticide contamination in India. J. *Water Resource and Protection, 2, 432–448.*

Albert, L. A., & Benítez, J. A. (2005). Impacto ambiental de los plaguicidas en los ecosistemas costeros. In Golfo de México Contaminación e Impacto Ambiental: Diagnóstico y Tendencias, 2nd ed.; Botello, A.V., Rendón von Osten, J., Gold Bouchot, G., Hernández, C.A., Eds.; Univ. Nal. Autón. de México, Instituto Nacional de Ecología:de Campeche, Mexico, pp. 157–176, ISBN 968-5722-37-4.

Albert, L. A. (2014). Los plaguicidas y sus riesgos para el ambiente. In Botello, A. V., Rendón von Osten, J., Benítez, J., Gold-Bouchot, G., Eds. Golfo de México Contaminación e Impacto Ambiental: Diagnóstico y Tendencias, 3rd ed.; La Universidad Nacional Autónoma de México (UNAM): de la Ciudad de México, Mexico, pp. 183–212, ISBN 978-607-7887-71-3

Arbeli, Z. (2009). Biodegradación de compuestos orgánicos persistentes (COP): I. El Caso de los Bifenilos Policlorados (PCB). *Acta Biológica Colombiana, 14, 57–88.*

Arellano-Aguilar, O., Betancourt-Lozano, M., Aguilar-Zárate, G., & Ponce de León-Hill, C. (2017). Agrochemical loading in drains and rivers and its connection with pollution in coastal lagoons of the Mexican Pacific. *Environmental Monitoring and Assessment, 189, 270.*

Arisekar, U., Shakila, R. J., Jeyasekaran, G., Shalini, R., Kumar, P., Malani, A. H., & Rani, V. (2019). Accumulation of organochlorine and pyrethroid pesticide residues in fish, water, and sediments in the Thamirabarani river system of southern peninsular India. *Environmental Nanotechnology, Monitoring & Management, 11*, 100194.

Arisekar, U., Shakila, R. J., Shalini, R., & Jeyasekaran, G. (2021). Pesticides contamination in the Thamirabarani, a perennial river in peninsular India: The first report on ecotoxicological and human health risk assessment. *Chemosphere, 267*, 129251.

Bakore, N., John, P. J., & Bhatnagar, P. (2002). Evaluation of organochlorine insecticide residue levels in locally marketed vegetables of Jaipur City, Rajasthan, India. *Journal of Environmental Biology, 23(3), 247–252.*

Bakore, N., John, P. J., & Bhatnagar, P. (2004). Organochlorine pesticide residues in wheat and drinking water samples from Jaipur, Rajasthan, India. *Environmental Monitoring and Assessment, 98*, 381–389.

Carvalho, F. P., Gonzalez-Farias, F. Villeneuve, J. P., Cattini, C., Hernandez-Garza, M., & Mee, L. D. (2002). Distribution, fate and effects of pesticide residues in tropical coastal lagoons of Northwestern Mexico. *Environmental Toxicology, 23, 1257–1270.*

Environmental Protection Agency. (1976). Velsicol Chemical Co., consolidated heptachlor/chlordane hearing. *Federal Register, 41, 7552–7572.*

Environmental Protection Agency. (1987a). Heptachlor and Heptachlor Epoxide, Washington DC, Office of Drinking Water.

García-Gutiérrez, C.; Rodríguez-Meza, G.D. Problemática y riesgo ambiental por el uso de plaguicidas en Sinaloa. Ra Ximhai 2012, 8, 1–10.

Grimalt, J.; Fernández, P.; Berdie, L.; Vilanova, R.M.; Catalán, J.; Psenner, R.; Hofer, R.; Appleby, P.G.;Rosseland, B.O.; Lien, L.; et al. Selective trapping of organochlorine compounds in mountain lakes of temperate areas. *Environ. Sci. Technol. 2001, 35, 2690–2697*

Gupta M,Garg NK,Joshi H,Sharma MP (2012) Persistence and mobility of 2, 4-D in unsaturated soil zone under winter wheat crop in subtropical region of India. *Agriculture, Ecosystems Environment,146, 60–72.*

Heikkilä, P., & Kauppinen, T. (1992). Occupational exposure to carcinogens in Finland. *American Journal of Industrial Medicine, 21*(4), 467–480.

IARC. (1991a). Chlordane and heptachlor. In: IARC Monographs on the Evaluation of Carcinogenic Risks to Humans, Volume 53. Occupational exposures in insecticide application, and some pesticides. International Association for Research on Cancer, Lyon, France. pp. 115–178.

IPEN, POPs.Link, T. (2018). Persistent *organic pollutants (POPs) in India: Country situation report.*

John, P. J., Bakore, N., & Bhatnagar, P. (2001). Assessment of organochlorine pesticide residue levels in dairy milk and buffalo milk from Jaipur City, Rajasthan, India. *Environment International, 26*(4), 231–236.

Kamble, S. T., Ogg, C. L., Gold, R. E., & Vance, A. D. (1992). Exposure of applicators and residents to chlordane and heptachlor when used for subterranean termite control. *Archives of Environmental Contamination and Toxicology, 22*, 253–259.

Kassegne, A. B., Okonkwo, J. O., Berhanu, T., Daso, A. P., Olukunle, O. I., & Asfaw, S. L. (2020). Ecological risk assessment of organochlorine pesticides and polychlorinated biphenyls in water and surface sediment samples from Akaki River catchment, Central Ethiopia. *Emerging Contaminants, 6*, 396–404.

Kauppinen, T. P., Partanen, T. J., Nurminen, M. M., Nickels, J. I., Hernberg, S. G., Hakulinen, T. R., et al. (1986). Respiratory cancers and chemical exposures in the wood industry: A nested case-control study. *Occupational and Environmental Medicine, 43*(2), 84–90.

Kidd, H., & James, D. R. (1991). *The agrochemical handbook* (3rd ed.). The royal Society of Chemistry.

Krithiga, T., Sathish, S., Renita, A. A., Prabu, D., Lokesh, S., Geetha, R., et al. (2022). Persistent organic pollutants in water resources: Fate, occurrence, characterization and risk analysis. *Science of the Total Environment, 831*, 154808.

Kumari, K. (Ed.). (2021). *Persistent organic pollutants: Gaps in management and associated challenges* (1st ed.). CRC Press. https://doi.org/10.1201/9781003046806

Lee, S.-R. (1982) [Overall assessment of organochlorine insecticide residues in Korean foods.]Korean J. Food Sci. Technol., 14, 82–93

Leyva-Cardoso, D. O., Ponce-Vélez, G., Botello, A. V., & Díaz-González, G. (2003). Persistent organochlorine pesticides in coastal sediments from Petacalco Bay, Guerrero, México. *Bulletin of Environmental Contamination and Toxicology, 71, 1244–1251.*

Leyva-Morales, J. B., Valdez-Torres, J. B., Bastidas-Bastidas, P. J., Angulo-Escalante, M. A., Sarmiento-Sánchez, J. I., Barraza-Lobo, A. L., Olmeda-Rubio, C., Chaidez-Quiroz, C. (2017). Monitoring of pesticides residues in northwestern Mexico rivers. *Acta University, 27, 45–54.*

Lupi, L., Bedmar, F., Wunderlin, D. A. (2016). Miglioranza, K.S.B. Organochlorine pesticides in agricultural soils and associated biota. *Environmental Earth Sciences, 75, 519.*

Mathur, V., Bhatnagar, P., Sharma, R. G., Acharya, V., & Sexana, R. (2002). Breast cancer incidence and exposure to pesticides among women originating from Jaipur. *Environment International, 28*(5), 331–336.

Mathur, H. B., Agarwal, H. C., Johnson, S., & Saikia, N. (2005). *Analysis of pesticide residues in blood samples from villages of Punjab* (CSE Report, India, pp. 1–15).

Mekonen, S., Ibrahim, M., Astatkie, H., & Abreha, A. (2021). Exposure to organochlorine pesticides as a predictor to breast cancer: A case-control study among Ethiopian women. *PLoS One, 16*(9), e0257704.

Mesfin, K. (2019). Organochlorine residues determination in the air of Zeway Town, Ethiopia. *International Journal of Environmental Sciences & Natural Resources, 20*, 138–140. https://doi.org/10.19080/IJESNR.2019.20.556046

Mussalo-Rauhamaa, H. (1991). Partitioning and levels of neutral organochlorine compounds in human serum, blood cells, and adipose and liver tissue. *Science of the Total Environment, 103*(2–3), 159–175.

National Academies of Sciences, Engineering, and Medicine. (2003). *Gulf war and health: Volume 2: Insecticides and solvents*. The National Academies Press. https://doi.org/10.17226/10628

National Center for Biotechnology Information. (2023). PubChem Compound Summary for CID 3589, *Heptachlor*. Retrieved October 13, 2023 from https://pubchem.ncbi.nlm.nih.gov/compound/Heptachlor

NCBI, https://pubchem.ncbi.nlm.nih.gov/compound/Heptachlor accessed on 2023.

Pokethitiyook, P., & Poolpak, T. (2012). Heptachlor and its metabolite: Accumulation and degradation in sediment. *Pesticides-Recent Trends in Pesticide Residue Assay*, 217–252.

Pozo, K., Harner, T., Rudolph, A., Oyola, G., Estellano, V. H., Ahumada-Rudolph, R., et al. (2012). Survey of persistent organic pollutants (POPs) and polycyclic aromatic hydrocarbons (PAHs) in the atmosphere of rural, urban and industrial areas of Concepcion, Chile, using passive air samplers. *Atmospheric Pollution Research, 3*(4), 426–434.

Qing Li, Q., Loganath, A., Seng Chong, Y., Tan, J., & Philip Obbard, J. (2006). Persistent organic pollutants and adverse health effects in humans. *Journal of Toxicology and Environmental Health, Part A, 69*(21), 1987–2005.

Ramírez-Elías, M. A., Córdova-Quiroz, A. V., Cerón-Bretón, J. G., Cerón-Bretón, R. M., Rendón-von Osten, J., & Cortés-Simón, J. H. (2016). Dichloro-diphenyl-trichloroethane (DDT) and endosulfan in sediments of Sabancuy Lagoon, Campeche, Mexico. *Open Journal of Ecology, 6*(1), 22–31.

She, J., Holden, A., Sharp, M., Tanner, M., Williams-Derry, C., & Hooper, K. (2007). Polybrominated diphenyl ethers (PBDEs) and polychlorinated biphenyls (PCBs) in breast milk from the Pacific Northwest. *Chemosphere, 67*(9), S307–S317.

Singh, K. P., Malik, A., & Sinha, S. (2007). Persistent organochlorine pesticide residues in soil and surface water of northern Indo-Gangetic alluvial plains. *Environmental Monitoring and Assessment, 125*, 147–155.

Smith, A. G. (1991). Chlorinated hydrocarbon insecticides. In W. J. Hayes Jr. & E. R. Laws Jr. (Eds.), *Handbook of pesticide toxicology* (Vol. 2). Academic.

Solomon, G. M., & Weiss, P. M. (2002). Chemical contaminants in breast milk: Time trends and regional variability. *Environmental Health Perspectives, 110*(6), A339–A347.

Srimurali, S., Govindaraj, S., Krishna Kumar, S., & Babu Rajendran, R. (2015). Distribution of organochlorine pesticides in atmospheric air of Tamilnadu, Southern India. *International journal of Environmental Science and Technology, 12*, 1957–1964.

Toxics link. (2018). Persistent Organic Pollutants (POPs) in India: Country Situation Report. https://toxicslink.org/wp-content/uploads/2022/08/country%20situation%20report_new%20size.pdf (Acessed in December 2023).

Villanueva, F.S.; Páez-Osuna, F. Contaminantes críticos: Metales. In Golfo de México, Contaminación e Impacto Ambiental: Diagnostico y Tendencias; Botello, A.V., Rojas-Galaviz, J.L., Benítez, J., Zárate-Lomelí, D., Eds.; EPOMEX Serie Científica 5; Universidad Autónoma de Campeche: México D.F., Mexico, 1995; pp. 681–710

United States. Environmental Protection Agency. Office of Drinking Water. (1987). *Health advisories for 16 pesticides*. (87, No. 200176. The Office.

USEPA. (1986). Guidelines for carcinogen risk assessment. pp. 38.

Waliszewski, S. M., Caba, M., Goméz-Arroyo, S., Villalobos-Pietrini, R., Mártinez, A., Valencia-Quintana, R., Lozano-Flores, M. E., Regalado-Torres, M. A. (2013). Niveles de plaguicidas

organoclorados en habitantes de México. *Revista Internacional de Contaminación Ambiental, 29, 121–131.*
Westbom, R., Hussen, A., Megersa, N., Retta, N., Mathiasson, L., & Björklund, E. (2008). Assessment of organochlorine pesticide pollution in Upper Awash Ethiopian state farm soils using selective pressurised liquid extraction. *Chemosphere, 72*(8), 1181–1187.
World Health Organization. (1984a). *Environmental health criteria 38: Heptachlor*. World Health Organization.
World Health Organization. (1984b). *Health promotion: A discussion document on the concept and principles: Summary report of the Working Group on Concept and principles of health promotion, Copenhagen, 9–13 July 1984* (No. ICP/HSR 602 (m01)). WHO Regional Office for Europe.
World Health Organization. (2003). *Health risks of persistent organic pollutants from long-range transboundary air pollution.*
Wurl, O., & Obbard, J. P. (2005). Organochlorine pesticides, polychlorinated biphenyls and polybrominated diphenyl ethers in Singapore's coastal marine sediments. *Chemosphere, 58*(7), 925–933.

Hexachlorobenzene

Kanchan Kumari and Senerita Swamy

1 Introduction

In order to treat seeds, particularly for the prevention of wheat bunt, Hexachlorobenzene (HCB) was introduced in 1945. HCB is also a by-product of the production of industrial chemicals such pentachlorbenzene, perchloroethylene, trichloroethylene, and carbon tetrachloride. It is a recognized contaminant in a number of pesticide formulations, such as pentachlorophenol and dicloram, and it could be present in additional formulations as well. It does not emerge in the environment naturally. It is generated as a by-product during the production of other chemicals, in the waste streams of wood preservation and chloralkaline facilities, and while burning municipal garbage. Among all environmental contaminants, HCB is one that has received the most attention from researchers. Due to its persistent presence in the environment, several research investigations have been conducted, including toxicological in vitro and in vivo studies, tests for its detection in biological and environmental samples, and studies of its transport through the atmosphere, water, soil, and sediments. HCB was first used as an agricultural pesticide in 1945 and was outlawed in Europe in 1981 (European Food Safety Authority, 2018). HCB was frequently employed as a pesticide to prevent the fungus in the seeds of crops such as wheat, sorghum, and onions.

K. Kumari (✉)
CSIR-National Environmental Engineering Research Institute (NEERI), Kolkata Zonal Center, Kolkata, West Bengal, India
e-mail: k_kumari@neeri.res.in

S. Swamy
CSIR-National Environmental Engineering Research Institute (NEERI), Nagpur, India

K. Kumari, *Pollutants of Global Concern*, Emerging Contaminants and Associated Treatment Technologies, https://doi.org/10.1007/978-3-031-50996-4_7

HCB was produced on a global scale between 1000 and 2000 tonnes annually in the middle of the 1970s (Barber et al., 2005). With an annual production of 10,000 tonnes, HCB production peaks between late 1970s to early 1980s (Barber et al., 2005). Due to global restrictions on its use, production has since decreased. Due to HCB's high stability and persistence (Table 1), it can be released into the environment through seed dispersal, waste stockpiling, and disposal. When too much of a source material is used to make chemicals, it can also be released into the environment. HCB waste production in past is also a problem.

2 Technical Details of HCB

Table 1 Physico-chemical properties of HCB

Details	Description (HCB)
Structure	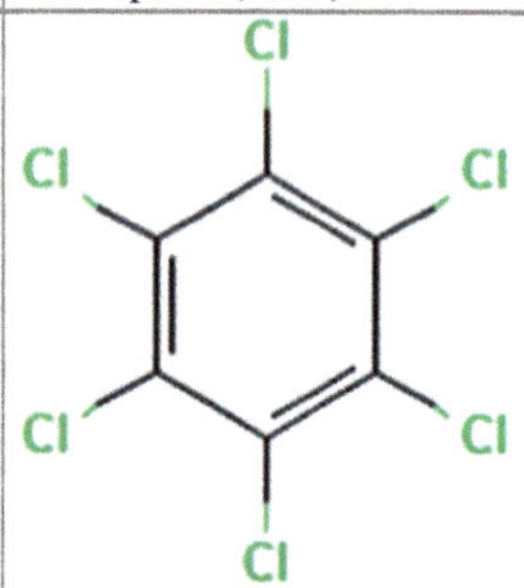
IUPAC name	1,2,3,4,5,6-HCB
Molecular formula	C_6Cl_6
CAS no.	118-74-1
Molecular weight	284.8 g/mol
Color/form	White crystal
Odor	Slight aromatic odor
Boiling point	325 °C
Melting point	231 °C (ICSC)
Flash point	468 °F (NTP, 1992)
Solubility	Sparingly soluble in cold alcohol; soluble in benzene, chloroform, and ether
Density	2.044 at 75.2 °F (USCG, 1999)
Vapor pressure	1 mm Hg at 237.9 °F; 0.0000109 mm Hg at 68 °F (NTP, 1992)
Henry's law constant	5.8×10^{-4} atm-cu m/Mol at 25 °C
Stability/shelf life	Very stable, even to acids and bases.
Decomposition	Upon heating for decomposition, toxic hydrogen chloride fumes come out
Safety and hazards	Potentially carcinogenic and cause organ harm as a result of prolonged or repetitive exposure. Highly harmful to aquatic life
Toxicity	Toxic

3 Status of Works and Usage

According to a review report by Environment Australia's Environment Protection Group, there are 8255 tonnes of crystalline HCB and 898 tonnes of tar-based solid waste that have been contaminated with HCB in Australia (Tong and Yuan, 2012). Formerly, legislators considered HCB present in pesticides only. As a result, they did not originally set a maximum value for HCB's unintentional trace contaminants (UTC). According to the European Commission, HCB may also be found as an impurity in a variety of materials, including various plastics and some inks, coatings, paints, toners, wood, and textiles. Surprisingly, HCB is also produced as a by-product of several production operations. The stakeholders were uncertain because of this discovery. The European Commission had proposed a model regulation amending EU Regulation 2019/1021 Part A, Annex I in order to make this clearer and make enforcement easier. The goal was to control HCB as a UTC in substances, mixtures, and even finished goods. As a result, the limit value for HCB was set at 10 mg/kg (0.001% by weight). This HCB POP amendment was made with input from the Competent Authorities expert group on POPs (POP CA meeting) and stakeholders. Similar to other product environmental regulations, the European Commission publishes POP amendments in the Official Journal of the European Union (OJEU). Table 2 summarizes listing of HCB by Stockholm Convention and Table 3 describes its effects, toxicity, and persistence. In India, HCB was never registered as pesticide but banned for manufacture, use and import since March, 2014 (Kumari, 2021).

4 Degradation and Environmental Fate

HCB mostly persists in the vapor phase when it is discharged into the atmosphere, and it degrades very slowly. According to estimates, the half-life of HCB in the atmosphere varies greatly depending on the region, ranging from 0.63 years in subtropical and tropical areas to 1.94 years in temperate and boreal areas to 6.28 years in polar areas (Wania & Mackay, 1995). An estimated half-life of 1.69 years was calculated using a measured hydroxyl rate constant 2.7 × 10–14 $cm^3/mol\ s^{-1}$ (Brubaker & Hites, 1998). From the temperate to the polar areas, long-distance

Table 2 Uses, sources, and regulation (Kumari, 2021)

Identified POP	Uses and sources	Annex	Year of listing by SC
HCB	Until 1965, HCB was frequently used as a pesticide to guard against fungus in the seeds of grains such as wheat, sorghum, and onions. In addition, it was used to produce synthetic rubber, ammunition, and fireworks	A and C	2001

Table 3 Effects, toxicity, and persistence of HCB

HCB	Data
Effects	A number of harmful health consequences linked to HCBs have been established. It has been demonstrated to produce a variety of significant non-cancer health impacts in animals, including immune system, reproductive, neurological, endocrine, and other health problems. Studies on people provide evidence for HCB impacts that may be both cancer-causing and not. HCBs may have a number of interconnected health impacts. The effects of changes in one system on other systems of the body may be profound
Toxicity	The National Toxicology Programme (NTP) stated in 2021 that the detrimental impacts of HCB are comparable to those of 2,3,7,8 tetrachlorodebenzodioxin (TCDD). As a result, HCB was omitted from the toxic equivalency factor approach for assessing adverse potency. The liver is the primary target organ of intoxication. According to studies conducted on people, HCB can harm the hepatic, thyroid, neurological system, kidneys, bones, blood, immunological, and endocrine systems (toxicological profile for HCB; 2012)
Persistence/fate	Due to its chemical stability and resistance to deterioration, HCB is one of the most enduring environmental contaminants. HCB mostly persists in the vapor phase when it is discharged into the atmosphere, and it degrades very slowly

global transit is feasible (Barber et al., 2005). HCB can be physically removed from the atmosphere via dry deposition, washout from snowfall or rain, or both. HCB is persistent in water and not readily decomposed by either biotic or abiotic processes. HCB is reported to have a half-life of between 2.7 and 5.7 years in the surface water and between 5.3 and 11.4 years in the groundwater (Howard, 1991). HCB will volatilize from the water column, but the process is slowed down by the compound's high adsorption to particles and organic materials in the water. HCB can volatilize from the soil surface if it is discharged into the soil, although it will be highly adsorbed to organic materials and is typically thought to be immobile with regard to leaching. According to Beck and Hansen (1974), it has a half-life value in soils that can be anywhere between 3 and 6 years. From all the studies, the average half-life of HCB is nearly 9 years, while the temporal patterns of HCB concentration in the environment differ based on the media and research site, as well as the time period studied. The loadings in treated and background soils, sediments, and seas make up the majority of estimates of the current environmental burden, which range between 20 and 60,000 million pounds (Barber et al., 2005). According to O'Neil et al. (2006), HCB has a moderate vapor pressure of 1.09×10^{-5} mmHg and a very low solubility of 0.005815 mg/L in water. HCB can exist in the vapor phase as well as in conjunction with particles, if it is discharged into the atmosphere (Eisenreich et al., 1981); however, monitoring studies have shown that the vapor phase typically predominates (Ballschmiter & Wittlinger, 1991; Bidleman et al., 1989; Lane et al., 1992). HCB's net residence period in the atmosphere is considerably shorter than a year and is dependent on physical translocation rather than chemical change (Ballschmiter, 2013). The atmospheric processes may function on a hemispheric scale over great distances (Montmessin et al., 2004).

HCB is predicted to volatilize at a considerable rate when released into surface water, and volatilization can be a substantial transfer route, according to the Henry's law constant value for this substance is 5.8×10^{-4} atm-m^3/mol (Hulscher et al., 1992). Due to the high organic carbon partition coefficient (K_{oc}) value, adsorption of HCB on sediment or suspended particle matter is also anticipated to be substantial if discharged into water (Sittig, 1981). The rate of volatilization is slowed down by this propensity to partition to sediment and adsorb to suspended particles in the water column. HCB may accumulate in the bottom sediments of large aquatic systems because of its high adsorption to soil particles and sediments. HCB is anticipated to remain immobile in soil and unlikely to leak into groundwater due to its strong sorption qualities (Swann et al., 1983). HCB should volatilize at a modest pace when released into damp soil, according to the Henry's law constant value. Volatilization may be a key process for removal of HCB emitted to soils, according to several studies. The catfish, the species at the top of the food chain, accumulated more than ten times as much as the species at the next trophic level such as snails and mosquitofish, and these species accumulated 1.5 to two times as much as the species at the bottom of the food chain such as the daphnids and the algae. The authors concluded that the food web main producers were additionally subjected to bio magnification (Ballschmiter, 2013) (Fig. 1).

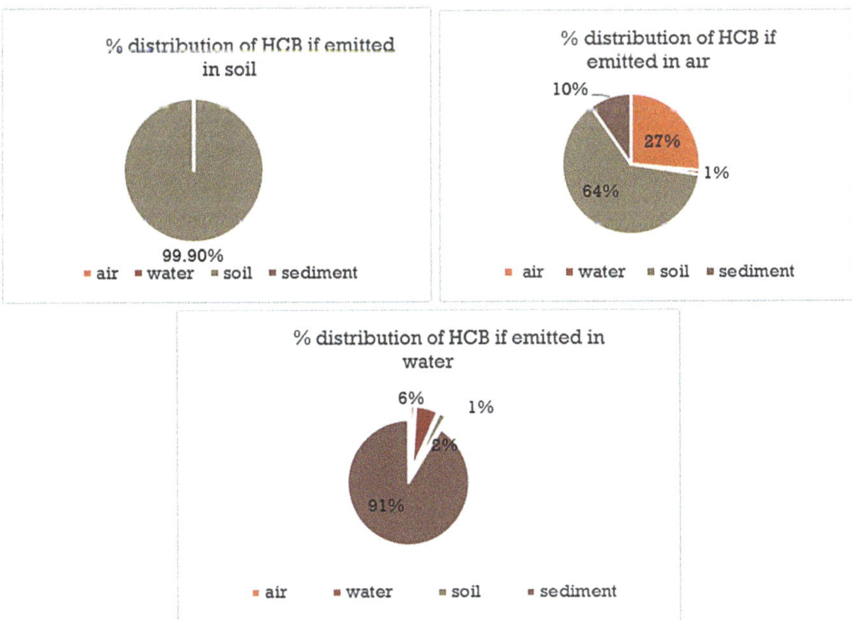

Fig. 1 Distribution of HCB in soil, water, and air

5 In-House (Countrywide) Scientific Work Done on the Chemical

A number of chlorinated pesticides, including pentachloronitrobenzene (PCNB), pentachlorophenol, chlorothalonil (TPN), dicloran, trimethyl 2,3,5,6-terephthalate (TCTP), dacthal, and picloram, have historically produced HCB as a by-product during synthesis (Fig. 2). These pesticides are a major source of HCB emissions. They are typically discovered in these pesticides as an impurity (Ritter et al., 1995). However, due to restrictions and bans against the use of HCB, current uses of such pesticides have been abolished. HCB is released into the environment when pesticides with HCB are used on lawns, gardens, or other types of vegetation. Bailey (2001) calculated the level of HCB contamination in several pesticides.

Tables 4, 5, and 6 summarize the levels of HCB prevailing in water, soil, and human breast milk samples collected from different parts of India.

The levels of HCB in the sediments and water samples of the Gomti River were between BDL and 0.98 ng g^{-1} dry weight and BDL and 38.36 ng l^{-1}, respectively. The Gomti River's sediments contain HCB residues at levels that are significantly lower than the Yangtse River's (0.14–3.96 ng g^{-1}) (Xu & Zhang, 2000). The results from the study done by Guzzella et al. (2005) may be representative of the HCB levels produced in the industrial and crowded city of Kolkata (with a population of 14.5 million). The limited sources and volatile nature of HCB may be reflected in the residues found at the other sites. According to Devanathan et al. (2009), the pollution of HCB is typically small and quite equal among the sites New Delhi (3.2 ng/g), Mumbai (1.7 ng/g), and Kolkata (4.4 ng/g). This is possibly because India generated very less HCB. As HCB is primarily advised for treatment of seed, large-scale treatment methods like spraying and dusting are not used. The presence of HCB in breast milk samples may therefore be the result of pollution from other pesticide composition that exist either as an impurity or as a by-product of different chlorination procedures and industrial product ignition. Breivik et al. (2004) astutely observed that when emerging countries enhance their chemical and metal manufacturing and handling procedures, HCB emissions can be anticipated to decline. In developed economies, far lower concentrations—typically 50 pg/m^3—have been observed (Bailey, 2001). In India, HCB was never authorized as a pesticide. However, from 1995 to 1997, 42,612 metric tonne of HCB of technical grade was produced (Ministry of Chemicals and Fertilisers, 2000).

6 International Data Versus The Indian Scenario

HCB emissions into the environment (Barber et al., 2005; Bailey, 2001), environmental destiny and levels (Barber et al., 2005; Wang et al., 2010), as well as toxicity and human exposures (Wang et al., 2010) have all been examined by a number of researchers. Bitterfeld in Germany (Vrana et al., 2001), Orica stockpile site in

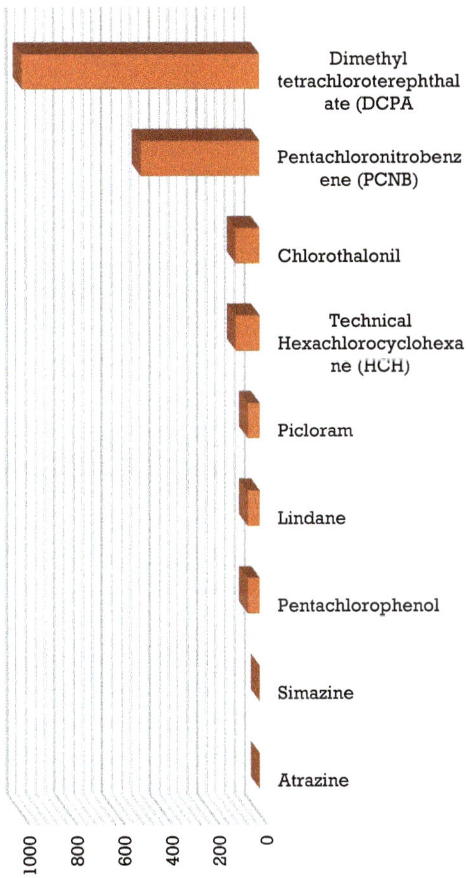

Fig. 2 Concentration of HCB in pesticide (in ppm)

Table 4 Summary of HCB concentration (ng/g) in water samples in India

Site	Sample type	Concentration	Reference
Koshi River	River water (ng/L)	9.2 ng/l	Ghimire et al. (2020)
Gomti River	Surface water (ng/L)	6.40 ± 10.45	Malik et al. (2009)
Indian Ocean	Sea water (pg/L)	0.94–13	Huang et al. (2014)

Table 5 Summary of HCB concentration (ng/g) in sediment samples in India

Site	Sample type	Observed range	Reference
Babughat, Hugli estuary	Sediment	0.98 ng/g dry wt	Guzzella et al. (2005)
Kachubaria	Sediment	0.09 ng/g dry wt	Guzzella et al. (2005)
Baj Baj	Sediment	0.11 ng/g dry wt	Guzzella et al. (2005)
Diamond harbor	Sediment	>0.05 ng/g	Guzzella et al. (2005)
Bay of Bengal	Sediment	0.05 to 1.39 ng g^{-1}	Sarkar et al. (2008)

Table 6 Residual HCB concentration in human breast milk in India

Site	Sample type	Observed range	Reference
Chennai (city)	Human breast milk	3.2–7.6 ng/g lipid wt.	Subramanian et al. (2007)
New Delhi	Human breast milk	0.40–19 ng/g lipid wt.	Devanathan et al. (2009)
Mumbai	Human breast milk	0.40–7.0 ng/g lipid wt	Devanathan et al. (2009)
Kolkata	Human breast milk	1.3–9.3 ng/g lipid wt	Devanathan et al. (2009)

Australia (Brown, 2009), Sao Paulo State's coastal plain in Brazil (Do Nascimento et al., 2004), and the Niagara River watershed in the United States (USEPA, 2004) are a few examples of contaminations.

The final National Priority List for the US Environmental Protection Agency's (EPA) Superfund Program includes 44 hazardous waste sites where HCB is a pollutant of concern (NPL). According to several studies, the levels of HCB in the air in North America and Northern Europe have significantly decreased during the previous four decades. The peak level of HCB is associated with 1960, with significantly lower levels present in the 1970s and early 1980s. Tian et al. (2021) evaluated HCB's historical and contemporary geographic environment. In nine cities of the Pearl River Delta (PRD), where historical OCP use was the most extensive in China, an air monitoring campaign was conducted in 2018. During the sample period, the PRD's average HCB concentration was 81 ± 39 pg/m^3. Its level in winter (105 ± 35 pg/m^3) was double than summer (58 ± 27 pg/m^3), with a statistical seasonal variation. In Zhaoqing, the wintertime HCB concentration peaked at 138 ± 38 pg/m^3. HCB was found in all samples of maternal blood in Arctic Canada and also in higher concentrations in the blood of Inuit women than in that of local Caucasian women. The trend in cord blood plasma concentrations was similar (Walker et al., 2003). In New York State and Finland, populations of women who consumed tainted local fish had higher levels of HCB in their breast milk (Kostyniak et al., 1999; Greizerstein et al., 1999; Fitzgerald et al., 2001; Damgaard et al., 2006). All blood samples taken from expectant women in a Californian farming community contained HCB (Fenster et al., 2006). To identify the cause of the high and rising concentration of HCB, the food of the Inuit people in Greenland was investigated. The concentration of hexachlorobenzene in blood of these Greenland Arctic populations were found to be correlated with consumption of foods containing seal and whale (Deutch et al., 2004, 2006)

7 Adverse Effects of HCB

HCB is poisonous to aquatic life, terrestrial vegetation, terrestrial animals, and humans. It has been classified as Group 2 B carcinogen by the International Agency for Research on Cancer (IARC), meaning it may cause cancer in people, and it also appears to encourage tumor growth. HCB may harm the liver, immunological system, thyroid, kidneys, and central nervous system in addition to the growing fetus. The effects are primarily noticeable on the liver and nervous system. HCB poisoning frequently manifests as porphyria. High or prolonged exposure may harm the nervous system and lead to symptoms such as irritation, coordination issues, muscle weakness, tremor, and/or a pins-and-needles sensation on the skin. Animals exposed to HCB orally have been thoroughly investigated, and the whole spectrum of end goals has been evaluated. The toxicity of HCB in animals after cutaneous exposure has not been researched at all; only the immunological effects of inhalation exposure have been investigated in animals.

I. *Carcinogenic Potential* – After HCB's inclusion in the third Annual Report on Carcinogens, more rat studies have come to light. Females who were exposed to HCB during pregnancy and thereafter consumed it for up to 130 weeks developed benign liver tumors, men developed benign parathyroid gland tumors, and both sexes developed benign adrenal gland tumors. No epidemiological research had examined the connection between human cancer and exposure to HCB particularly in the Third Annual Report on Carcinogens. Numerous case-control studies, particularly focusing on breast cancer, have since been published. According to the International Agency for Research on Cancer, there is insufficient proof that HCB causes cancer in humans (WHO, IARC, 2001).
II. *Neurobehavioral Effects* – There is no conclusive proof that HCB exposure through inhalation causes neurotoxic consequences in either humans or animals. Studies of the southeast Turkish citizens who ate contaminated bread provide evidence of neurotoxicity in people after oral exposure to HCB. The exposed people experienced myotonia, paresthesia, neuritis, and muscle wasting. The breastfed infants of HCB-exposed mothers in Turkey who were under 24 months old displayed convulsions, tremors, and gradual weakening before they passed away. Serious neurological abnormalities were present in a portion of the exposed population that was monitored by multiple researchers for the long-term effects of this incident. Both in the presence and in the absence of histological evidence of changes in the neurological structures, HCB is neu-

rotoxic in adult animals. The central nervous system (CNS) is not overly damaged by HCB at low dosages, but at large levels, CNS diseases may be the main symptom. The level or severity of the CNS effects is dose and time dependant and includes electrophysiological abnormalities, tremors, seizures, muscle weakness, lethargy, and unsteady gait.

III. *Reproductive Effects* – Research data is not available on the effects of HCB on reproduction in either humans or animals.

8 Management and Remedial Measures

Although HCB production for commercial purposes has been outlawed globally for decades, it is still produced as an industrial by-product (Jiang et al., 2018). Around the world, including in the USA, Germany, Australia, Japan, and Brazil, HCB pollution in soil poses a serious problem (Jiang et al., 2018). By 1988, China had produced 79,278 t of HCB cumulatively (Leading Group for National Implementation Plan Compiling of China, 2007). In Anhui Province, China, the mean soil concentration of HCB was 8.01 ng g^{-1}. HCB is present in most soils (Wang et al., 2010; Song et al., 2017). Soil remediation from HCB contamination is a significant and ongoing challenge. Remediation entails the removal of HCB from a variety of areas, including air, water, soil, and sediments. As a result, HCB was widely found in soils and sediments (Fig. 1). The majority of adopted remediation techniques involve removing HCB from soils. Due to its prolonged half-life (>6 years), which affects the stability of the soil, HCB was able to remain in soils for a long time. The traces of HCB from soils and sediments can be removed using a variety of techniques. Both of these treatments lower the level of chemical congeners that the HCB produces. Toxic substances are converted into nontoxic substances like water and carbon dioxide through the treatment process known as biodegradation (Mori & Kondo, 2002). The microbial population decreased after the HCB was broken down because they had completely used it as a food source. The type of microorganism chosen will determine how much biodegradation occurs. Both aerobic and anaerobic environments are suitable for bioremediation. Microorganisms in the soils and sediments will break down HCB in the absence of oxygen. Chlorinated compounds will be broken down by microorganisms into carbon dioxide and water in aerobic conditions with sufficient oxygen (Kumar et al., 2013).

A promising environmentally friendly method to get rid of POPs is mechanochemical degradation (MCD). One of the important factors in initiating and escalating the de chlorination and degradation of POPs that was previously thought to be mechanically induced by free-radical attack. Using both experimental analysis and quantum chemical calculations, Deng et al. (2018) investigated the mechanochemical production of free radicals and their roles in the reclamation of (HCB)-contaminated soil. The production of chlorinated phenoxy radicals (CB-O%) in the milling process and their importance in the de chlorination of HCB were both shown to depend on the results of electron spin resonance (ESR) and X-ray photoelectron spectra.

Literature on the large-scale and pilot remediation of HCB-contaminated soils, sediments, and waters is scarce. Recent reports on notable pilot scale electrokinetic remediations focused on the treatment of old sediment that had been contaminated with HCB for years (Li et al., 2010). Sediments totaling 0.5 m^3 were treated at a constant voltage while having HCB concentrations of several tens of mg/kg. Hydroxypropyl-β-cyclodextrin (HPCD) was added to anodic solution to improve the remediation process. Significant HCB movement was seen across the sediments after 100 days of processing, but only very little was actually removed. A total of 563 kWh/m^3 was determined to be the electric energy consumption. Even with HPCD's help, it is challenging to use the EK technique to remove old HCB in real sediments on a large scale. Even though HCB use has been prohibited for many years, cleaning up the contamination is still a problem for the environment (Chen et al., 2021).

9 Conclusion

HCB is a hydrophobic, toxic, and persistent pollutant in the environment. Despite the substance's use being prohibited for many years, HCB contamination clean-up continues to be a problem for the environment. All administration routes for HCBs are regarded as hazardous. The most sensitive indications of these chemical substances have reportedly been found to be hematotoxicity and immunotoxicity. HCB detected in water soil and sediment in India is quite less, as India produced relatively less amount of HCB. This might be due to the fact that, HCB is primarily recommended for seed treatment, large-scale treatment procedures like as spraying and dusting are rarely employed. Despite the fact that HCB has been detected in the human breast milk compositions that may be present as an impurity or as a consequence of industrial combustion. Some studies claims that as growing nations improve their chemical and metal manufacturing and handling procedures, HCB emissions may decline. Researchers should consider the limitations of current technology in order to choose the best technology for HCB remediation and disposal. Additional data is thus required for estimating the equivalence factors between the consequences and their toxic potential in order to permit the normalization of measures across various research.

References

Bailey, R. E. (2001). Global hexachlorobenzene emissions. *Chemosphere, 43*(2), 167–182.

Ballschmiter, K. (2013). Department of analytical chemistry. In *Organic micropollutants in the aquatic environment: Proceedings of the sixth European symposium, held in Lisbon, Portugal, May 22–24, 1990* (Vol. 13152, p. 260). Springer.

Ballschmiter, K., & Wittlinger, R. (1991). Interhemisphere exchange of hexachlorocyclohexanes, hexachlorobenzene, polychlorobiphenyls, and 1, 1, 1-trichloro-2, 2-bis (p-chlorophenyl) ethane in the lower troposphere. *Environmental Science & Technology, 25*(6), 1103–1111.

Barber, J. L., Sweetman, A. J., Van Wijk, D., & Jones, K. C. (2005). Hexachlorobenzene in the global environment: Emissions, levels, distribution, trends and processes. *Science of the Total Environment, 349*(1–3), 1–44.

Beck, J., & Hansen, K. E. (1974). The degradation of quintozene, pentachlorobenzene, hexachlorobenzene and pentachloroaniline in soil. *Pesticide Science, 5*(1), 41–48.

Bidleman, T. F., Patton, G. W., Walla, M. D., Hargrave, B. T., Vass, W. P., Erickson, P., ... & Gregor, D. J. (1989). Toxaphene and other organochlorines in Arctic Ocean fauna: evidence for atmospheric delivery. *Arctic*, 307–313.

Breivik, K., Alcock, R., Li, Y. F., Bailey, R. E., Fiedler, H., & Pacyna, J. M. (2004). Primary sources of selected POPs: regional and global scale emission inventories. *Environmental Pollution, 128*(1–2), 3–16.

Brown, P. (2009). Toxic waste in our midst: Towards an interdisciplinary analysis. *Journal of Environmental Management, 90*(4), 1559–1566.

Brubaker, W. W., & Hites, R. A. (1998). OH reaction kinetics of gas-phase α-and γ-hexachlorocyclohexane and hexachlorobenzene. *Environmental Science & Technology, 32*(6), 766–769.

Chen, Q., Duan, G., Liu, Q., Ye, Q., Sharma, R. P., Chen, Y., et al. (2021). Estimating crown width in degraded forest: A two-level nonlinear mixed-effects crown width model for Dacrydium pierrei and Podocarpus imbricatus in tropical China. *Forest Ecology and Management, 497*, 119486.

Damgaard, I. N., Skakkebæk, N. E., Toppari, J., Virtanen, H. E., Shen, H., Schramm, K. W., et al. (2006). Persistent pesticides in human breast milk and cryptorchidism. *Environmental Health Perspectives, 114*(7), 1133–1138.

Deng, S., Feng, N., Kang, S., Zhu, J., Yu, B., Chen, J., & Xie, X. (2018). Mechanochemical formation of chlorinated phenoxy radicals and their roles in the remediation of hexachlorobenzene contaminated soil. *Journal of Hazardous Materials, 352*, 172–181.

Deutch, B., Pedersen, H. S., & Hansen, J. C. (2004). Dietary composition in Greenland 2000, plasma fatty acids and persistent organic pollutants. *Science of the Total Environment, 331*(1–3), 177–188.

Deutch, B., Dyerberg, J., & Pedersen, H. S., Asmund, G., Moller, P., & Hansen, J. C. (2006). Dietary composition and contaminants in north Greenland, in the 1970s and 2004. *Science of the Total Environment, 370*(2–3), 372–381.

Devanathan, G., Subramanian, A., Someya, M., Sudaryanto, A., Isobe, T., Takahashi, S., ... & Tanabe, S. (2009). Persistent organochlorines in human breast milk from major metropolitan cities in India. *Environmental Pollution, 157*(1), 148–154.

Do Nascimento, N. R., Nicola, S. M. C., Rezende, M. O. D. O., Oliveira, T. A., & Öberg, G. (2004). Pollution by hexachlorobenzene and pentachlorophenol in the coastal plain of São Paulo state, Brazil. *Geoderma, 121*(3–4), 221–232.

Eisenreich, S. J., Looney, B. B., & Thornton, J. D. (1981). Airborne organic contaminants in the Great Lakes ecosystem. *Environmental Science & Technology, 15*(1), 30–38.

Fenster, L., Eskenazi, B., Anderson, M., Bradman, A., Harley, K., Hernandez, H., ... & Barr, D. B. (2006). Association of in utero organochlorine pesticide exposure and fetal growth and length of gestation in an agricultural population. *Environmental Health Perspectives, 114*(4), 597–602.

Fitzgerald, E. F., Hwang, S. A., Deres, D. A., Bush, B., Cook, K., & Worswick, P. (2001). The association between local fish consumption and DDE, mirex, and HCB concentrations in the breast milk of Mohawk women at Akwesasne. *Journal of Exposure Science & Environmental Epidemiology, 11*(5), 381–388.

Ghimire, P. S., Qu, B., Wang, X., Tripathee, L., Paudyal, R., Sharma, C. M., & Kang, S. (2020). Nutrients and organic carbons in river waters of the Third Pole. In *Water quality in the Third Pole* (pp. 179–209). Elsevier.

Greizerstein, H. B., Stinson, C., Mendola, P., Buck, G. M., Kostyniak, P. J., & Vena, J. E. (1999). Comparison of PCB congeners and pesticide levels between serum and milk from lactating women. *Environmental Research, 80*(3), 280–286.

Guzzella, L., Roscioli, C., Vigano, L., Saha, M., Sarkar, S. K., & Bhattacharya, A. (2005). Evaluation of the concentration of HCH, DDT, HCB, PCB and PAH in the sediments along the lower stretch of Hugli estuary, West Bengal, northeast India. *Environment International, 31*(4), 523–534.

Howard, P. H. (Ed.). (1991). *Handbook of environmental degradation rates*. CRC Press.

Huang, Y., Li, J., Xu, Y., Xu, W., Cheng, Z., Liu, J., ... & Zhang, G. (2014). Polychlorinated biphenyls (PCBs) and hexachlorobenzene (HCB) in the equatorial Indian Ocean: Temporal trend, continental outflow and air–water exchange. *Marine Pollution Bulletin, 80*(1–2), 194–199.

Hulscher, T. E. T., Van Der Velde, L. E., & Bruggeman, W. A. (1992). Temperature dependence of Henry's law constants for selected chlorobenzenes, polychlorinated biphenyls and polycyclic aromatic hydrocarbons. *Environmental Toxicology and Chemistry: An International Journal, 11*(11), 1595–1603.

Indian Ministry of Chemicals and Fertilisers (MC&F). (2000). Information compiled by Plant Protection Adviser, Department of Agriculture and Cooperation, Ministry of Agriculture and Cooperation, New Delhi.

Jiang, Y., Shang, Y., Yu, S., & Liu, J. (2018). Dechlorination of hexachlorobenzene in contaminated soils using a nanometallic Al/CaO dispersion mixture: Optimization through response surface methodology. *International Journal of Environmental Research and Public Health, 15*(5), 872.

Kostyniak, P. J., Stinson, C., Greizerstein, H. B., Vena, J., Buck, G., & Mendola, P. (1999). Relation of Lake Ontario fish consumption, lifetime lactation, and parity to breast milk polychlorobiphenyl and pesticide concentrations. *Environmental Research, 80*(2), S166–S174.

Kumar, J. M., Kumar, S. D., Kubendran, D., & Kalaichelvan, P. T. (2013). Hexachlorobenzene-sources, remediation and future prospects. *International Journal of Current Research and Review, 5*(1), 1.

Kumari, K. (Ed.). (2021). *Persistent organic pollutants: Gaps in management and associated challenges* (1st ed.). CRC Press. https://doi.org/10.1201/9781003046806

Lane, D. A., Johnson, N. D., Hanley, M. J. J., Schroeder, W. H., & Ord, D. T. (1992). Gas-and particle-phase concentrations of. alpha.-hexachlorocyclohexane,. gamma.-hexachlorocyclohexane, and hexachlorobenzene in Ontario air. *Environmental Science & Technology, 26*(1), 126–133.

Li, T., Yuan, S., Wan, J., & Lu, X. (2010). Hydroxypropyl-β-cyclodextrin enhanced electrokinetic remediation of sediment contaminated with HCB and heavy metals. *Journal of Hazardous Materials, 176*(1–3), 306–312.

Malik, A., Ojha, P., & Singh, K. P. (2009). Levels and distribution of persistent organochlorine pesticide residues in water and sediments of Gomti River (India)—a tributary of the Ganges River. *Environmental Monitoring and Assessment, 148*, 421–435.

Montmessin, F., Forget, F., Rannou, P., Cabane, M., & Haberle, R. M. (2004). Origin and role of water ice clouds in the Martian water cycle as inferred from a general circulation model. *Journal of Geophysical Research: Planets, 109*(E10).

Mori, T., & Kondo, R. (2002). Degradation of 2, 7-dichlorodibenzo-p-dioxin by wood-rotting fungi, screened by dioxin degrading ability. *FEMS Microbiology Letters, 213*(1), 127–131.

National Toxicology Program, Institute of Environmental Health Sciences, National Institutes of Health (NTP). (1992). *National Toxicology Program Chemical Repository Database*. Research Triangle Park, North Carolina: NTP.

O'Neill, S. M., Ylitalo, G. M., West, J. E., Bolton, J., Sloan, C. A., & Krahn, M. M. (2006). Regional patterns of persistent organic pollutants in 5 Pacific salmon species (Oncorhynchus spp) and their contributions to contaminant levels in northern and southern resident killer whales (Orcinus orca). In: Proc 2006 Southern Resident Killer Whale Symp, NOAA- Fisheries, Northwest Fisheries Science Center, Seattle, WA. www.wdfw.wa.gov/publications/01034/wdfw01034.pdf

Ritter, L., Solomon, K. R., Forget, J., Stemeroff, M., & O'Leary, C. (1995). A review of selected persistent organic pollutants. International Programme on Chemical Safety (IPCS). *PCS/95.39. Geneva: World Health Organization, 65, 66.*

Sarkar, S. K., Binelli, A., Riva, C., Parolini, M., Chatterjee, M., Bhattacharya, A. K., ... & Satpathy, K. K. (2008). Organochlorine pesticide residues in sediment cores of Sunderban wetland, northeastern part of Bay of Bengal, India, and their ecotoxicological significance. *Archives of Environmental Contamination and Toxicology, 55*, 358–371.

Sittig, M. (1981). *Handbook of toxic and hazardous chemicals.*

Song, Y., Bian, Y., Wang, F., Herzberger, A., Yang, X., Gu, C., & Jiang, X. (2017). Effects of biochar on dechlorination of hexachlorobenzene and the bacterial community in paddy soil. *Chemosphere, 186*, 116–123.

Subramanian, A., Ohtake, M., Kunisue, T., & Tanabe, S. (2007). High levels of organochlorines in mothers' milk from Chennai (Madras) city, *India. Chemosphere, 68*(5), 928–939.

Swann, R. L., Laskowski, D. A., McCall, P. J., Vander Kuy, K., & Dishburger, H. J. (1983). A rapid method for the estimation of the environmental parameters octanol/water partition coefficient, soil sorption constant, water to air ratio, and water solubility. In *Residue reviews: Residues of pesticides and other contaminants in the total environment* (pp. 17–28). Springer.

Tian, Z., Zhao, H., Peter, K. T., Gonzalez, M., Wetzel, J., Wu, C., et al. (2021). A ubiquitous tire rubber–derived chemical induces acute mortality in coho salmon. *Science, 371*(6525), 185–189.

Tong, M., & Yuan, S. (2012). Physiochemical technologies for HCB remediation and disposal: A review. *Journal of Hazardous Materials, 229–230*, 1–14, ISSN 0304-3894. https://doi.org/10.1016/j.jhazmat.2012.05.092.

US EPA/NYSDEC. (2004). *Reduction of toxic loadings to the Niagara River from hazardous waste sites in the United States.* June 2004.

U.S. Coast Guard. (1999). *Chemical Hazard Response Information System (CHRIS) - Hazardous Chemical Data.* Commandant Instruction 16465.12C. Washington, D.C.: U.S. Government Printing Office.

Vrana, B., Paschke, A., Popp, P., & Schuurman, G. (2001). Use of semi-permeable membranes: Determination of bioavailable organic waterbourne contaminants in the industrial region of Bitterfeld, Saxony-Anhalt, Germany. *Environmental Science and Pollution Research, 8*, 27–34.

Walker, J. B., Seddon, L., McMullen, E., Houseman, J., Tofflemire, K., Corriveau, A., et al. (2003). Organochlorine levels in maternal and umbilical cord blood plasma in Arctic Canada. *Science of the Total Environment, 302*(1–3), 27–52.

Wang, G., Lu, Y., Han, J., Luo, W., Shi, Y., Wang, T., & Sun, Y. (2010). Hexachlorobenzene sources, levels and human exposure in the environment of China. *Environment International, 36*(1), 122–130.

Wania, F., & Mackay, D. (1995). A global distribution model for persistent organic chemicals. *Science of the Total Environment, 160*, 211–232.

World Health Organization. (2001). *IARC monographs on the evaluation of carcinogenic risks to humans. Volume 79: Some thyrotropic agents.*

Xu, Y., & Zhang, W. X. (2000). Subcolloidal Fe/Ag particles for reductive dehalogenation of chlorinated benzenes. *Industrial and Engineering Chemistry Research, 39*(7), 2238–2244.

Mirex

Kanchan Kumari and Deepak Marathe

1 Introduction

Historically, the chemical mirex has been used as both an insecticide and a fire retardant. It has also been used as a fire-resistant compound in materials including plastics, rubber, and electrical items (Hunt, 2007). Mirex was first used to fight fire ants, other ant species, and termites (Dhang, 2023). Despite its many uses, there are rising worries about its possible effects on the environment and human health. Although mirex does not appear to affect people right away when directly exposed to them, laboratory tests on animals have led to its designation as a potential human carcinogen (Mahire et al., 2023; Fernandes et al., 2023).

This suggests a potential link between prolonged mirex exposure and an increased risk of cancer in humans (Mahire et al., 2023). Furthermore, mirex's toxicity extends to various plant species, aquatic life forms such as fish and crustaceans, highlighting its potential to disrupt ecosystems and endanger wildlife (Khan & Law, 2005). One of the most notable characteristics of mirex is its remarkable stability and persistence in the environment (Waters et al., 1977). With a half life of up to 10 years, mirex's longevity contributes to its lasting effects on the ecosystems it enters (Zitko, 2003). This attribute, combined with its ability to accumulate in the food chain, particularly in meat, fish, and wild game, underscores the importance of understanding its implications for both human and environmental health (SC, 2001). As awareness of mirex's potential negative impacts has grown, regulatory measures have

K. Kumari (✉)
CSIR-National Environmental Engineering Research Institute (NEERI),
Kolkata Zonal Center, Kolkata, West Bengal, India
e-mail: k_kumari@neeri.res.in

D. Marathe
CSIR-National Environmental Engineering Research Institute (NEERI), Nagpur, India

K. Kumari, *Pollutants of Global Concern*, Emerging Contaminants and Associated Treatment Technologies, https://doi.org/10.1007/978-3-031-50996-4_8

been implemented in many countries to restrict or ban its use. This approach aims to encourage the adoption of safer alternatives for pest management and fire protection, ultimately promoting the well-being of both people and the planet.

2 Technical Details of Mirex

Mirex is a polychlorinated pesticide that belongs to the class of chemicals known as organochlorines (León-Santiesteban & Rodríguez-Vázquez, 2017). Various physicochemical properties of the Mirex are shown in Table 1. It has the chemical formula $C_{10}H_8Cl_{12}$ and is classified as a hexachlorocyclopentadiene derivative. Mirex has a unique structure consisting of a hexachlorocyclopentadiene ring with additional chlorine atoms attached to various positions on the ring. The arrangement of these chlorine atoms gives mirex its insecticidal properties. The molecular weight of mirex is approximately 545.45 g/mol. Mirex is a white crystalline solid with a melting point of around 505 °C (941 °F). It is practically insoluble in water, which contributes to its persistence in the environment. One of the defining characteristics of mirex is its high environmental persistence. It has a long half-life of up to 10 years, meaning it takes a significant amount of time for half of the compound to degrade in the environment. It was historically used as an insecticide to control fire ants, other ant species, and termites. Its ability to disrupt insect nervous systems contributed to its effectiveness as a pesticide. As an important fire retardants, mirex has the ability to suppress combustion and reduce flammability making it useful in improving the fire safety of these products. Furthermore, its toxic effects on aquatic life and potential for bioaccumulation in the food chain are concerning from environmental perspective. Due to its persistence, toxicity, and potential for negative impacts on human health and the environment, the use of mirex has been restricted or banned in many countries. Regulatory measures have been put in place to control its production, distribution, and application.

3 Status of Usage Within the Country and Beyond

Mirex, though was historically used in various countries due to its chemical resilience and insecticidal ability, (Tschinkel, 2013) at present it is banned or restricted in many countries, including the United States, Canada, European Union member states, Australia, and others. Regulatory agencies in these countries have taken action to limit its production, distribution, and use to mitigate the potential risks associated with its application (Pope, 2014). The restrictions and bans on mirex have spurred the development and adoption of safer alternatives for pest control and fire retardation. These alternatives aim to provide effective solutions while minimizing the negative impacts on the environment and human health (Marshall, 1982). The Environmental Protection Agency (EPA) canceled most mirex uses in the 1980s and 1990s. Mirex was banned in Canada in the 1970s due to concerns about its

Table 1 Physicochemical properties of mirex (PubChem)

Details	Description (mirex)
Structure	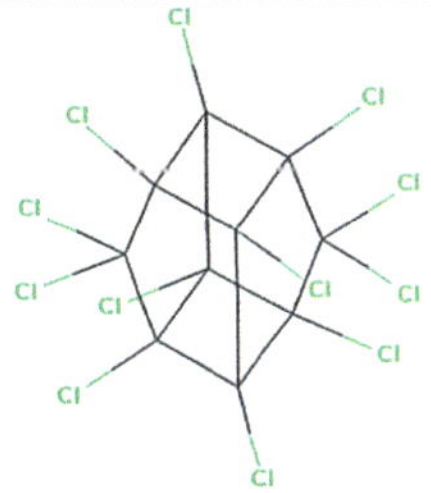
IUPAC name	1,2,3,4,5,5,6,7,8,9,10,10-dodecachloropentacyclo[5.3.0.$0^{2,6}$.$0^{3,9}$.$0^{4,8}$] decane
Molecular formula	$C_{10}Cl_{12}$
CAS no.	2385-85-5
Molecular weight	545.5 g/mol
Color/form	White crystalline solid
Odor	Odorless
Boiling point	485.0 °C
Melting point	485.0 °C
Solubility	In water, 0.085 mg/L at 25 °C
Vapor pressure	8.00e-07 mmHg
Octanol/water partition coefficient	Log kOW = 6.89
Henry's law constant	8.11e-04 atm-m^3/mole
Stability/shelf life	Very stable at normal temperatures
Decomposition	Decomposes above 500 °C to give hexachlorobenzene
Corrosivity	Practically non-corrosive to metals
Safety and hazards	Irritant, health hazard, and environmental hazard
Toxicity	Toxic

environmental persistence and toxicity. Mirex is banned for use within the European Union due to its persistent nature, potential for long-range transport, and potential health risks. Mirex usage has also been banned in Australia. In addition, many other countries around the world have restricted or banned the use of mirex due to similar concerns (Lallas, 2001). It is important to note that regulatory information can change over time as new research findings emerge and as countries update their policies. Therefore, for the most current and accurate information about the status of mirex usage in a specific country, it is recommended to consult with the country's environmental protection agency or relevant regulatory authorities. In India, Mirex was used as a pesticide in-order to control fire ants and other ants, and thus, based upon its physicochemical properties, it was listed in the category of Annex A of the Stockholm convention in the year 2001 (Table 2). In addition, it has been banned for production, distribution, and its use in India (Lallas 2001).

Known among the classic organochlorines, mirex's chronic effects of biota accumulation and environmental persistence played a key role for it being categorized

Table 2 Status of usage within the country (Stockholm Convention) (Kumari, 2021)

Identified POP	Uses and sources	Annex	Year of listing by SC	Regulation in India
Mirex	Pesticide used to control fire ants and other ants	A	2001	Never registered but it is banned form manufacture, distribution, use and production

under the POPs (Jyoti et al., 2023; Zitko, 2003). Numerous studies have shown that despite their ban globally and their applications being almost phased out, residual concentrations can be detected in environmental as well as in ecological compartments (Girones et al., 2020; Isogai et al., 2018). Regulations have been put in place to control its production, distribution, and application (Su et al., 2012). Safer alternatives for pest control and fire retardation have been developed to mitigate the potential negative impacts of mirex on ecosystems and human health. Over time, concerns about mirex's negative impact on the environment and human health led to restrictions and bans in many countries (Su et al., 2012; Shen et al., 2010). The main reasons for these actions include persistence, bioaccumulation, toxicity, and human health risks.

4 Effects, Toxicity, and Persistence of Mirex

Environmental Impact: Mirex is considered as a persistent organic pollutant. Its high stability and resistance to degradation contribute to its long-lasting presence in the environment. This persistence can lead to bioaccumulation in various organisms, especially in the fatty tissues of animals (Fisher, 1999; Zitko, 2003).

Wildlife Impact: Mirex has been found to be toxic to a range of species, including aquatic life like fish and crustaceans, as well as certain plant species. This can disrupt ecosystems and food chains, potentially leading to population imbalances and biodiversity loss (Bevenue et al., 1975; Kendall, 1982; Imanishi et al., 2004; Capella et al., 2023).

Human Health Impact: Direct exposure to mirex does not appear to cause immediate harm to humans. However, mirex has been classified as a possible human carcinogen based on studies in laboratory animals, suggesting a potential link between exposure and an increased risk of cancer. Additionally, the accumulation of mirex in the food chain, particularly in fatty animal tissues, raises concerns about human exposure through consumption of contaminated food (Rawn et al., 2023; Zhou et al., 2014).

Toxicity

Acute Toxicity: Acute toxicity refers to the immediate harmful effects that can occur shortly after exposure to a substance. While mirex is not highly acutely toxic to humans, it can cause harm to aquatic organisms and wildlife. It affects the nervous systems of insects, which was one of the reasons for its use as an insecticide (Ens, 2023; Puertas et al., 2010; Ren et al., 2011).

Chronic Toxicity: Chronic toxicity refers to the long-term, cumulative effects of exposure to a substance. Mirex's classification as a possible human carcinogen is based on chronic toxicity studies conducted on animals. Chronic exposure to mirex may increase the risk of cancer over time (Waters et al., 1977; Rawn et al., 2023; Vimalkumar et al., 2022; Prabhu & Lakshmipraba, 2022; Reshi et al., 2022; Martínez et al., 2022).

Environmental Persistence: Mirex is highly persistent in the environment due to its chemical stability and resistance to degradation. This means that it can remain in the environment for an extended period. Its half-life, which is the time it takes for half of the substance to degrade, can be up to 10 years or even longer (Waters et al., 1977; Zitko, 2003; Mehendale et al., 1972; Mackay, 1979; Li et al., 2023).

Bioaccumulation

Mirex's persistence contributes to its potential for bioaccumulation in the food chain. Organisms at higher trophic levels (predators) can accumulate higher levels of mirex as they consume other organisms containing the compound. This bioaccumulation increases the potential for adverse effects in animals and humans at the top of the food chain (Mahire et al., 2023; Peivasteh-Roudsari et al., 2023; Kelly et al., 2004; Capella et al., 2023; Rawn et al., 2023).

Mirex is highly bioaccumulative and amount depends upon the concentration and duration of exposure. There is evidence of accumulation of mirex in aquatic and terrestrial food chains to harmful levels. After six applications of mirex bait at 1.4 kg/ha, high mirex levels were found in some species; turtle fat contained 24.8 mg mirex/kg, kingfishers, 1.9 mg/kg, coyote fat, 6 mg/kg, opossum fat, 9.5 mg/kg, and raccoon fat, 73.9 mg/kg (Fig. 1).

5 International Data Versus Indian Scenario

Mirex is a persistent organic pollutant that can be found in various environmental compartments, including soil, water, and biological organisms. Mirex can accumulate in soil due to its persistence and low solubility in water (Assey & Mogusu, 2023). It can persist in soil for a long time and has the potential to move within the soil profile. This accumulation can lead to increased concentrations in areas where mirex was historically used or where waste containing the compound was disposed of. Mirex can be transported through water bodies due to its low water solubility (Jones & Hodges, 1974; Waters et al., 1977; Takkar et al., 2022). Studies have reported residual quantities of mirex in the Tamiraparani River Basin where mirex concentrations were found to be 0.01–0.7 ng/L (pre-monsoon), 0.01–0.47 ng/L (monsoon), 0.01–0.17 ng/L (post- monsoon) and 0.01–0.15 ng/L (summer) respectively (Agarwal et al., 2015). A similar study by Patil et al. (2015) put forward the possibility of mirex contamination in the waters of River Kaveri (Cauvery) where it

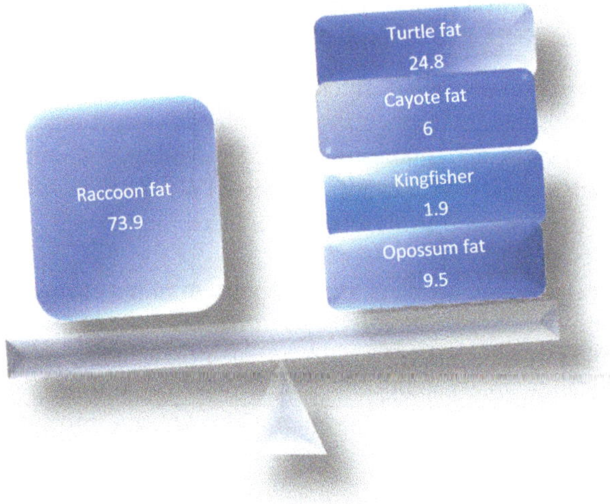

Fig. 1 Mirex concentration in different species (mg/kg) (El-Bayomey et al., 2002)

was found to be upto 17,500 ng/L in water samples and 190 ng/g in the muscle tissues of fish. It can also attach to suspended particles and sediments, potentially leading to sediment contamination as reported by Guzzella et al. (2005) along the Hugli Estuary and by Kumarasamy et al. (2012) in the Tamiraparani River Basin. The sediments of the coastal estuarine areas of Hugli accounted mirex concentrations of <10 ng/g (dry weight). In aquatic ecosystems, mirex can bio accumulate in fish and other organisms, particularly in areas with historical mirex usage or runoff from contaminated soils. A study of marine species especially of bivalves indicated a concentration of mirex below 0.18 ng/g collectively with endosulfan and chlordanes (Girones et al., 2020). Atmospheric monitoring have provided evidences of the possibility of mirex transport and contamination of the studied localities. Study conducted by Isogai et al. (2018) in the West African countries of Togo, Benin, Nigeria and Cameroon inferred the presence of mirex, though low in comparison to other OCPs, with a range of 0.1 to 3.3 pg/m^3. A similar study conducted by Srimurali et al. (2015) reported mirex concentrations to be 4,609 pg/m^3 in air samples from Tamil Nadu. The OCP concentrations including mirex were found to the highest during the monsoon season. A research on avian species showed the presence of mirex quantities in eggs collected from the wetland site of Barberspan Lake, Vaal River and western Cape province in South Africa. The bird species included Cattle egret, African darter, Reed cormorant, African sacred ibis, Crowned plover, Little grebe, Whitefronted plover and Kelp gull. Mirex concentrations reported in the eggs of these bird species were 0.68, 2.0, 1.5, 0.3, 0.56, 0.49, 0.32 and 0.81 (ng/g, ww) respectively (Bouwman et al., 2008). Laboratory studies have shown that rats exposed to mirex can accumulate the compound in their tissues (Reagan & Mulcahy, 2019; Kaur et al., 2019). Mirex has been found to accumulate in the fatty tissues of animals, including rodents. Chronic exposure to mirex in animal studies has raised

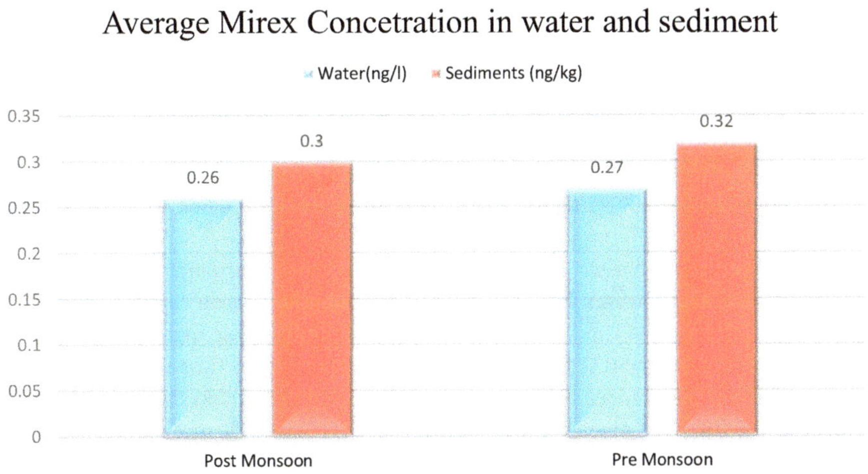

Fig. 2 Average concentrations of mirex in water and sediments during pre-monsoon and post-monsoon (Mujtaba Baqar et al., 2018)

concerns about potential carcinogenic effects and other health impacts. Mirex can also accumulate in humans through the consumption of contaminated food, particularly fatty animal products. This bioaccumulation in the food chain can lead to measurable concentrations of mirex in human tissues, including blood. Human exposure to mirex primarily occurs through the consumption of contaminated food. Once ingested, mirex can be absorbed into the bloodstream. Mirex and its metabolites have been detected in human blood samples in various studies. However, levels of exposure can vary based on dietary habits, geographic location, and other factors. It is noteworthy that Mirex was never produced or registered as a pesticide in India as indicated in the Persistent Peril report of 2004. Furthermore, there is a lack of available information regarding the current stockpiles of Mirex in India. However, one study by Mujtaba Baqar et al. (2018) reported the descriptive statistical of OCPs in water and sediments (during pre-monsoon and post monsoon) and it is given in Fig. 2.

There isn't much information accessible for the chemical mirex at the international level that can be applicable to the Indian situation based on the facts of relevance of scientific data (available with convention) as per the Indian condition.

6 Adverse Effects of Mirex

Limited research has been conducted on human exposures to mirex, and its impact on human health remains insufficiently documented. However, studies on animals have unveiled several adverse effects resulting from dietary exposure to mirex.

Liver Hypertrophy: Mirex induces liver hypertrophy after prolonged, low-dose exposure in rats and has harmful effects on developing fetuses, including the development of cataracts (WHO, Environment Health Criteria 44; (WHO, Environment Health Criteria 44; Waters et al., 1977; Kim et al., 1985). Additionally, mirex has been associated with immune system suppression (Colborn & Clement, 1992), and it negatively affects the germination and emergence of various plant species, demonstrating its hazardous nature on multiple systems, including test animals (TSMP, Environment Canada).

Carcinogenic Potential: Based upon the facts of International Agency for Research on Cancer, mirex has been listed as a potential human carcinogen due to proof/indications of its cancer-causing ability in animals. Male and female rats fed diets containing 50 and100 ppm of mirex for long periods of time-developed damage to the liver and cancer of the hepatocellular system (Ulland et al., 1977). The National Institutes of Health conducted a two-year research that clearly demonstrated carcinogenic activity in male and female rats, as evidenced by a rise in the occurrence of many aberrant cellular states (Toxicology and carcinogenesis investigations). The hepatocytes of hens given mirex at 10 ppm and higher for 12 to 16 weeks showed structural alterations, cell abnormalities, and cells dying. Rats given mirex at 1 to 100 ppm for 2–4 weeks experienced liver hypertrophy (Davison et al., 1976) (WHO Environmental Health Criteria, 1984). Mirex causes kidney lesions and gill damage in fish when given to them.

Hormonal Effects on Reproduction: Mirex is known to disrupt hormones in the reproductive system. In female rats, doses of 0.4–50 mg suppressed ovulation induced by PMS (Fuller & Draper, 1975). Pregnant rats exposed to 25-ppm mirex had smaller litters, lower pup survival rates, and a 33% incidence of cataracts in pups compared to 0% in controls. Fetal mortality and abnormalities, such as edema and ectopic gonads, were observed in pregnant rats exposed to mirex (US EPA, 1992). Additionally, fetuses of mothers exposed to mirex at specific doses exhibited scoliosis, runts, short tails, cleft palates, and heart malformations.

These findings underscore the potential health risks associated with mirex exposure, particularly its carcinogenic properties and disruptions to reproductive hormones in animals. However, further research is needed to fully understand the extent of these effects on human health.

7 Remedial Action Taken for Management of POP

In accordance with the authority granted by clause (a) of sub-section (1) of Section 18 read with subsection (2) of Section 27 and Section 28 of the Insecticides Act, 1968 (46 of 1968), the Ministry of Agriculture issued a notification on March 27, 2014, known as S.O. 910(E), ordering the ban on the production, formulation, distribution, stocking, and use of mirex. There are exceptions, though, and mirex may be imported and used in the amounts necessary for research and development operations with the previous consent of the Central Insecticides Board and Registration Committee established under the Insecticide Act of 1968.

8 Recommendations

The ban imposed on this legacy persistent pollutant reflects upon the fact that no further effective evaluation is required. However, periodical monitoring to keep mirex under critical evaluation is a need to actually eliminate its impacts imparted due to its previous usages. Pest control techniques must be approached from a holistic standpoint, taking into account not just their immediate efficacy but also any possible negative effects they may have on ecosystems and human health. Global regulatory organizations responded to the possible problems linked with mirex as scientific understanding and environmental consciousness increased. It became essential to outlaw or severely restrict its usage in order to preserve ecosystems, human health, and a sustainable balance in nature. The trajectory of mirex from a commonly used pesticide to a material that is banned illustrates the intricate relationship that exists between environmental stewardship, technological innovation, and human needs.

References

Agarwal, A., Prajapati, R., Singh, O. P., Raza, S. K., & Thakur, L. K. (2015). Pesticide residue in water—a challenging task in India. *Environmental Monitoring and Assessment, 187*, 1–21.

Assey, G. E., & Mogusu, E. (2023). Effects of persistent organic pollutants on environment, health and mountains: A review. *Journal of Environmental Engineering and Science, 40*(XXXX), 1–8.

Baqar, M., Sadef, Y., Ahmad, S. R., Mahmood, A., Li, J., & Zhang, G. (2018). Organochlorine pesticides across the tributaries of River Ravi, Pakistan: Human health risk assessment through dermal exposure, ecological risks, source fingerprints and spatio-temporal distribution. *Science of the Total Environment, 618*, 291–305.

Bevenue, A., Ogata, J. N., Tengan, L. S., & Hylin, J. W. (1975). Mirex residues in wildlife and soils, Hawaiian pineapple growing areas – 1972–74. *Pesticides Monitoring Journal, 9*(3), 141–149.

Bouwman, H., Polder, A., Venter, B., & Skaare, J. U. (2008). Organochlorine contaminants in cormorant, darter, egret, and ibis eggs from South Africa. *Chemosphere, 71*(2), 227–241.

Capella, R., Guida, Y., Loretto, D., Weksler, M., & Meire, R. O. (2023). Occurrence of legacy organochlorine pesticides in small mammals from two mountainous National Parks in south eastern Brazil. *Emerging Contaminants, 9*(2), 100211.

Colborn, T., Clement C. eds. (1992). Chemically induced alterations in sexual and functional development: The wildlife/human connection. Princeton NJ: Princeton Scientific Publishing, 365–394.

Davison, K. L., Mollenhauer, H. H., Younger, R. L., & Cox, J. H. (1976). Mirex-induced hepatic changes in chickens, Japanese quail and rats. *Archives of Environmental Contamination and Toxicology,4*: 469–482.

Dhang, P. (2023). Role of insect bait in sustainable management of urban pests. In *Urban pest management: An environmental perspective* (pp. 62–79). GB: CABI.

Ens, J. (2023). *Corpus-based symbolic music generation: Data, representation, models, evaluation.*

EL-Bayomey, A. A., Somak, I. W., & Branch, S. (2002). Embryotoxicity of the pesticide Mirex In vitro. Teratogenesis, *Carcinogenesis, and Mutagenesis, 22*, 239–249.

Fernandes, I. D. A. A., Maciel, G. M., Bortolini, D. G., Pedro, A. C., Rubio, F. T. V., de Carvalho, K. Q., & Haminiuk, C. W. I. (2023). The bitter side of teas: Pesticide residues and their impact on human health. *Food and Chemical Toxicology*, 113955.

Fisher, B. E. (1999). Most unwanted. *Environmental Health Perspectives, 107*(1), A18–A23.

Fuller, G. B., & Draper, S. W. (1975). Effect of mirex on induced ovulation in immature rats. *Proceedings of the Society for Experimental Biology and Medicine, 148*(2), 414–417.

Girones, L., Oliva, A. L., Marcovecchio, J. E., & Arias, A. H. (2020). Spatial distribution and ecological risk assessment of residual organochlorine pesticides (OCPs) in South American marine environments. *Current Environmental Health Reports, 7*, 147–160. http://toxicslink.org/docs/06029_per_peril.pd

Guzzella, L., Roscioli, C., Vigano, L., Saha, M., Sarkar, S. K., & Bhattacharya, A. (2005). Evaluation of the concentration of HCH, DDT, HCB, PCB and PAH in the sediments along the lower stretch of Hugli estuary, West Bengal, northeast India. *Environment International, 31*(4), 523–534.

Hunt, G. T. (2007). Persistent organic pollutants as compounds of emerging concern. In *Microconstituents conference 2007* (pp. 224–239). Water Environment Federation.

Imanishi, K., Kawakami, M., Shimada, A., Chikaishi, K., Kimura, Y., Kajiwara, N., et al. (2004). *Detection of pesticides unregistered in Japan, toxaphene and mirex, in the cetaceans from Japanese coastal waters* (Vol. 15, p. 429).

Isogai, N., Hogarh, J. N., Seike, N., Kobara, Y., Oyediran, F., Wirmvem, M. J., et al. (2018). Atmospheric monitoring of organochlorine pesticides across some West African countries. *Environmental Science and Pollution Research, 25*, 31828–31835.

Jones, A. S., & Hodges, C. S. (1974). Persistence of mirex and its effects on soil microorganisms. *Journal of Agricultural and Food Chemistry, 22*(3), 435–439.

Jyoti, Yadav, N., Thakur, I. S., & Srivastava, S. (2023). Omics approach in nano-bioremediation of persistent organic pollutants. *Genomics Approach to Bioremediation: Principles, Tools, and Emerging Technologies*, 271–293.

Kaur, N., Singh, P., Bedi, J. S., & Gupta, A. (2019). Studies on persistent organic pollutants residue in water, sediment and fish tissues of River Sutlej, India. *Journal of Environmental Biology, 40*(2), 258–264.

Kelly, B. C., Gobas, F. A., & McLachlan, M. S. (2004). Intestinal absorption and biomagnification of organic contaminants in fish, wildlife, and humans. *Environmental Toxicology and Chemistry: An International Journal, 23*(10), 2324–2336.

Kendall, R. J. (1982). Wildlife toxicology. *Environmental Science & Technology, 16*(8), 448A–453A.

Khan, M. Z., & Law, F. C. (2005). Adverse effects of pesticides and related chemicals on enzyme and hormone systems of fish, amphibians and reptiles: A review. *Proceedings of the Pakistan Academy of Sciences, 42*(4), 315–323.

Kim, H. T., Kim, K. S., Kim, J. S., & Stone, W. B. (1985). Levels of polychlorinated biphenyls (PCBs), DDE, and mirex in waterfowl collected in New York State, 1981–1982. *Archives of Environmental Contamination and Toxicology, 14*, 13–18.

Kumarasamy, P., Govindaraj, S., Vignesh, S., Rajendran, R. B., & James, R. A. (2012). Anthropogenic nexus on organochlorine pesticide pollution: a case study with Tamiraparani river basin, South India. *Environmental Monitoring and Assessment, 184*, 3861–3873.

Kumari, K. (Ed.). (2021). *Persistent organic pollutants: Gaps in management and associated challenges* (1st ed.). CRC Press. https://doi.org/10.1201/9781003046806

Lallas, P. L. (2001). The Stockholm convention on persistent organic pollutants. *American Journal of International Law, 95*(3), 692–708.

León-Santiesteban, H. H., & Rodríguez-Vázquez, R. (2017). Fungal degradation of organochlorine pesticides. *Microbe-Induced Degradation of Pesticides*, 131–149.

Li, L., Chen, C., Li, D., Breivik, K., Abbasi, G., & Li, Y. F. (2023). What do we know about the production and release of persistent organic pollutants in the global environment? *Environmental Science: Advances, 2*(1), 55–68.

Mackay, D. (1979). Finding fugacity feasible. *Environmental Science & Technology, 13*(10), 1218–1223.

Mahire, S., Tiwana, A. S., Khan, A., Nalawade, P. M., Bandekar, G., Trehan, N., et al. (2023). Accumulation and effects of persistent organic pollutants and biogeographical solutions: Appraisal of global environment. *Arabian Journal of Geosciences, 16*(10), 570.

Marshall, E. (1982). Mississippi Inc., Pesticide Manufacturer: The state wants to attack fire ants with a cheap version of Mirex; Safer alternatives may be held back as a result. *Science, 218*(4572), 548–550.

Martínez, C., Arroyo, A. M., Alemán, D. B., García, A. G., Caba, M., Garcidueñas, A. L. C., et al. (2022). Persistent organic compounds in human milk and evaluation of the effectiveness of the Stockholm convention in Mexico. *Environmental Advances, 8*, 100190.

Mehendale, H. M., Fishbein, L., Fields, M., & Matthews, H. B. (1972). Fate of mirex-14 C in the rat and plants. *Bulletin of Environmental Contamination and Toxicology, 8*, 200–207.

Mirex. https://pubchem.ncbi.nlm.nih.gov/compound/Mirex

Patil, N. N., Selvaraj, K. K., Krishnamoorthy, V., Elaiyaraja, A., & Ramaswamy, B. R. (2015). Organochlorine pesticide contamination in the Kaveri (Cauvery) river, India: a review on distribution profile, status, and trends. *Water Challenges and Solutions on a Global Scale*, 115–128.

Peivasteh-Roudsari, L., Barzegar-Bafrouei, R., Sharifi, K. A., Azimisalim, S., Karami, M., Abedinzadeh, S., et al. (2023). Origin, dietary exposure, and toxicity of endocrine-disrupting food chemical contaminants: A comprehensive review. *Heliyon, 9*, e18140.

Pope. (2014). *Mirex, encyclopedia of toxicology* (3rd ed., pp. 345–346., ISBN:9780123864550). Academic. https://doi.org/10.1016/B978-0-12-386454-3.00165-2

Prabhu, R. N., & Lakshmipraba, J. (2022). Persistent organic pollutants (Part I): The "dirty dozen"–sources and adverse effects. *Organic Pollutants: Toxicity and Solutions*, 1–27.

PubChem. https://pubchem.ncbi.nlm.nih.gov/compound/Mirex

Puertas, R., Lopez-Espinosa, M. J., Cruz, F., Ramos, R., Freire, C., Pérez-García, M., et al. (2010). Prenatal exposure to mirex impairs neurodevelopment at age of 4 years. *Neurotoxicology, 31*(1), 154–160.

Rawn, D. F., Quade, S. C., Corrigan, C., Ménard, C., Sun, W. F., Breton, F., et al. (2023). Differences in mirex [dechlorane] and dechlorane plus [syn-and anti-] concentrations observed in Canadian human milk. *Chemosphere, 316*, 137784.

Reagan, T. E., & Mulcahy, M. M. (2019). Interaction of cultural, biological, and varietal controls for management of stalk borers in Louisiana sugarcane. *Insects, 10*(9), 305.

Ren, B., Yang, Y., & Wang, Z. (2011). Notice of retraction: The health implications of Mirex in the environment. In *2011 5th international conference on bioinformatics and biomedical engineering* (pp. 1–4). IEEE.

Reshi, M. S., Mustafa, R. A., Javaid, D., & Haque, S. (2022). Pesticide toxicity associated with infertility. In *Oxidative stress and toxicity in reproductive biology and medicine: A comprehensive update on male infertility* (Vol. II, pp. 59–69). Springer.

SC, Stockholmn Convention. (2001). https://chm.pops.int/TheConvention/ThePOPs/ListingofPOPs/tabid/2509/Default.aspx

Shen, L., Reiner, E. J., MacPherson, K. A., Kolic, T. M., Sverko, E. D., Helm, P. A., et al. (2010). Identification and screening analysis of halogenated norbornene flame retardants in the Laurentian Great Lakes: Dechloranes 602, 603, and 604. *Environmental Science & Technology, 44*(2), 760–766.

Srimurali, S., Govindaraj, S., Krishna Kumar, S., & Babu Rajendran, R. (2015). Distribution of organochlorine pesticides in atmospheric air of Tamilnadu, southern India. *International Journal of Environmental Science and Technology, 12*, 1957–1964.

Stockholm Convention. https://chm.pops.int/TheConvention/ThePOPs/The12InitialPOPs/tabid/296/Default.aspx

Su, N. Y., Lagnaoui, A., Wang, Q., Li, X., & Tan, S. (2012). A demonstration project of Stockholm POPs convention to replace chlordane and mirex with IPM for termite control in China. *Journal of Integrated Pest Management, 3*(4), D1–D8.

Takkar, S., Shandilya, C., Agrahari, R., Chaurasia, A., Vishwakarma, K., Mohapatra, S., et al. (2022). Green technology: Phytoremediation for pesticide pollution. In *Phytoremediation technology for the removal of heavy metals and other contaminants from soil and water* (pp. 353–375). Elsevier.

Toxics Link. (2004). *Persistent Peril report*. Report prepared by Ravi Aggarwal & Papiya Sarkar.

Tschinkel, W. R. (2013). *The fire ants*. Harvard University Press.

Ulland, B. M., Page, N. P., Squire, R. A., Weisburger, E. K., & Cypher, R. L. (1977). A carcinogenicity assay of mirex in Charles River CD rats. *Journal of the National Cancer Institute, 58*(1), 133–140.

USEPA (United States Environmental Protection Agency). (1992). https://iris.epa.gov/static/pdfs/0251_summary.pdf

Vimalkumar, K., Sangeetha, S., Felix, L., Kay, P., & Pugazhendhi, A. (2022). A systematic review on toxicity assessment of persistent emerging pollutants (EPs) and associated microplastics (MPs) in the environment using the Hydra animal model. *Comparative Biochemistry and Physiology Part C: Toxicology & Pharmacology, 256*, 109320.

Waters, E. M., Huff, J. E., & Gerstner, H. B. (1977). Mirex. An overview. *Environmental Research, 14*(2), 212–222.

World Health Organization (WHO). (1984). *World Health Organization environmental health criteria 36: Fluorine and fluorides*. World Health Organization.

Zhou, S. N., Siddique, S., Lavoie, L., Takser, L., Abdelouahab, N., & Zhu, J. (2014). Hexachloronorbornene-based flame retardants in humans: Levels in maternal serum and milk. *Environment International, 66*, 11–17.

Zitko, V. (2003). Chlorinated pesticides: Aldrin, DDT, endrin, dieldrin, mirex. *Persistent Organic Pollutants*, 47–90.

Toxaphene

Kanchan Kumari and Shaheen Akhtar

1 Introduction

Toxaphene first appeared as a versatile insecticide in the 1940s and quickly acquired global recognition due to its potency against a variety of pests. However, due to its complex chemical composition and resistance to breakdown, it remained in the environment, causing environmental and health issues.

Toxaphene is not naturally occurring and are produced instead using technical camphene and chlorine gas (Vetter & Scherer, 1998). During the beginning of the 2000s, global efforts have been made to stop the manufacture and usage of toxaphene. Consequently, the Stockholm Convention aimed to restrict the manufacture and use of specific chemical compounds among its signatory governments, listed toxaphene as a persistent organic pollutant (POP) (Stockholm Convention, 2008). Toxaphene manufacturing and usage are strictly prohibited under the convention's provisions, with no exceptions. By the end of June 2010, more than 150 nations have joined the 2005-enacted pact. There are evidences that PCCs (polychlorinated camphene) congeners may be an unintentional outcome of chlorination-based industrial processes, such as those used in paper and pulp (EPA, 2005).

Cotton, cereal grains, fruits, nuts, and vegetables were all treated with this pesticide. It has also been used to keep ticks and mites at bay in animals. Counted among the most widely used insecticide in the USA in 1975, its persistence in soil was estimated to be around 12 years. Food is a common source where exposure to toxaphene is probable for humans. While the toxicity of toxaphene to humans is low, it has been identified as a probable human carcinogen based on its effects on

K. Kumari (✉) · S. Akhtar
CSIR-National Environmental Engineering Research Institute (NEERI), Kolkata Zonal Center, Kolkata, West Bengal, India
e-mail: k_kumari@neeri.res.in

K. Kumari, *Pollutants of Global Concern*, Emerging Contaminants and Associated Treatment Technologies, https://doi.org/10.1007/978-3-031-50996-4_9

laboratory animals. It is extremely harmful to fish; toxaphene exposed brook trout lost 46% of their weight and had worse egg viability after 90 days, while long-term exposure to doses of 0.5 micrograms per liter of water resulted in 0% egg viability (Stockholm Convention, 2001).

2 Technical Details of Toxaphene

According to the Agency for Toxic Substances and Disease Registry (ATSDR), toxaphene is a complex combination of about 670 compounds. This material is a waxy yellow-to-amber solid with a peculiar turpentine-like odor. Toxaphene's odor threshold has been established to be 0.14 ppm (parts per million). Toxaphene's typical chemical formula is $C_{10}H_{10}C_{l}8$, and its average molecular weight is 411.8 g/mol. Furthermore, at 77 °F, toxaphene has a vapor pressure of about 0.4 mm Hg and a log K_{ow} (octanol/water partition coefficient) of 2.474 (Table 1). Despite being easily

Table 1 Physico-chemical properties of toxaphene

Details	Description (toxaphene)
Structure	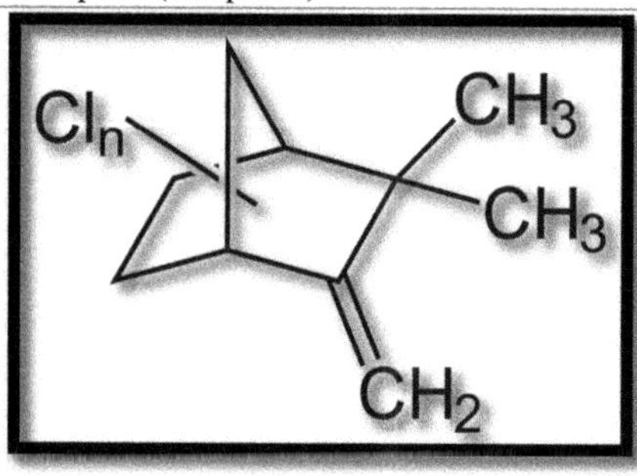
IUPAC name	1,4,5,6,7,7-hexachloro-2,2-bis(chloromethyl)-3-methylidenebicyclo[2.2.1] heptane
Molecular formula	$C_{10}H_8Cl_8$
Molecular weight	411.8 g/mol
Color/form	Yellow-to-amber waxy solid
Odor	Chlorine- and camphor-like odor
Melting point	65–90 °C
Flash point	84 °F
Solubility	Less than 1 mg/mL at 66 °F
Vapor pressure	0.4 mm Hg at 77 °F
Safety and hazards	Acute toxic, irritant, health hazard, and environmental hazard

soluble in aromatic hydrocarbons like benzene and xylene, it is largely insoluble in water (ATSDR, 1996).

3 Status of Works and Usage in India

Toxaphene had been primarily utilized as a pesticide, particularly for controlling pests on cotton. Additionally, it is found to manage poultry and livestock pests, as well as on various other field crops, as stated in the ATSDR report of 1996. However, in 1982, the Environmental Protection Agency (EPA) took the decision to cancel the approval of toxaphene as a pesticide for the majority of applications. The sole exclusions were its continuing use on livestock for urgent cases and for managing insects in Puerto Rico and the Virgin Islands' banana and pineapple crops. Following that, in 1990, all other applications of toxaphene were likewise discontinued (ATSDR, 1996). Though listed in the POPs Annex A by the Stockholm Convention in 2001 (Kumari, 2021), toxaphene was banned in India since 1989. This ban was strictly on its import, use, and/or manufacture (PIB, 2023). Since then, India has neither imported nor produced toxaphene. Therefore, no information on that has been published. Additionally, given that toxaphene was never registered in India, there are no reports of any stocks of the substance (Toxics Link, 2018).

4 International Data for Toxaphene Residues

Even though toxaphene was heavily used to prevent pesticides in crop plants as well in livestock, the majority of its applications were revoked in the United States in 1982. The Environmental Protection Agency (EPA) initiated steps to cancel toxaphene's registration for most pesticide purposes, allowing it to be used only in emergency cases to control insects on specific crops and livestock. Subsequently, in 1990, all remaining uses of toxaphene were completely discontinued (EPA, 1987). Despite this, they continue to have large atmospheric quantities and circulate throughout the global ecosystem. Various organochlorine pesticides are identified as pollutants of priority in national and worldwide screening programs because of their toxicity. All of these efforts have advocated for a better comprehension of their origin and fate, particularly the necessity to differentiate between atmospheric load caused by local reemissions and burdens caused by long-distance transboundary transport from countries that continue to use OC pesticides (Harner et al., 2001).

Assessing the formerly used organochlorines pesticides (OCPs) in agricultural soil, a study in Alabama (USA) showed the highest concentrations in toxaphene residue among all the other OCPs with a value of 285 ± 390 ng/g (dry weight of soil) (Harner et al., 1999). Another study documented 2500 ng/g (dry weight) of toxaphene in soil collected from cotton fields of Georgia and South Carolina (USA) (Kannan et al., 2003). Toxaphene measurements in coastal locations on the Pacific

Ocean's coast (Nicaragua) revealed that environmental concentrations are notably high in the typical cotton producing zone. Toxaphene residues in soils reached 44μg/g (dry weight), while concentrations in lagoon sediments reached 6.9μg/g (dry weight) around the mouths of rivers running through the agricultural zone. Concentrations in clam (*Anadara* spp.) soft tissues were found to be as high as 1.6μg/g (dry weight) in aquatic biota measurements. Consumption of shellfish, particularly clams from highly polluted locations, may expose the population to an unacceptable level of toxaphene, 30μg/d per person, through diet (Carvalho et al., 2003). Similarly, studies conducted in the cities of Mexico (Chiapas, Veracruz, and Tabasco) detected toxaphene levels in soil (0.066–69 ng/g) and in air (6.2–230 pg/m^3). A drop in Parlar congeners 39 and 42 was seen in toxaphene chromatographic profiles in soil and air, indicating that soil emissions of ancient residues are the main source of toxaphene in the atmosphere (Wong et al., 2008). In Lake Superior, parallel air and water samples were collected in August 1996 and May 1997. There was however not much variations noted in concentrations of toxaphene in Lake Superior (average ~ 918,218 pg/L), where the concentrations in air quality in May (124.6 pg/m^3) were lower than in August (2810 pg/m^3). Parlar 26 and 50, two obstinate congeners, were also identified (Jantunen & Bidleman, 2003). Harnet et al. (2001) recorded the toxaphene concentrations in Alabama (USA). Assessing the soil-air equilibrium state and identifying the chemicals that contributed significantly to the reported air loads, toxaphene was responsible for up to 50% of the air load that was detected, with soils in Alabama emitting between 3000 and 11,000 kg of toxaphene per year.

Consequently, the inference drawn from the traces of organochlorine pesticides still detected in the environmental compartments, where some of these OCPs had been banned from usage years ago, has strongly been related to the hypothesis that pesticides used in agriculture and for similar such purposes around the world are being volatilized and are likely contributing to the long-range movement of these chemicals (Kannan et al., 2003).

Moreover, toxaphene formulations has been seen to be stable in water and soil majorly thus various studies has assessed its complications in different organisms especially that in aquatic species (Figs. 1 and 2).

5 Adverse Effects of Toxaphene

Toxaphene's main impact is to stimulate the central nervous system (CNS), which can lead to convulsive seizures. Acute inhalation exposure studies in humans and animals are not available. Acute oral toxaphene exposure in humans, on the other hand, can cause CNS activation and convulsive seizures, with the approximate amount required to elicit nonfatal convulsions being about 10 mg/kg of body weight/day (mg/kg/day). Studies on animals have also revealed that acute oral exposure to toxaphene causes liver, kidney, and CNS effects, based on brief oral examination in rats (ATSDR, 1996; RTECS, 1993). Toxaphene exposure for humans is majorly via

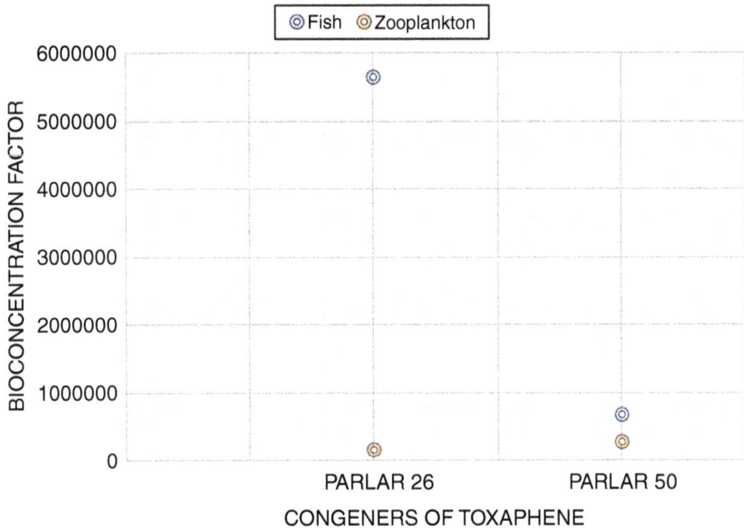

Fig. 1 Congeners of toxaphene and their consequent BCFs (bioconcentration factors) in aquatic species (mg/kg fresh weight). (Jongbloed et al., 2000)

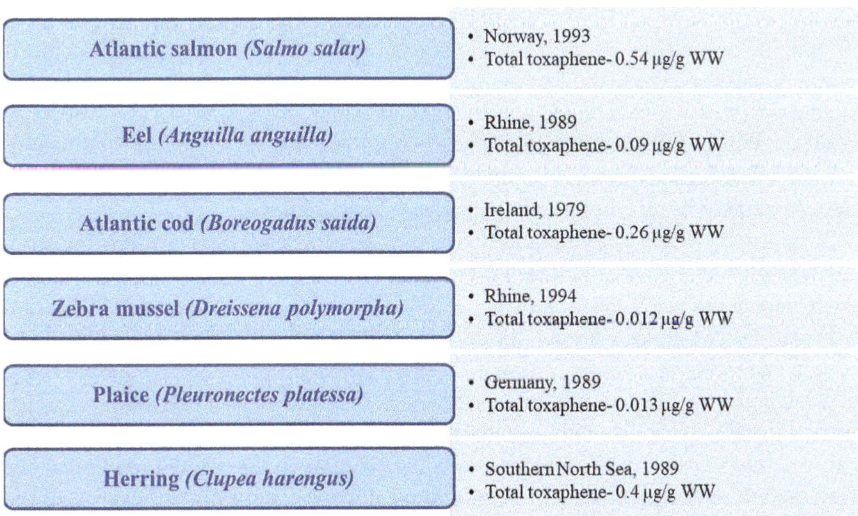

Fig. 2 Residues of toxaphene in freshwater aquatic species. (Jongbloed et al., 2000) [***WW* - wet weight]

inhalation, which is closely linked to reversible respiratory toxicity in terms of long-term effects. Chronic oral exposure to toxaphene, on the other hand, has had a wide range of effects on organs such as the liver, adrenal, kidney, thyroid glands and spleen, the central nervous system, and also the immune system in animals. Toxaphene use has been linked to an increase in thyroid gland and liver malignancies in animal studies (ATSDR, 1996; NTP, 1979). Toxaphene has been categorized by the EPA as a Group B2, potential human carcinogen (IRIS, 1999). An individual's exposure to toxaphene and its by-products can be assessed using blood, urine, breast milk, and body tissue analyses; however, due to the substance's fast clearance from the body, these approaches are only efficient for identifying recent exposures (ATSDR, 1996).

6 Management and Remedial Measures

Toxaphene management and remedial measure needs a critical perspective. An approach to on-site toxaphene treatment could include facilitating dechlorination under conditions favorable to anaerobic microbial metabolism, as studied by Mirsatari et al. (1987), who found that soil and sediment had a dissipation half-life of around 3 weeks and 1 week, respectively. Other laboratory experiments revealed that toxaphene was reductively dechlorinated with Fe (II) protoporphyrin and the redox pair of Fe (II)/Fe (III) seen in marshy sediments of anoxic salt. Reduction happens in sewage sludge and irrigation drainage ditches as well, boosting the creation of products being less chlorinated and hence more volatile. Dehydrochlorination can also happen, especially in alkaline environments, resulting in reduced molecular weight compounds. Two potential environmental effects of toxaphene dechlorination processes include the formation of a residue that may volatilize from the soil more quickly than the technical material and a residue that may be more susceptible to subsequent chemical and/or biological breakdown in a variety of soil conditions. For the on-site remediation of soils polluted with toxaphene and probably other halogenated chemicals, both approaches may be helpful.

Recent advancements include extensive papers and data sheets on non-combustion methods for POP treatment developed by the UNEP (United Nations Environment Programme) and the IHPA (International HCH and Pesticides Association) (Rubin & Burhan, 2006). Bio-stimulation with amendments has been used in anaerobic of soil and sediment. It entails the use of biological additives such as dried and powdered animal blood known as blood meal as a nutrition, and a pH buffer mostly a phosphate buffer, to stimulate the breakdown of toxaphene by anaerobic microbes (Allen et al., 2002). Anaerobic bioremediation using blood meal has been used to remediate toxaphene-contaminated low-strength waste. Several animal dip vat locations have employed this method to treat toxaphene. In 2004, full-scale implementation cleaning costs ranged from $98 to $296 per cubic yard. The Environmental Response Team (ERT) of the US EPA created the technique. This technique is not copyrighted and is freely available to the public (Rubin & Burhan, 2006).

Another approach for the similar objective is DARAMEND®, which is also an amendment-enhanced technology mainly utilized to remediate toxaphene-contaminated low-strength wastes. The DARAMEND® can be used both ex situ and in situ. This approach entails creating anoxic and oxic conditions sequentially (Phillips et al., 2001).

7 Conclusion

Toxaphene is a combination of different congener mostly composed of chlorinated bornanes, bornenes, bornadienes, camphenes, and dihydrocamphenes. Being a non-systemic contact insecticide, it was first used in cotton and other crops in 1949. Toxaphene emissions were mostly caused by its usage as a contact insecticide and acaricide in agriculture, including the cotton growing industry, the production of crops such as grains, nuts, cereals, soybeans, vegetables and fruits, the control of pests especially the likes of mites in cattle, and its use as a piscicide. An important dispersion mechanism of it is atmospheric transport and subsequent deposition. It has thus been found all over the world, including the Arctic. Residues have been found in biota such as fish, molluscs, etc. Since global emissions are crucial to evaluate because of their persistence and potential for long-distance atmospheric movement, toxaphene's utilization was projected to be 1,330,000 tonnes in 1950 and 1993 (Jongbloed et al., 2000). Toxaphene usage and production in India have not been adequately documented because it was banned for such purposes relatively early. On July 25, 1989, the Ministry of Agriculture issued a notification, S. O. 569 (E), formed an expert committee and proceeded (in consultation with the Registration Committee) to pass the order to completely ban toxaphene from production, use, import, export, manufacturing, and distribution (Toxics Link, 2018).

Previously, occupational toxaphene exposure risks were greatest for toxaphene manufacturers, on-field pesticide distributors, cotton farmers, etc. The studies conducted to assess its concentrations in the various environmental compartments and its subsequent effects on biota have enabled the approach for its mitigation. These approaches have in turn contributed in the understanding of their origins and fates, particularly the loads on different environmental matrices due to re-emissions and loads caused by long-distance transboundary transport from countries still having significant stockpiles of toxaphene.

Toxaphene's legacy thus serves as a reminder of the importance of responsible chemical usage and proactive efforts to safeguard our globe and future generations from the detrimental effects of persistent organic pollutants as we face environmental concerns.

References

Allen Harry, L., Mandel Robert, M., Tom, D., Dan, S., Michael, T., Crouse Daniel, G., … & Douglas, H. (2002). Anaerobic bioremediation of toxaphene-contaminated soil: A practical solution. In *17th World congress of soil science*, Bangkok (Thailand), 14–21 August 2002.

ATSDR. (1996). *Toxicological profile for Toxaphene*. Agency for Toxic Substances and Disease Registry. http://www.atsdr.cdc.gov/toxprofiles/tp94.pdf

Carvalho, F. P., Montenegro-Guillén, S., Villeneuve, J. P., Cattini, C., Tolosa, I., Bartocci, J., et al. (2003). Toxaphene residues from cotton fields in soils and in the coastal environment of Nicaragua. *Chemosphere, 53*(6), 627–636.

EPA. (2005). *Toxic chemical release inventory reporting forms and instructions: Revised 2004 version*. Section 313 of the Emergency Planning and Community Right-to-Know Act (Title III of the Superfund Amendments and Reauthorization Act of 1986). U.S. Environmental Protection Agency. Office of Environmental Information EPA260B05001.

Harner, T., Wideman, J. L., Jantunen, L. M. M., Bidleman, T. F., & Parkhurst, W. J. (1999). Residues of organochlorine pesticides in Alabama soils. *Environmental Pollution, 106*(3), 323–332.

Harner, T., Bidleman, T. F., Jantunen, L. M., & Mackay, D. (2001). Soil—Air exchange model of persistent pesticides in the United States cotton belt. *Environmental Toxicology and Chemistry: An International Journal, 20*(7), 1612–1621.

Jantunen, L. M., & Bidleman, T. F. (2003). Air—Water gas exchange of toxaphene in Lake Superior. *Environmental Toxicology and Chemistry: An International Journal, 22*(6), 1229–1237.

Jongbloed, R. H., Visschedijk, A. J. H., Van Dokkum, H. P., & Laane, R. W. P. M. (2000). *Toxaphene: An analysis of possible problems in the aquatic environment*.

Kannan, K., Battula, S., Loganathan, B. G., Hong, C. S., Lam, W. H., Villeneuve, D. L., et al. (2003). Trace organic contaminants, including toxaphene and trifluralin, in cotton field soils from Georgia and South Carolina, USA. *Archives of Environmental Contamination and Toxicology, 45*, 0030–0036.

Kumari, K. (Ed.). (2021). *Persistent organic pollutants: Gaps in management and associated challenges* (1st ed.). CRC Press. https://doi.org/10.1201/9781003046806

Mirsatari, S. G., McChesney, M. H., Craigmill, A. C., Winterlin, W. L., & Seiber, J. N. (1987). Anaerobic microbial dechlorination: An approach to on-site treatment of toxaphene-contaminated soil. *Journal of Environmental Science & Health Part B, 22*(6), 663–690.

National Toxicology Program. (1979). Bioassay of toxaphene for possible carcinogenicity. *National Cancer Institute Carcinogenesis Technical Report Series, 37*, 1–104.

Phillips, T., Bell, G., Raymond, D., Shaw, K., & Seech, A. (2001). DARAMEND® technology for in situ bioremediation of soil containing organochlorine pesticides. In *6th International HCH and Pesticides Forum*.

PIB. (2023). *Banning of pesticides*. Press Information Bureau. Ministry of Agriculture & Farmers Welfare. https://pib.gov.in/PressReleaseIframePage.aspx?PRID=1896140.

Rubin, E., & Burhan, Y. (2006). Noncombustion technologies for remediation of persistent organic pollutants in stockpiles and soil. *Remediation Journal: The Journal of Environmental Cleanup Costs, Technologies & Techniques, 16*(4), 23–42.

Stockholm Convention. (2001). *Listing of POPs in Stockholm Convention*. Retrieved September 7, 2022, from https://chm.pops.int/TheConvention/ThePOPs/The12InitialPOPs/tabid/296/Default.aspx

Stockholm Convention. (2008). *What are POPs?* Stockholm Convention. http://chm.pops.int/Convention/ThePOPs/tabid/673/language/en-US/Default.aspx

Toxics Link. (2018). Retrieved in October 18, 2022, from https://ipen.org/sites/default/files/documents/POPs%20Country%20Situation%20report.pdf

U.S. Department of Health and Human Services. (1993). *Registry of Toxic Effects of Chemical Substances (RTECS, online database)*. National Toxicology Information Program, National Library of Medicine, Bethesda, MD.

U.S. Environmental Protection Agency. (1999). *Integrated Risk Information System (IRIS) on Toxaphene*. National Centre for Environmental Assessment, Office of Research and Development, Washington, DC.

United States. Environmental Protection Agency. Office of Drinking Water. (1987). *Health advisories for 16 pesticides* (Vol. 87, No. 200176). The Office.

Vetter, W., & Scherer, G. (1998). Variety, structures, GC properties and persistence of compounds of technical toxaphene (CTTs). *Chemosphere, 37*(9–12), 2525–2543.
Wong, F., Alegria, H. A., Jantunen, L. M., Bidleman, T. F., Salvador-Figueroa, M., Gold-Bouchot, G., et al. (2008). Organochlorine pesticides in soils and air of southern Mexico: Chemical profiles and potential for soil emissions. *Atmospheric Environment, 42*(33), 7737–7745.

Polychlorinated Biphenyls (PCBs)

Kanchan Kumari and Senerita Swamy

1 Introduction

PCBs, also known as polychlorinated biphenyls, are part of the broad category of organic chemicals synthesized by humans called chlorinated hydrocarbons. There are many different industrial and commercial uses for this class of chlorinated organic chemicals. PCBs are used in technical mixtures rather than as single compounds. Tri-, tetra-, and pentachlorinated biphenyls make up 70% of all PCBs produced worldwide, with trichlorinated biphenyls serving as the dominant homolog (Breivik et al., 2002). Pentachlorobiphenyl was primarily used as an additive, whereas trichlorobiphenyl was primarily used in electric power capacitors and transformers (You et al., 2011). In addition to being used as electric transformer and capacitor additives, these substances are also used in the manufacturing of plastics, paint, and carbonless copy paper. Thirteen of the 209 different PCB types show toxicity similar to that of dioxin. Their environmental persistence is correlated with the level of chlorination, and their half-lives range can be upto ten days to one and half years. Biphenyl is converted into polychlorinated biphenyls (PCBs) by replacing one to ten hydrogen atoms with chlorine atoms. Mono-chlorobiphenyl 3, di-12, tri-24, tetra-42, penta-46, hexa-42, hepta-24, octa-12, nona-3, and decachlorobiphenyl 1 are the isomers that make up each homolog group. There are a total of 209 potential PCB congeners. The PCB congeners, also known as chlorobiphenyls or CBs, are typically identified with numerals ranging from 1 (mono-Cl) to 209

K. Kumari (✉)
CSIR-National Environmental Engineering Research Institute (NEERI), Kolkata Zonal Center, Kolkata, West Bengal, India
e-mail: k_kumari@neeri.res.in

S. Swamy
CSIR-National Environmental Engineering Research Institute (NEERI), Nagpur, India

K. Kumari, *Pollutants of Global Concern*, Emerging Contaminants and Associated Treatment Technologies, https://doi.org/10.1007/978-3-031-50996-4_10

(deca-Cl) (Table 1). The chemical stability of PCBs includes resistance to both strong acids and bases as well as resistance to oxidation and reduction. They are also non-flammable, heat stable, and soluble in a variety of organic solvents with low electrical conductivity. An estimated 1.2–1.5 million metric tonnes of PCBs have been manufactured since the industrial manufacturing of PCBs commenced in 1929. In most nations, production came to an end in the 1970s and 1980s after reaching its peak in the late 1960s. Even in 2006, it was uncertain whether production had entirely ceased in all nations. According to Reddy et al. (2019) approximately one third of all PCBs manufactured in the United States (US) are thought to have been released into the environment. After officials learned about PCBs' harmful effects on the environment due to their persistence, bioaccumulative properties, and toxicity, PCB production was stopped worldwide sometime between the late 1970s and the early 1980s. Aroclor (USA), Kanechlor (Japan), Chlophen (Germany), Fenclor (Italy), and other commercial PCB products are manufactured under these names, among others. The toxicology of PCBs has attracted a lot of attention recently, especially with regard to the congeners. On May 22, 2001, the Stockholm Convention classified PCBs as persistent organic pollutants (POPs) in response to growing concerns regarding the risks associated with their use.

2 Technical Details of PCB

Table 1 Physicochemical properties of PCB

Details	Description (PCB)
Structure	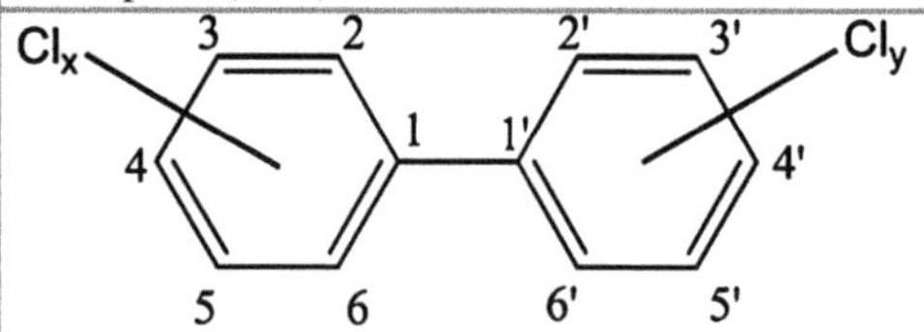 Since structural isomers are feasible at each degree of chlorination (homolog), there exists 209 possible PCBs (known as congeners). A limited number of PCBs also exist as enantiomers, with the link between the phenyl rings acting as the chiral center. When three or four chlorine atoms are arranged in ortho positions in these compounds, steric hindrance prevents free rotation
IUPAC	PCB congeners are commonly referred to by two names: an IUPAC name, such as 2,2′,5-trichlorobiphenyl, and a numbering system, such as PCB 18, or #18, developed by Ballschmitter and Zell and accepted by the IUPAC. PCB 1 is 2-monochlorobiphenyl, and PCB 209 is decachlorobiphenyl
Molecular formula	$C_{12}H_{10-x}Cl_x$
CAS No.	1336-36-3
Molecular weight	Variable
Color/form	Commercial mixtures are mostly viscous liquids; however, compounds containing highest chlorine content are solids at room temperature

(continued)

Table 1 (continued)

Details	Description (PCB)
Solubility	Soluble in a wide range of organic solvents, at 25 °C, the solubility of PCBs in water ranges from 2×10^{-6} mol m^{-3} for decachlorobiphenyl to 29 mol m^{-3} for monochlorobiphenyl
Conductivity	Low electrical conductivity
Vapor pressure	1.2 pa for monochlorinated congeners to 7×10^{-6} pa for decachlorobiphenyl at 25 °C
Octanol/water partition coefficient	4.2 °C (monochlorobiphenyl) to 8.3 °C (decachlorobiphenyl) at 25 °C

3 Status of Works and Usage in India

PCBs were used in industrial applications but were never produced in India. Since 1966, PCBs have been found in environmental matrices, and their route of exposure in soil includes the deposition of industrial applications, incinerators, and biomass combustions. Due to their characteristics, PCBs adsorb to particle of soil and linger there for a longer period. India is required to uphold the goals of minimizing and eventually getting rid of these pollutants as a signatory to the Stockholm Convention (Table 2), and it is encouraged to carry out POPs research. In India, some research has been done on POPs like the intentionally released insecticide DDT, but little has been done regarding the accidentally released POPs like PCBs and dioxins.

Table 2 Uses, sources, and regulation in India (Kumari, 2021)

Identified POP	Uses and sources	Annex	Year of listing by SC	Regulation in India
PCBs	In addition to transformer, capacitors, hydraulic fluids, heat transfer flame retardants, inks, adhesives, microencapsulated dyes for carbonless copying paper, pesticide extenders, paints, plasticizers, slide mounting mediums for microscopes, polyolefin catalyst carriers, surface coatings, wire insulators, and metal coatings, PCBs were also used in these applications. PCBs have mostly been used as insulating dielectric materials in electrical machinery like capacitors and transformers.	A and C	2001	PCB manufacture and new applications are prohibited, and parties to the Stockholm Convention are required to stop using PCB in equipment by 2025 and to guarantee that PCB-contaminated equipment and liquids are disposed of in a sustainable way by 2028.

Table 3 Effects, toxicity, and persistence of PCBs

DDT	Data of India
Effects	Humans have mainly been exposed to relatively high levels of PCBs while working in establishments that produced and utilized PCBs and PCB-containing equipment on a large scale. Chloracne (a rare form of acneiform dermatosis brought on by exposure to halogenated aromatic compounds) and related dermal lesions, as well as respiratory issues, are just a few of the wide range of effects that occupational exposure to PCBs can have (Emmett, 1985; Lawton et al., 1985; Safe, 1993). Low-level environmental PCB exposure may have harmful effects on human health, but these effects are complex and still need to be confirmed (Safe, 2007).
Toxicity	PCBs have a significant impact on all ecosystem components as well as individual organisms. Due to its carcinogenicity and various toxicity reported in brain cells, reproductive system immune system, hepatic system, and cardiovascular system, PCBs have been shown to potentially be detrimental to human health.
Persistence/ fate	PCBs are among the most evident and pervasive environmental contaminants due to their chemical stability, which additionally renders it environmental persistent nature. The ability of PCBs to biodegrade itself in the environment is quite slow as compared to other organic compound. PCBs are easily transported from nearby or regional contamination sites to remote areas due to their low vapor pressure and low water solubility in combination with water, air, and soil transport processes.

Studies on the assessment of PCBs in environmental matrices have been conducted in India (Kumar et al., 2013; Chakraborty et al., 2018). However, there are not many publications on the evaluation of the human health risk caused by PCBs. In particular, PCBs in soils and the risk they pose were detected in North India, but very few research was done in Central part of India. So, in the industrialized city of Korba in Central India, Kumar et al. (2018) identified 28 PCBs, which includes 12 dioxin-like PCBs, in residential land and their potential health risks. This initial data was compared to results from other research, that can be helpful in determining the hazards that industrial pollutants pose to the general public.

Burning biomass and emissions from the wood processing, paint and dyeing, chemical, and electronic recycling industries are two potential sources of PCBs in India. The Yamuna River receives domestic and industrial wastewater from factories, including those that produce PVC (polyvinyl chloride), paint, textile, iron, and paper, as well as off-gassing from older equipment. The PCBs, which are frequently used in industrial production, might also be found in the waste stream from electronic devices. An estimated 150,000 t of electronic waste were produced in India in 2005, and the majority of this waste is recycled informally in New Delhi, which is a potential source of PCBs (Amoyaw-Osei et al. 2011). Table 3 summarizes the effects, toxicity, and persistence of PCBs.

4 Degradation and Environmental Fate

PCBs are persistent environmental contaminants owing to their chemical stability. A significant issue arises when these unfavorable characteristics are coupled with PCBs' low water solubility because these chemicals may accumulate through the food web and reach humans through aquatic species such as fish, shrimp, and crab. PCBs can, however, be dehalogenated in freshwater and in estuarine sediments, which has attracted a lot of attention (Furukawa, 2000; Girones et al., 2021; Zanaroli et al., 2015; Berkaw et al., 1996). The great deal of interest in the discovery is due to the possibility that dehalogenation will detoxify these toxic substances (Bedard et al., 1997). Two other significant routes for PCB loss include volatilization and colloidal adsorption to dissolved organic matter, which reduces bioavailability. PCBs in soil or aquatic environments can only be broken down by the biodegradation process. PCBs are expected to break down biologically to release CO_2, chlorine, and water. The biphenyl ring is subjected to chlorine removal (Fig. 1) in this process, which is followed by ring breakage and oxidation of the resultant product (Boyle et al., 1992). Numerous investigations have demonstrated that different PCB dechlorination pathways seem to be responsible for microbial dechlorination (Bedard, 2008; Khalid et al., 2021), but for maximum chlorine removal, two or more such dechlorination processes must work in concert (Quensen et al., 1990). According to a theory put forth by Bedard et al. (1997), the various dechlorination processes that have been described are all caused by discrete dechlorinating microorganisms that contain dehalogenases with various regiospecificities.

5 International Scientific Work on PCBs

The ecosystem only receives PCBs through human activity. Accordingly, regions with a high concentration of PCB contamination are often industrial. Because PCBs are so pervasive in the ecosystem as a whole, they are carried by the wind and the water. In the soil, water column, sediments, and air particulates, PCBs adhere

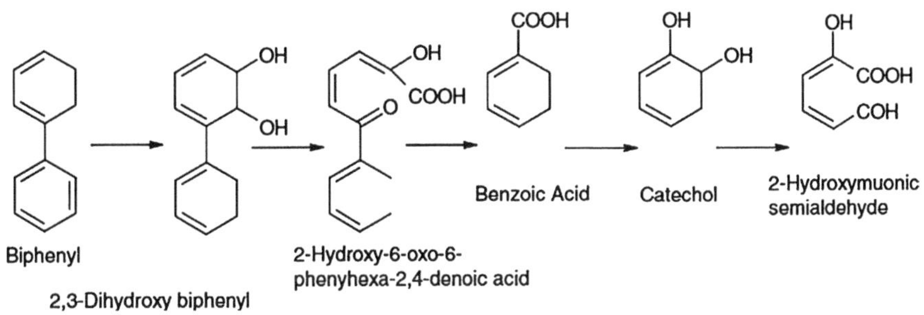

Fig. 1 PCB dechlorination pathways

tenaciously to organic particles that are present. PCBs are predominantly accumulated in sediments and soils as a result of atmospheric deposition of particle-bound PCBs and sedimentation from water. Waste streams introduced into receiving waters serve as the main pathway for PCB transmission via aquatic systems. According to Swackhamer and Armstrong (1986), environmental cycling is currently the main cause of PCB emission into surface water. The most significant levels of PCB-153 were found in children aged 6–9 years (7.3 ng/g lipid), 12–19 years (5.4 ng/g lipid), and adults over 20 years (24.2 ng/g) between 2005 and 2007 according to an investigation done in California and Ohio (Whitehead et al., 2010). As of right now, the threshold values of total PCB per gram of maternal plasma lipids recorded in international literature for infants under 3 years old, nursing mothers, and women of reproductive age were between 700 and 1000 ng.

Polychlorinated biphenyls (PCBs) were outlawed in Lebanon in 1982 and 1997, respectively, and the Stockholm Convention on persistent organic pollutants (POPs) was ratified in 2003. A nationwide inventory of PCBs, which were mostly found in insulating oils and equipment in power plants, is part of a nationwide Implementation Plan for POPs that has been developed and revised multiple times. Additionally, sediments from the Port of Tripoli, Lebanon's second-largest seaport, show high PCB concentrations. Studies on PCB biomonitoring in humans have also been conducted, and various researches found detectable PCB levels in human serum (Du et al., 2020; Schettgen et al., 2018; Polachova et al., 2021; Chang et al., 2020; Bandow et al., 2020). These levels were somewhat greater than those noted by the National Health and Nutrition Examination Survey, although being much below those seen in other nations. Municipal sewage systems continue to operate as continuous secondary sources for pollution in receiving aquatic life notwithstanding the prohibition on new PCB manufacturing and commercial usage. By strategically collecting the freely dissolved and PCBs associated with biosolids, Needham and Ghosh (2019) reported a 16-month monitoring to ascertain the inputs, fate, and export of PCBs throughout a municipal waste collection as well as treatment system in Baltimore city. A total of 170 g/day of PCBs were identified to be entering the treatment facility, from this, 100 g/day left the facility with the biosolids and approximately 5.2 g/day freely dissolved PCBs was released in the waste water. The net loss of 68 g/day due to volatilization and biodegradation occur in the facility. The amount of freely dissolved PCBs in the treated waste water detected was significant factor in the dissolved quantity in the river and to be an order of concentration greater than the water quality standards for the safety of human health through fish intake. The biosolids had an average PCB content of 760 g/kg and were being fed to the ground as fertilizer. A rise in PCB levels in the earthworm *E. fetida* caused by the addition of biosolids to clean cultivable land was caused by a six times increase, which could be removed by adding 1% by weight of activated carbon. In a study, Saktrakulkla et al. (2020) used the Diet History Questionnaire II to determine the dietary exposure to PCBs in women and children participating in the study and found that dietry exposure was also comparable to Airborne Exposures to Semivolatile Organic Pollutants (AESOP). Salmon and canned tuna had the greatest concentrations and egg showed the lowest PCB concentration, although seafood only makes up a small portion of total exposure. In this study group, other animal derived proteins are more significant dietary route of PCB exposure. In most meals, the scientists discovered strong correlations between PCBs' congener distributions and aroclors as well as indications of a drop in PCB concentrations over the last 20 years. They also reported PCBs production by-products, including PCB11, in catfish and tilapia.

6 In-House (Countrywide) Scientific Work Done on the Chemical

In the 1970s and 1980s, majority of the developed nations outlawed the manufacture and utilization of PCBs. However, the production and usage of PCBs continued in developing nations like India, particularly in the shipbreaking industries, which is a major source of PCBs around the Indian Ocean. According to Hess et al. (2001), it can produce 0.25–0.8 metric tonnes of PCBs per demolished merchant ship. With 38% of all global shipbreaking operations, India is the leading shipbreaking nation in the world in terms of volume. According to Andeobu (2021), India produces roughly 146,000 tonnes of garbage annually from obsolete or damaged electrical and electronic equipment. For persistent organic chemicals, soils serve as a significant reservoir and sink (Valle et al., 2005; Wild & Jones, 1995). According to a study by Devi et al. (2014), among the 25 PCBs analyzed in the Manipur region, hepta-PCB, penta-PCB, tri-PCB, and tetra-PCB congeners were relatively prevalent in soil and air. Compared to PCB concentrations found in air samples, PCB concentrations in soil were two to three times greater. High PCB concentrations were also found in the air in cities. The content of PCBs in Manipur's soil is a major factor in regulating PCB levels in the atmosphere. The stated TEQ level in Delhi's agricultural soil was lower than the PTEQ level in the soil sample. The sources of the PCB that has been detected at the current study site may include local emissions and long-range atmospheric transport, primarily from southern and northern India and to a lesser extent till Myanmar. In a similar study, Kumar et al. (2012) provided information on the concentrations of PCBs in surface water from various sources in the NCR region. The majority of PCBs in this study region were found in PCB congeners with three to five chlorines. In NCR surface waters, the tetra-PCB predominated the PCB homologs on average. Twelve dioxin-like PCBs make up 12% of the total 28 PCBs in existence. However, PCB levels at some areas were greater than recommended levels and Criteria Continuous Concentration (CCC), which suggests potential chronic toxicological effects on continuous contact. The average concentrations were lower than recommended values. In a recent research, PCB contamination and sources of ambient air, road dust, and sediments in Raipur, India were assessed from 2008 to 2015. Road dust, ambient air (related to particle pollution), and sediment samples had total PCB values in 2008 that ranged from 186 to 645 pg/m^3, 102 to 537 ng/g, and 241 to 538 ng/g, respectively. The two most common chemical substances found were 2- and 4-chlorobiphenyl (PCB-1 and PCB-3, respectively). The PCBs in the silt were found to be migrating substantially vertically (Nayak et al., 2021). According to the statistics, India has to put more efforts to meet the Stockholm Convention's deadlines for eliminating PCBs from equipment by 2028 and phasing out their usage in equipment by 2025 (Ripley, 2017).

7 Adverse Effects of PCBs

PCBs are categorized as potential carcinogenic which belong to Group 2A by the International Agency for Research on Cancer (IARC). The brain, endocrine, reproductive, cardiovascular and the immunological systems are just a few of the human systems that PCBs impact.

I. *Carcinogenic Potential:* PCBs have been linked to the emergence of a number of cancer types (ATSDR, 2000; Liu et al., 2010; Lauby-Secretan et al., 2016). According to statistics from Serpukhov, the incidence of cancer was 335 cases per 100,000 people in a district where PCBs were present at a concentration of 1200–30,000 lg/kg dw soil, compared to 165 cases per 100,000 people in a district where PCB concentrations were lower (Orlinskii et al., 2001). According to medical statistics from Dzerzhinsk, Russia, there were 462.6 cancer cases per 100,000 residents in 2009 (Dolgova & Denisenko, 2010). According to epidemiological studies conducted in Kazakhstan, Ablaketka (35,700–1,720,000 lg PCB/kg dw soil) has 749.8 incidences of malignant neoplasm per 100,000 residents (Ishankulov, 2008).

II. *Neurobehavioral Effects:* Neurotoxicity and abnormalities of neurodevelopment are linked to PCB exposure. According to epidemiological research, children who are exposed to PCBs are more likely to experience neuropsychiatric problems (Klocke et al., 2020). Many axially chiral neurotoxic PCB congeners have cellular targets that are affected by PCB neurotoxicity (Uwimana et al., 2019). Multiple neurologic illnesses, such as Parkinson's disease, Alzheimer's disease, dementia, and amyotrophic lateral sclerosis (ALS), can develop in humans after exposure to PCBs. PCBs are converted to phenols through the microsomal monooxygenase system, which is catalyzed by CYP450. These phenols can undergo further hydroxylation to become catechols. These intermediates, which are electrophilic by nature, can form covalent bonds with nucleophilic biological macromolecules (such as protein, RNA, and DNA) and, after doing so, break and repair DNA strands, which adds to PCBs' toxicity. Additionally, arene oxide intermediates can combine with glutathione to generate conjugates, which then go on to form methyl sulphonyl metabolites through additional metabolism. These metabolites' binding to certain proteins may contribute to the negative effects of PCBs. Some proteins may play a role in the negative effects of PCBs, according to a theory that suggests hydroxylated PCB metabolites can also contribute to PCBs' toxicity (Hofvander, 2006).

III. *Reproductive Effects:* The health of human reproduction can also be impacted by PCBs. The Inuit population residing in Greenland, the urban populations from Warsaw in Poland, Kharkiv in Ukraine and the Swedish fishermen and their wives are just a few of the four diverse populations that Aakash et al. (2021) reported the adverse effects of PCBs on reproductive health in humans. These populations demonstrate the wide variances in PCB exposure levels caused by individual differences in dietary contamination. According to this study's findings, male and female reproductive functioning in Arctic and European populations may be altered by PCB-153 at the present levels of PCB exposure in these communities (Toft, 2014). However, the specific pathway through which PCBs might have negative consequences is still not entirely recognized (Portigal et al., 2002). PCBs have been reported in fetal cord blood, breast milk, uterine muscle, ovarian tissue, follicular and amniotic fluid, and uterine and amniotic fluid (Huang et al., 2019; Bloom et al., 2017). Infertility can result from ovarian damage caused by a woman's exposure to PCBs, which can also cause aberrant hormone levels, early ovarian failure, and other reproductive issues (Patel et al., 2015). The beginning of menopause sooner, decreased menstrual function, increased incidence of miscarriage, and a longer gestation period are additional negative consequences in women related to PCB exposure (Montano et al., 2022; Toft et al., 2010; Grindler et al., 2015). Contrary to this evidence, a recent study examined the relationship between serum PCBs' concentration and a variety of female reproductive health outcomes and found no association between serum PCB concentration and infertility, only a relationship between PCBs' concentration and the decline in the number of pregnancies (Neblett et al., 2020). Additionally, a significant number of studies that linked PCB exposure during pregnancy to reduced cognitive abilities in children did not account for several potential interfering elements, such as concurrent exposure to other hazardous substances including Hg, Pb, and iodine (Kalloo et al., 2020, 2021; Brucker-Davis et al., 2015; Kim et al., 2018).

IV. *Other Effects:* Other negative health consequences from PCB exposure have been hypothesized or verified by occupational and epidemiologic research. Cardiovascular, gastrointestinal, immune, and musculoskeletal systems may all be impacted by these health impacts. Immune system impacts in populations exposed to PCBs have been shown (Montano et al., 2022). These effects include changes in the ratio of helper to killer (CD4+/CD8+) T-cells, drops in IgA and IgM antibody levels, drops in monocyte and granulocyte counts, and drops in natural killer cell count. According to Guo et al. (1995), the immunosuppressive effects of PCB exposure were linked to an increased frequency of chronic respiratory infections and improved response to mitogens in the Yusho and Yu-Cheng populations. Workers who manufacture transformers and electrical equipment and exposed to different PCB-containing combinations have been reported significant appetite loss. Workers exposed to PCBs also report experiencing epigastric pain and discomfort, nausea, and an aversion to fatty meals (Carlson et al., 2022; Emmett et al., 1988). According to research (Robertson & Ludewig,

2011), a number of PCB metabolites cause gene mutations, chromosomal breakage, chromosome loss, and polyploidization in cells in culture. The study also offered the first proof that a PCB congener is mutagenic in vivo.

8 Management and Remedial Measures

A study of PCB elimination operations worldwide was conducted by UNEP in 2016, and an update was issued in 2019. The study highlighted issues that were stated, including incomplete PCB inventories in several places and significant ambiguities regarding existing supplies and the scope of ESM. According to the research, "the majority of countries are currently not on track to achieve the ESM of PCBs by 2028." The PCBs that have already been eradicated, according to UNEP, were probably "low-hanging fruit." PCBs, which are still used in open applications and are rarely documented even in nations that report ESM, provide a significant concern. While there have been worldwide inventories of PCB usage and emissions, estimates for specific nations based on bottom-up methods are rare and frequently indicate patterns that diverge from the global inventory. Parmar and Qureshi (2023) estimated middle (low-high) emissions between 1950 and 2100 in their assessment based on mass balance model. Transformers have utilized up to 7296 tonnes of PCBs. It is estimated that there are around 5000 (2400–9100) tonnes of PCBs imported as garbage. Transformer usage and disposal, industrial activities, and the disposal of imported garbage are all expected to produce a total of 13 (0.1–537) tonnes, 89.26 (0.5–178) tonnes, and 63 (3–910) tonnes of emissions, respectively, between 1950 and 2100. Depending on the emission scenarios, industrial emissions are growing in importance as sources of PCBs and may even supersede. The management of equipment containing PCBs is a problem in India, since there is a lack of information on PCBs in the electrical manufacturing sector. PCB-contaminated trash should not be disposed off until all used oil is sold to local businesses for various purposes. The choices for disposing of PCBs are often limited. Figure 2 represents the different guidelines and standards set for PCBs by different regulatory authorities. Despite being covered by the Hazardous Waste Management Rules, handling of PCBs falls short of meeting the Convention's requirements. There is a lack of qualified personnel to oversee the application of the relevant laws and regulations. Better enforcement of the laws governing the management of hazardous waste at shipbreaking yards is required. Effective institutional setup and a thorough legal framework are key components of effective management and control of PCBs. The establishment of an efficient management system based on rational decision-making should be prioritized in light of the current scenario in India. The current laws and statutes need to be reformed to reflect the unique constraints with PCB management in the nation. Establishment and implementation of PCBs' rules are absolutely necessary. After the Convention's ratification and legitimate enforcement, the Parties were given a lot of time. It is essential for the authorities to make informed recommendations. Any PCB-related issues cannot be resolved

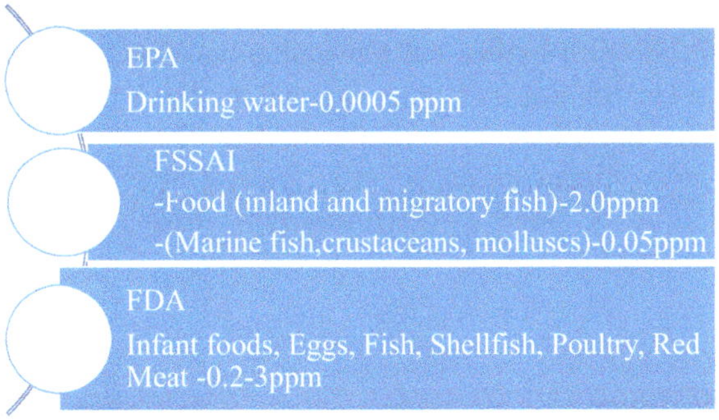

Fig. 2 Guidelines and standards sets by various regulatory bodies for PCBs

successfully if only government administrative agencies and research institutions are engaged in the decision-making process. According to Van et al. (2004), PCBs can be emitted into the atmosphere by burning PCB-containing waste, leaking from landfills containing PCB-containing items, and disposing of industrial waste. In the USA, landfill disposal or incineration is the most popular remediation techniques for PCB-contaminated soil or sediments (Gomes et al., 2013). The use of other techniques including biological, chemical, physical, and thermal procedures is also common in remediation of PCB. The most common remediation techniques are phytoremediation, microbiological degradation, chemical reagent dehalogenation, and activated carbon removal of PCBs. Recent studies on remediation of PCBs have revealed new insights about nanoscale zero-valent iron based reductive dehalogenation and biofilm covered activated carbon, supercritical water oxidation, ultrasonic radiation, bimetallic systems, electro kinetic remediation, and nZVI particles in combination with a second metal. Some of these techniques still need further investigation because they are in the early stages of development. Although several potential technologies have been proposed, there is no one well-developed technology. Therefore, it may be an excellent strategy to use a combination of methods for the removal of PCBs. A holistic treatment strategy may be the future of remediation of PCBs. Since effective remediation of PCBs depends on a number of factors, including the acceptance of the remediation approach, the feasibility of the approach, the overall cost-effectiveness, and its effects on the environment and human health. Since none of these goals have been met till now, remediation of PCBs is still a futuristic approach.

9 Conclusion

Numerous studies have shown the presence and geographic distribution of PCBs in different media; however, more research is needed to understand the interactions and transformations of PCBs in different matrices. Future studies should place more emphasis on the overall environmental pressure than on particular matrix like the atmosphere, soil, water, and organisms. A more thorough investigation is necessary on the distribution and transportation of PCBs across national as well as global levels. Studies on human external exposure and its connection to health concerns have been done. As a result, the majority of research on external exposure might underestimate the danger of PCB exposure to humans. However, additional research has to be done to determine how these matrices can more accurately reflect on humans.

References

Akash, M. S. H., Rehman, K., & Hashmi, M. Z. (Eds.). (2021). *Endocrine disrupting chemicals-induced metabolic disorders and treatment strategies*. Springer.

Amoyaw-Osei, Y., Agyekum, O., Mueller, E., Fasko, R., Schluep, M. (2011). *Ghana E-Waste country assessment*. Secretariat of the Basel Convention. Accessed at www.ewasteguide.info/Amoyaw-Osei_2011_GreenAd-Empa. Date Accessed 20th July, 2016

Andeobu, L., Wibowo, S., & Grandhi, S. (2021). A systematic review of e-waste generation and environmental management of Asia Pacific countries. *International Journal of Environmental Research and Public Health, 18*(17), 9051.

ATSDR, A. (2000). *Toxicological profile for polychlorinated biphenyls (PCBs)*. Agency for Toxic Substances and Disease Registry.

Bandow, N., Conrad, A., Kolossa-Gehring, M., Murawski, A., & Sawal, G. (2020). Polychlorinated biphenyls (PCB) and organochlorine pesticides (OCP) in blood plasma–Results of the German environmental survey for children and adolescents 2014–2017 (GerES V). *International Journal of Hygiene and Environmental Health, 224*, 113426.

Bedard, D. L. (2008). A case study for microbial biodegradation: Anaerobic bacterial reductive dechlorination of polychlorinated biphenyls—From sediment to defined medium. *Annual Review of Microbiology, 62*, 253–270.

Bedard, D. L., Van Dort, H. M., May, R. J., & Smullen, L. A. (1997). Enrichment of microorganisms that sequentially meta, para-dechlorin-ate the residue of aroclor 1260 in Housatonic River sediment. *Environmental Science & Technology, 31*(11), 3308–3313.

Berkaw, M., Sowers, K. R., & May, H. D. (1996). Anaerobic ortho dechlorination of polychlorinated biphenyls by estuarine sediments from Baltimore Harbor. *Applied and Environmental Microbiology, 62*(7), 2534–2539.

Bloom, M. S., Fujimoto, V. Y., Storm, R., Zhang, L., Butts, C. D., Sollohub, D., & Jansing, R. L. (2017). Persistent organic pollutants (POPs) in human follicular fluid and in vitro fertilization outcomes, a pilot study. *Reproductive Toxicology, 67*, 165–173.

Boyle, A. W., Silvin, C. J., Hassett, J. P., Nakas, J. P., & Tanenbaum, S. W. (1992). Bacterial PCB biodegradation. *Biodegradation, 3*, 285–298.

Breivik, K., Sweetman, A., Pacyna, J. M., & Jones, K. C. (2002). Towards a global historical emission inventory for selected PCB congeners—A mass balance approach: 2. Emissions. *Science of the Total Environment, 290*(1–3), 199–224.

Brucker-Davis, F., Ganier-Chauliac, F., Gal, J., Panaïa-Ferrari, P., Pacini, P., Fénichel, P., & Hiéronimus, S. (2015). Neurotoxicant exposure during pregnancy is a confounder for assessment of iodine supplementation on neurodevelopment outcome. *Neurotoxicology and Teratology, 51*, 45–51.

Carlson, L. M., Christensen, K., Sagiv, S. K., Rajan, P., Klocke, C. R., Lein, P. J., et al. (2022). A systematic evidence map for the evaluation of noncancer health effects and exposures to polychlorinated biphenyl mixtures. *Environmental Research, 202*, 115148.

Chakraborty, P., Selvaraj, S., Nakamura, M., Prithiviraj, B., Cincinelli, A., & Bang, J. J. (2018). PCBs and PCDD/Fs in soil from informal e-waste recycling sites and open dumpsites in India: Levels, congener profiles and health risk assessment. *Science of the Total Environment, 621*, 930–938, ISSN 0048-9697,. https://doi.org/10.1016/j.scitotenv.2017.11.083

Chang, C. J., Terrell, M. L., Marcus, M., Marder, M. E., Panuwet, P., Ryan, P. B., ... & Barr, D. B. (2020). Serum concentrations of polybrominated biphenyls (PBBs), polychlorinated biphenyls (PCBs) and polybrominated diphenyl ethers (PBDEs) in the Michigan PBB Registry 40 years after the PBB contamination incident. *Environment International, 137*, 105526.

Devi, N., Yadav, I., Qi, S., Chakraborty, P., & Dan, Y. (2014). Distribution and risk assessment of polychlorinated biphenyls (PCBs) in the remote air and soil of Manipur, India. *Environmental Geology, 72*, 3955. https://doi.org/10.1007/s12665-014-3284-8

Dolgova, S. V., & Denisenko, A. N. (2010). Cancer morbidity in the Nizhny Novgorod region. Oncology service state of art and the ways for its improvement. *Meditsinskiy al'manakh, 3*, 42–45.

Du, S., Rodenburg, L., Patterson, N., Chu, C., Riker, C. D., Yu, C. H., & Fan, Z. T. (2020). Concentration of polychlorinated biphenyls in serum from New Jersey biomonitoring study: 2016–2018. *Chemosphere, 261*, 127730.

Emmett, E. A. (1985). Polychlorinated biphenyl exposure and effects in transformer repair workers. *Environmental Health Perspectives, 60*, 185–192.

Emmett, E. A., Maroni, M., Jefferys, J., Schmith, J., Levin, B. K., & Alvares, A. (1988). Studies of transformer repair workers exposed to PCBs: II. Results of clinical laboratory investigations. *American Journal of Industrial Medicine, 14*(1), 47–62.

Furukawa, K. (2000). Biochemical and genetic bases of microbial degradation of polychlorinated biphenyls (PCBs). *The Journal of General and Applied Microbiology, 46*(6), 283–296.

Girones, L., Oliva, A. L., Negrin, V. L., Marcovecchio, J. E., & Arias, A. H. (2021). Persistent organic pollutants (POPs) in coastal wetlands: A review of their occurrences, toxic effects, and biogeochemical cycling. *Marine Pollution Bulletin, 172*, 112864.

Gomes, H. I., Dias-Ferreira, C., & Ribeiro, A. B. (2013). Overview of in situ and ex situ remediation technologies for PCB-contaminated soils and sediments and obstacles for full-scale application. *Science of the Total Environment, 445*, 237–260.

Grindler, N. M., Allsworth, J. E., Macones, G. A., Kannan, K., Roehl, K. A., & Cooper, A. R. (2015). Persistent organic pollutants and early menopause in US women. *PLoS One, 10*(1), e0116057.

Guo, Y. L., Lambert, G. H., & Hsu, C. C. (1995). Growth abnormalities in the population exposed in utero and early postnatally to polychlorinated biphenyls and dibenzofurans. *Environmental Health Perspectives, 103*(suppl 6), 117–122.

Hess, W. R., Rocap, G., Ting, C. S., Larimer, F., Stilwagen, S., Lamerdin, J., & Chisholm, S. W. (2001). The photosynthetic apparatus of Prochlorococcus: Insights through comparative genomics. *Photosynthesis Research, 70*, 53–71.

Hofvander, L. (2006). *Polychlorinated biphenyls and their metabolites in human blood: Method development, identification and quantification*. Doctoral dissertation, Institutionen för miljökemi.

Huang, W., He, Y., Xiao, J., Huang, Y., Li, A., He, M., & Wu, K. (2019). Risk of breast cancer and adipose tissue concentrations of polychlorinated biphenyls and organochlorine pesticides: A hospital-based case-control study in Chinese women. *Environmental Science and Pollution Research, 26*, 32128–32136.

Ishankulov, M. (2008). PCB-contaminated areas in Kazakhstan and analysis of PCB impact on human health experience. In *The fate of persistent organic pollutants in the environment* (pp. 387–403). Springer.

Kalloo, G., Wellenius, G. A., McCandless, L., Calafat, A. M., Sjodin, A., Romano, M. E., et al. (2020). Exposures to chemical mixtures during pregnancy and neonatal outcomes: The HOME study. *Environment International, 134*, 105219.

Kalloo, G., Wellenius, G. A., McCandless, L., Calafat, A. M., Sjodin, A., Sullivan, A. J., et al. (2021). Chemical mixture exposures during pregnancy and cognitive abilities in school-aged children. *Environmental Research, 197*, 111027.

Khalid, F., Hashmi, M. Z., Jamil, N., Qadir, A., & Ali, M. I. (2021). Microbial and enzymatic degradation of PCBs from e-waste-contaminated sites: A review. *Environmental Science and Pollution Research, 28*, 10474–10487.

Kim, A. W., Vane, C. H., Moss-Hayes, V., Engelhart, S. E., & Kemp, A. C. (2018). PAH, PCB, TPH and mercury in surface sediments of the Delaware River Estuary and Delmarva Peninsula, USA. *Marine Pollution Bulletin, 129*(2), 835–845.

Klocke, C., Sethi, S., & Lein, P. J. (2020). The developmental neurotoxicity of legacy vs. contemporary polychlorinated biphenyls (PCBs): Similarities and differences. *Environmental Science and Pollution Research, 27*, 8885–8896.

Kumar, B., Singh, S. K., Mishra, M., Kumar, S., & Sharma, C. S. (2012). Assessment of polychlorinated biphenyls and organochlorine pesticides in water samples from the Yamuna River. *Journal of Xenobiotics, 2*(1), e6.

Kumar, B., Kumar, S., & Sharma, C. S. (2013). Ecotoxicological risk assessment of polychlorinated biphenyls (PCBs) in bank sediments from along the Yamuna River in Delhi, India. *Human and Ecological Risk Assessment: An International Journal, 19*(6), 1477–1487.

Kumar, K. K., Devi, B. U., & Neeraja, P. (2018). Elucidation of endocrine—Disrupting polychlorinated biphenyls binding potency with steroidogenic genes: Integration of in silico methods and ensemble docking approaches. *Ecotoxicology and Environmental Safety, 165*, 194–201.

Kumari, K. (Ed.). (2021). *Persistent organic pollutants: Gaps in management and associated challenges* (1st ed.). CRC Press. https://doi.org/10.1201/9781003046806

Lauby-Secretan, B., Loomis, D., Baan, R., El Ghissassi, F., Bouvard, V., Benbrahim-Tallaa, L., et al. (2016). Use of mechanistic data in the IARC evaluations of the carcinogenicity of polychlorinated biphenyls and related compounds. *Environmental Science and Pollution Research, 23*, 2220–2229.

Lawton, R. W., Ross, M. R., Feingold, J., & Brown, J. F., Jr. (1985). Effects of PCB exposure on biochemical and hematological findings in capacitor workers. *Environmental Health Perspectives, 60*, 165–184.

Liu, S., Li, S., & Du, Y. (2010). Polychlorinated biphenyls (PCBs) enhance metastatic properties of breast cancer cells by activating Rho-associated kinase (ROCK). *PLoS One, 5*(6), e11272.

Montano, L., Pironti, C., Pinto, G., Ricciardi, M., Buono, A., Brogna, C., et al. (2022). Polychlorinated biphenyls (PCBs) in the environment: Occupational and exposure events, effects on human health and fertility. *Toxics, 10*(7), 365.

Neblett, M. F., Curtis, S. W., Gerkowicz, S. A., Spencer, J. B., Terrell, M. L., Jiang, V. S., et al. (2020). Examining reproductive health outcomes in females exposed to polychlorinated biphenyl and polybrominated biphenyl. *Scientific Reports, 10*(1), 3314.

Needham, T. P., & Ghosh, U. (2019). Four decades since the ban, old urban wastewater treatment plant remains a dominant source of PCBs to the environment. *Environmental Pollution, 246*, 390–397.

Nayak, Y., Sahu, Y. K., Patel, K. S., Sharma, S., Hung, C. C., Martín-Ramos, P., & Yurdakul, S. (2021). Distribution and sources of polychlorinated biphenyls in air, dust, and sediment from India. *Journal of Hazardous, Toxic, and Radioactive Waste, 25*(1), 05020004.

Orlinskii, D., Priputina, I., Popova, A., Shalanda, A., Tsongas, T., Hinman, G., & Butcher, W. (2001). Influence of environmental contamination with PCBs on human health. *Environmental Geochemistry and Health, 23*, 313–328.

Parmar, J., & Qureshi, A. (2023). Accounting of the use and emissions of polychlorinated biphenyl compounds (PCBs) in India, 1951–2100. *Environmental Science & Technology, 57*(12), 4763–4774.

Patel, K. S., Ramteke, S., Sahu, B. L., Nayak, Y., Sharma, S., & Hung, C. C. (2015). Polychlorinated biphenyls contamination of sludge in India. *American Journal of Analytical Chemistry, 6*(11), 867.

Polachova, A., Gramblicka, T., Bechynska, K., Parizek, O., Parizkova, D., Dvorakova, D., ... & Pulkrabova, J. (2021). Biomonitoring of 89 POPs in blood serum samples of Czech city policemen. *Environmental Pollution, 291*, 118140.

Portigal, C. L., Cowell, S. P., Fedoruk, M. N., Butler, C. M., Rennie, P. S., & Nelson, C. C. (2002). Polychlorinated biphenyls interfere with androgen-induced transcriptional activation and hormone binding. *Toxicology and Applied Pharmacology, 179*(3), 185–194.

Quensen, J. F., III, Boyd, S. A., & Tiedje, J. M. (1990). Dechlorination of four commercial polychlorinated biphenyl mixtures (Aroclors) by anaerobic microorganisms from sediments. *Applied and Environmental Microbiology, 56*(8), 2360–2369.

Reddy, A. V. B., Moniruzzaman, M., & Aminabhavi, T. M. (2019). Polychlorinated biphenyls (PCBs) in the environment: Recent updates on sampling, pretreatment, cleanup technologies and their analysis. *Chemical Engineering Journal, 358*, 1186-1207. ISSN 1385–8947.

Ripley, K. (2017). The 2017 BRS COPs: Steps Forward, But Challenges Persist. *Environmental Policy and Law, 47*, 66.

Robertson, L. W., & Ludewig, G. (2011). Polychlorinated biphenyl (PCB) carcinogenicity with special emphasis on airborne PCBs. *Gefahrstoffe, Reinhaltung der Luft= Air quality control/ Herausgeber, BIA und KRdL im VDI und DIN, 71*(1-2), 25.

Safe, S. (1993). Toxicology, structure-function relationship, and human and environmental health impacts of polychlorinated biphenyls: Progress and problems. *Environmental Health Perspectives, 100*, 259–268.

Safe, S. H. (2007). Environmental and occupational medicine. In W. N. Rom (Ed.), *Chapter 77: Polychlorinated biphenyls* (4th ed.). Lippincott Williams & Wilkins.

Saktrakulkla, P., Lan, T., Hua, J., Marek, R. F., Thorne, P. S., & Hornbuckle, K. C. (2020). Polychlorinated biphenyls in food. *Environmental Science & Technology, 54*(18), 11443–11452.

Schettgen, T., Alt, A., Schikowsky, C., Esser, A., & Kraus, T. (2018). Human biomonitoring of polychlorinated biphenyls (PCBs) in plasma of former underground miners in Germany–A case-control study. *International Journal of Hygiene and Environmental Health, 221*(7), 1007–1011.

Swackhamer, D. L., & Armstrong, D. E. (1986). Estimation of the atmospheric and nonatmospheric contributions and losses of polychlorinated biphenyls to Lake Michigan on the basis of sediment records of remote lakes. *Environmental Science & Technology, 20*(9), 879–883.

Toft, G. (2014). Persistent organochlorine pollutants and human reproductive health. *Danish Medical Journal, 61*(11), B4967.

Toft, G., Thulstrup, A. M., Jönsson, B. A., Pedersen, H. S., Ludwicki, J. K., Zvezday, V., & Bonde, J. P. (2010). Fetal loss and maternal serum levels of 2, 2′, 4, 4′, 5, 5′-hexachlorbiphenyl (CB-153) and 1, 1-dichloro-2, 2-bis (p-chlorophenyl) ethylene (p, p′-DDE) exposure: A cohort study in Greenland and two European populations. *Environmental Health, 9*, 1–11.

Uwimana, E., Cagle, B., Yeung, C., Li, X., Patterson, E. V., Doorn, J. A., & Lehmler, H. J. (2019). Atropselective oxidation of 2, 2′, 3, 3′, 4, 6′-hexachlorobiphenyl (PCB 132) to hydroxylated metabolites by human liver microsomes and its implications for PCB 132 neurotoxicity. *Toxicological Sciences, 171*(2), 406–420.

Valle Dalla, M., Marcomini, A., Jones, K. C., & Sweetman, A. J. (2005). Reconstruction of historical trends of PCDD/Fs and PCBs in the Venice Lagoon, Italy. *Environment International, 31*(7), 1047–1052.

Van Gerven, T., Geysen, D., & Vandecasteele, C. (2004). Estimation of the contribution of a municipal waste incinerator to the overall emission and human intake of PCBs in Wilrijk, Flanders. *Chemosphere, 54*(9), 1303–1308.

Whitehead, A., Triant, D. A., Champlin, D., & Nacci, D. (2010). Comparative transcriptomics implicates mechanisms of evolved pollution tolerance in a killifish population. *Molecular Ecology, 19*(23), 5186–5203.

Wild, S. R., & Jones, K. C. (1995). Polynuclear aromatic hydrocarbons in the United Kingdom environment: A preliminary source inventory and budget. *Environmental Pollution, 88*(1), 91–108.

You, H., Ding, J., Zhao, X. S., Li, Y. F., Liu, L. Y., Ma, W. L., et al. (2011). Spatial and seasonal variation of polychlorinated biphenyls in Songhua River, China. *Environmental Geochemistry and Health, 33*, 291–299.

Zanaroli, G., Negroni, A., Häggblom, M. M., & Fava, F. (2015). Microbial dehalogenation of organohalides in marine and estuarine environments. *Current Opinion in Biotechnology, 33*, 287–295.

Dioxins and Furans

Kanchan Kumari, Shaheen Akhtar, and K. P. Prathish

1 Introduction

Famously known as the critical persistent organic pollutants (CPCB, 2004), dioxins or polychlorinated dibenzo-para-dioxin (PCDDs) and furans or polychlorinated dibenzofuran (PCDFs) are counted among the unintentional category of POPs labeled by the Stockholm Convention in 2001 (Annex C) (Kumari, 2021). They are addressed so due to their unintentional production during manufacturing of chemicals, mainly pesticides and chlorinated compounds (UNEP, 1997). Both dioxins and furans are structurally related hydrocarbons; they extensively exhibit properties (Table 1) such as being poorly soluble in water, lipophilic, non-polarity and an increase in structure stability with the degree of chlorine substitution in their parent molecule.

Their major emission sources have been observed to be waste incinerators where municipal, hospital, and hazardous wastes are combusted at high temperatures, metal production industries where both ferrous and non-ferrous metal are processed, chemical synthesis where pesticides and chlorinated products are produced, automobiles, as well as uncontrolled combustion of waste, wood, coal, peat etc. Improper residue disposal from industries such as asphalt mixing plants, pulp and paper industries, cement kiln, and metal treatment and processing industries also contribute heavily to the emission of PCDD/Fs into the environment. Seventy-five

K. Kumari (✉) · S. Akhtar
CSIR-National Environmental Engineering Research Institute (NEERI), Kolkata Zonal Center, Kolkata, West Bengal, India
e-mail: k_kumari@neeri.res.in

K. P. Prathish
CSIR-National Institute for Interdisciplinary Science and Technology (NIIST), Thiruvananthapuram, Kerala, India

K. Kumari, *Pollutants of Global Concern*, Emerging Contaminants and Associated Treatment Technologies, https://doi.org/10.1007/978-3-031-50996-4_11

Table 1 The physicochemical properties of dioxins and furans

Details	Dioxins (PCDDs)	Furans (PCDFs)
Molecular formula	$C_{12}H_8O_2$	$C_{12}H_8O$
CAS No.	262-12-4	132-64-9
IUPAC name	Dibenzo-p-dioxin	Dibenzofuran
Molecular weight	184.19 g/mol	168.19 g/mol
Color/form	Crystals from methanol	White crystals
Odor	No perceptible odor	Weak, characteristic odor
Density	1.8 g/cm^3	1.0886 g/cm$^{3\ \text{at}\ 210\ °\text{F}}$
Vapor pressure	4.12e-04 mm Hg	0.00248 mm Hg at 25 °C
Octanol/water partition coefficient	4.3	4.12
Boiling point	932 °F	549 °F at 760 mm Hg
Melting point	252–253 °F	187–189 °F
Solubility	Less than 1 mg/mL at 64 °F	10,000 mg/mL
Stability/shelf life	Chemical is stable under normal laboratory conditions	Chemical is stable under normal laboratory conditions
Decomposition	Emits toxic fumes when heated to decomposition	Emits acrid smoke and irritating fumes when heated to decomposition

congeners are theoretically possible for dioxins out of which seven exhibits toxicity induced through AhR mechanism in humans (dioxin-like toxicity). Similarly, 135 congeners are possible for furans of which 10-exhibit dioxin like toxicity. All the 17 congeners are classified as probable human carcinogen by International Agency for Research on Cancer (IARC). Owing to its toxic nature, the Tolerable Daily Intake (TDI) of dioxin recommended by the WHO is 1–4 pgTEQ/kg body weight (Van Leeuwen et al., 2000).

Resistance to degradation and lipophilicity of dioxins has contributed to longer half-life and bioaccumulation in biological systems. A considerable concentration of PCDD/Fs has been found in various foodstuffs especially of animal origin which includes meat, fish, milk, and dairy products (Bocio & Domingo, 2005). Around 90% of human exposure to residual dioxins and furans are through their diet where a substantial amount of exposure is contributed by milk and dairy products (Mayilsamy et al., 2022). In 1973, Baughman and Meselson first reported the presence of congeners of dioxin and furan in human tissue samples which has eventually led to numerous studies coming afterward finding the occurrence of PCCD/Fs in various other tissues of biota: blood, serum, and milk samples. However, there lies significant challenge in the identification of these compounds at trace/ultra-trace levels and the subsequent human and ecological risk detection of their presence so as to provide effective remediation or control processes. This chapter focuses on the status update of dioxins and furans levels in various environmental, food, and biological matrices in Indian and international scenario. Intake risk assessment of these chemicals can be made from the data compiled herewith which is highly recommended due to the very high thermal, photolytic, and biological stability of PCCD/Fs especially of dioxins.

2 Technical Details of Dioxin and Furan

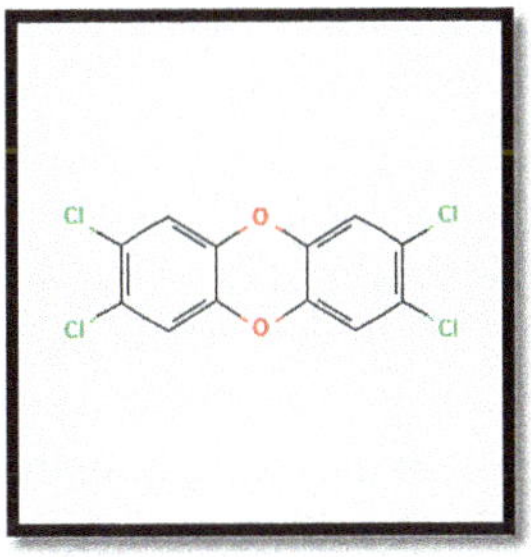

Structure of Dioxin

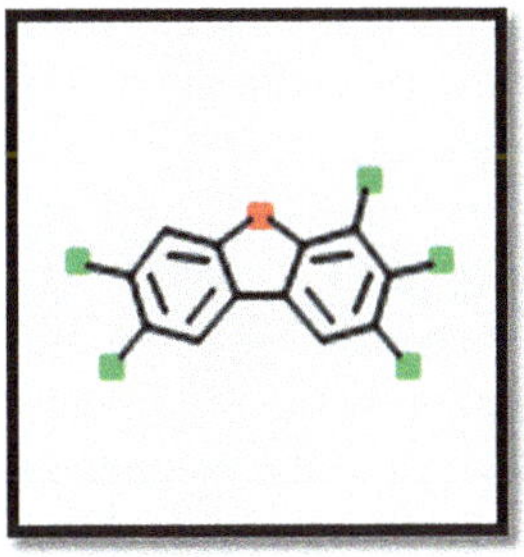

Structure of Furan

3 Status of Works and Usage in India

PCDD/Fs monitoring in the environmental, food, and human samples in the national territory started well before the Stockholm convention ratifications considering its critical effects on receiving community. The first study report came out in 1989 where the dioxins levels in human milk samples from the industrialized nations – Japan and the USA – were compared with the developing nation – India (Schecter et al., 1989). The observations suggested lowered levels in India and one of the reasons put forward was very low chemical production and consumption nationally. India underwent rapid changes in economical and industrial sectors in 1990–2000 period with large number of industrial hotspots emerging across the country. Since the generation of dioxins and furans can occur in many industrial sectors, their leaching in to the water streams and lands poses significant exposure risk to the various living beings such as dolphins, birds, and fishes. Senthil Kumar et al. (2001) collected and analyzed human tissues, fish muscles, lamb and chicken meat fat, liver and blubber of Ganges dolphins, and tissue samples from selected bird species, which were obtained from some of the hotspot sites in southern and eastern states of India (Senthil Kumar et al., 2001). The observed levels were in the order of: chicken < goat/lamb fat < fishes < dolphins < predatory birds, and the human tissue levels were observed to be less than the then contemporary levels reported at the USA, Canada, Germany, etc., while higher than China, Thailand, etc. The study pointed on the vegetarian diet as a possible reason for the lowered levels. Another study assessed butter as an indicator of regional persistent organic pollutant contamination and samples from 24 countries were studied. Sample from India reported an average level of 0.79 pgTEQ/gm lipid which was little above the study average but well below the European dairy products acceptable level (Santillo et al., 2003).

The dioxin levels in the soils of 13 dump yards in a south Indian city were studied by Minh et al. in 2003 and reported that, concentration ranged from 9.9 to 200 pgTEQ/gm which was higher than the then contemporary levels in Hanoi and

Vietnam (Minh et al., 2003). The study suggested that uncontrolled combustion of MSW has significant contribution to the observed high concentrations and it was not only attributed to atmospheric deposition. As a follow-up study, Kunisue et al. (2004) did an assessment of dioxins in human and bovine milk samples collected from the same study site (Kunisue et al., 2004). The observed concentrations in the human milk samples were in the range of 3.3–81 pgTEQ/gm lipid and for bovine milk, it was 4.1–14 pgTEQ/gm lipid. The study concluded that the residents in India who are residing near to dump yards are highly exposed to dioxins and related compounds and there is a need for introducing regulations to control the increasing levels. In a similar effort, Petrlik et al. (2005) studied the level of dioxins and furans in the eggs of free ranging hens in an industrial corridor in South India and medical waste incineration site in Northern India and found that the average levels were 13.91 pgTEQ/g and 19.80 pg/g, respectively, which was upto four times higher than the EU limit values (Petrlik et al., 2005). The study recommended the ratification of Stockholm Convention and the consideration and evaluation of the area as a potential hot spot. Breast muscles of common and jungle crow at the same dump yard site were also analyzed by Watanabe et al. (2005) and observed significantly higher levels of dioxins and furans in the range of 15–86 pgTEQ/gm lipid (Watanabe et al., 2005). Watanabe et al. in 2010 sampled and analyzed the livers of the free roaming pigs in the site as well in a follow-up study and found dioxins levels in the range of 8.9–350 pgTEQ/g fat (Watanabe et al., 2010). The toxicological aspects of dioxins and related compounds were discussed in both the studies which demanded more focus on regulations and monitoring for open dumping sites.

India ratified the Stockholm convention on POPs in 2006 and took steps to eliminate or control the dirty dozen compounds, which comprised dioxins and related compounds as well. A National Implementation Plan was also prepared in this regard which could act as a road map for the elimination of dl-POPs (dioxin-like POPs) from the Indian environment. A number of facility creation and monitoring programs were also been implemented during the period. The first in-lab characterization of 2,3,7,8 TCDD (tetrachlorodibenzodioxin) was also conducted as part of this program by Thacker et al. in 2007. Further investigations on the industrial and other releases of dioxins through different vectors across the nation have also been carried out during the NIP preparation period of 2007–2010 (Kashyap et al., 2008; Someya et al., 2010; Thacker et al., 2010).

The NIP estimated the annual release of PCCD/Fs in India, as 8656.55 gTEQ from the toolkit listed sources integrating all the vectors of emission. The percentage wise distribution of emission sources is provided in Fig. 1. As per the NIP, the major contributor was waste incineration sector (5777.97 gTEQ) followed by ferrous and non-ferrous metal production (1750.04 gTEQ), heat and power generation (504.15 gTEQ), production and use of chemicals and consumer goods (352.46 gTEQ), etc. Sixty-three percent of the annual release was through residues, followed by air (32.7%), products (3.6%), land (0.4%) and water (0.3%). India was not having any MSW or sewage sludge incinerators during the period and hence incineration sector accounted for hazardous waste and medical waste incinerators only. Drilling operations for oil and gas production, cleaning, emptying, and

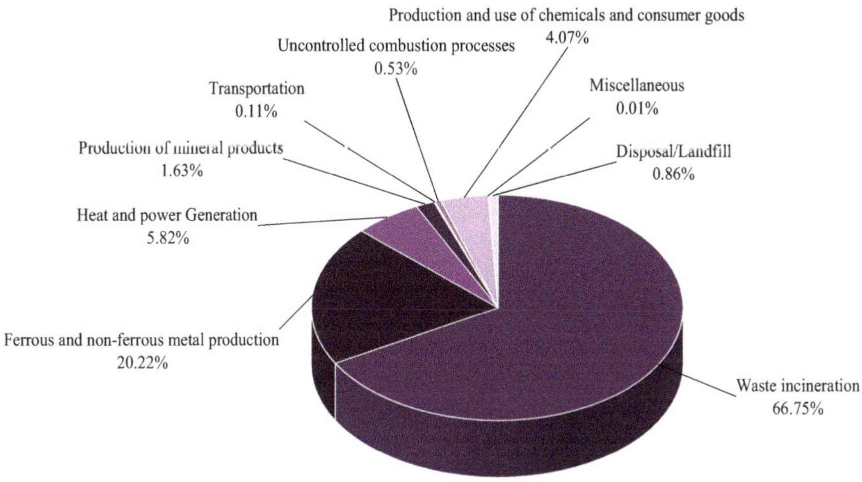

Fig. 1 Annual releases of PCDD/Fs from various point sources. (NIP, 2011)

maintenance of petroleum oil storage tanks, including ships, petrochemical processes, and pyrolytic operations, generate around 4.4 million tonnes of hazardous wastes in India every year. This very high activity rate made hazardous waste incineration the largest individual source with 60% contribution to annual emission where an estimated 5505.6 gTEQ was released from this sector alone. The NIP also presented a region-wise emission prospectus of country where the emissions were highest in the central and western region followed by northern and eastern region, and the southern region having the lowest emissions. The higher industrial presence in the central and western region must have been the reason for the highest contribution to dioxins emission of the region.

The NIP brought in conclusions for the reported emission levels of dioxins as well and the major ones were – many of the incinerators in operation were poorly designed or operated, unorganized operation of ferrous and non-ferrous industries and foundries and lack of public awareness in the waste management processes leading to uncontrolled open combustion. The gap analysis pointed out the requirements for introducing limit values for emissions and regular monitoring practices. With these the specific action plans to be implemented in the following years were proposed as follows:

1. Develop U-POP emission standards
2. Develop national standardized methods for their sampling and analysis
3. Developing coherent share of responsibilities between the authorities that register and monitor industrial processes
4. Develop a procedure for collecting, storing, transporting, eliminating, and decontaminating dioxins containing wastes
5. Financial facilitation and promotion of introducing BAT/BEP

Table 2 Residues of PCDDs/Fs in our daily food sources

Food source	Study site	Location	Dioxin concentration	Furan concentration	References
Eggs	Queen Mary's Hospital, Lucknow medical waste incinerator	Lucknow, Uttar Pradesh	19.80 pg/g	19.80 pg/g	Petrlik et al. (2005)
Eggs	Municipal corporation incineration sites	Western Zone of India	7.10 pg/g	0.39 pg/g	Kashyap et al. (2008)
Bovine milk	Dumping sites	Chennai, Tamil Nadu	2.8–6.6 pg/g	1.4–2.5 pg/g	Kunisue et al. (2004)
Bovine milk	Urban and rural area	Tamil Nadu	ND-6.676 pg TEQ/g	ND-6.676 pg TEQ/g	Mayilsamy et al. (2022)

Through the implementation of the action plan, NIP aimed significant reduction of dioxins emission from its major two sources – incineration (reduce to below 2 kgTEQ per annum) and ferrous and non-ferrous metal production (reduced to below 1 kgTEQ per annum) by 2016. There was a significant difference observed between the release estimates made based on the toolkit emission factors and measured emission factors which also needed further clarification. The NIP was submitted to Stockholm Convention in April 2011 with a commitment to strengthen policies, regulations, and frameworks to control the unintentional releases of POPs.

Following the NIP action plans India brought in the first regulatory limit for PCDD/Fs in stack emissions as 0.1 ng TEQ/Nm^3 (at 11% O_2) and was notified by the CPCB in the Bio Medical Waste Management Rule 2016 in March 28, 2016. Hazardous Waste Management Rules, 2016, also incorporated the leachable maximum levels for dioxins for the classification of hazardous waste as 1 μgTEQ/l. These available standards are at par with the currently available international regulations; however, India lacks regulations in direct exposure environmental media (ambient air, surface soil, sediment, etc.) and foods in domestic market. Also, the number of data available on the presence of these POPs in food is very low in the Indian subcontinent.

Number of studies on dioxins and related compounds significantly increased in the decade that followed the NIP submission, and the reported data are presented in Tables 2, 3, 4 and 5. In-house state of the art method development and validation has been reported during this period such as GC-triple quadrupole MS/MS which catalyzed the environmental and food monitoring in India (Ajay et al., 2022). The dioxins and related compounds contamination in the surface soils at the selected waste processing and dumping centers has been reported by Rajan et al. in 2021. These unintentional POPs can highly accumulate in soil (Chakraborty et al., 2016) due to their physiochemical properties and the organic content of the soil, and it is thus said that soil acts as a major global repository for these organic contaminants. This has incessantly increased with the anthropogenic effect of domestic and industrial land use pattern.

Table 3 PCCDs/Fs residues detected in different environmental compartments

Sample site	Location	Environmental matrix	Dioxin concentration	Furan concentration	References
E waste sites and dumpsites	New Delhi, Kolkata, Mumbai, Chennai	Soil	3.8 ± 3.7 ng/g	1.8 ± 2.1 ng/g	Chakraborty et al. (2018)
Waste incinerators and thermal processes	Central and western India	Air	0.0070–26.8140 ng TEQ/Nm3	0.0070–26.8140 ng TEQ/Nm3	Thacker et al. (2013)
Municipal dumpsites	Tamil Nadu	Soil	3.70–512 pg/g	0.26–236 pg/g	Rajan (2021)
Municipal dumpsites	Tamil Nadu	Ash	0.0081–1.2009 ng TEQ/g	0.0081–1.2009 ng TEQ/g	Mayilsamy et al. (2022)
North eastern coast of Bay of Bengal	Canning, West Bengal	Ambient Air	4.4–2490.9 fg/m^3	4.4–2490.9 fg/m^3	Kumar et al. (2012)
Urban site	Delhi	Ambient air	1090–4012 fg/m^3	620–5440 fg/m^3	Kumar et al. (2012)

Table 4 Residual PCDDs/Fs found in fauna

Sample	Location	Dioxin concentration	Furan concentration	References
Chicken	Tamil Nadu	11 pg/g	14 pg/g	Senthil Kumar et al. (2001)
Lamb	Tamil Nadu	13 pg/g	18 pg/g	Senthil Kumar et al. (2001)
Goat	Tamil Nadu	19 pg/g	23 pg/g	Senthil Kumar et al. (2001)
Birds	Tamil Nadu	24–1800 pg/g	19–860 pg/g	Senthil Kumar et al. (2001)
Fishes	West Bengal	310–740 pg/g	130–190 pg/g	Someya et al. (2010)
Dolphins	Chappra and Patna, Bihar	28–220 pg/g	22–200 pg/g	Senthil Kumar et al. (2001)
Fishes	Tamil Nadu, Bihar, West Bengal	9.5–82 pg/g	2.9–48 pg/g	Senthil Kumar et al. (2001)

Moreover, open burning of wastes is a serious problem that has been plaguing the Indian subcontinent since ages where household wastes of one kilogram can emit 4.28×10^{-4} mg/kg of PCDDs/Fs when burnt (Lemieux et al., 2004). In Asian developing countries, elevated levels of PCDDs/Fs and PCBs were detected due to uncontrolled open burning of municipal wastes (Rajan et al., 2021). A detailed appraisal and projection of dioxins and furans emission in India from the open burning of MSW was made by Kumari et al. in 2019. The study estimated the emissions for the year 2014 and 2016 and projected the emissions for the year 2021; 346 gTEQ

Table 5 Residual PCDDs/Fs found in human

Study site	Gender	Sample type	Dioxin concentration	Furan concentration	References
Coimbatore, Tamil Nadu	Female Male	Fat tissues	590 pg/g 440 pg/g	26 pg/g 33 pg/g	Senthil Kumar et al. (2001)
Chennai, Tamil Nadu	Female	Breast milk	290 pg/g	50 pg/g	Kunisue et al. (2004)
Kolkata, West Bengal	Female	Breast milk	610 ± 280 pg/g	44 ± 20 pg/g	Someya et al. (2010)

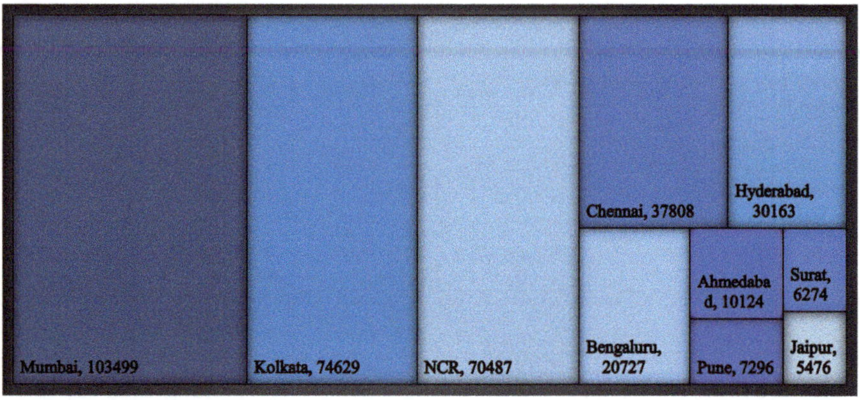

Fig. 2 Emission of PCDDs/Fs (μg TEQ/day) determined due to open burning of MSW based on the data of population and MSW generation for the year of 2014, 2016, and 2021 in the major ten metropolitan cities of India. (Kumari et al., 2019)

and 397 gTEQ were estimated as national inventory for the year 2014 and 2016, respectively, by the study and projected that the emissions will rise to 524 gTEQ by the year 2021. The detailed city wise and state wise emission levels are shown in Figs. 2 and 3, respectively (Kumari et al., 2019). Ajay et al. (2022) conducted first studies in Indian context on development of nation's default emission factors for dioxins and related compounds from open burning of MSW through simulated open burning studies. Simulated open burning test facility with provisions for weight change monitoring, temperature monitoring, air flow control, and iso-kinetic sampling ports was installed indigenously for conducting emission factor studies (Fig. 4). Both real dumpsite and simulated MSW samples were used in this study to simulate the national MSW composition and combustion conditions and reported the nation default emission factors as 167 μgTEQ/ton of MSW openly burned; 100 μgTEQ/ton (60%) is reported to be emitted into land emission vectors and 67 μgTEQ/ton (40%) released as air emissions.

PCDDs/Fs release into the environment occurs either directly via emissions or indirectly through volatization from water and land. Air currents also act as the transporting medium for the dispersal of dioxins and furans from fugitive and stack emissions of power plants, incinerators, etc. Chakraborty et al. (2016) conducted

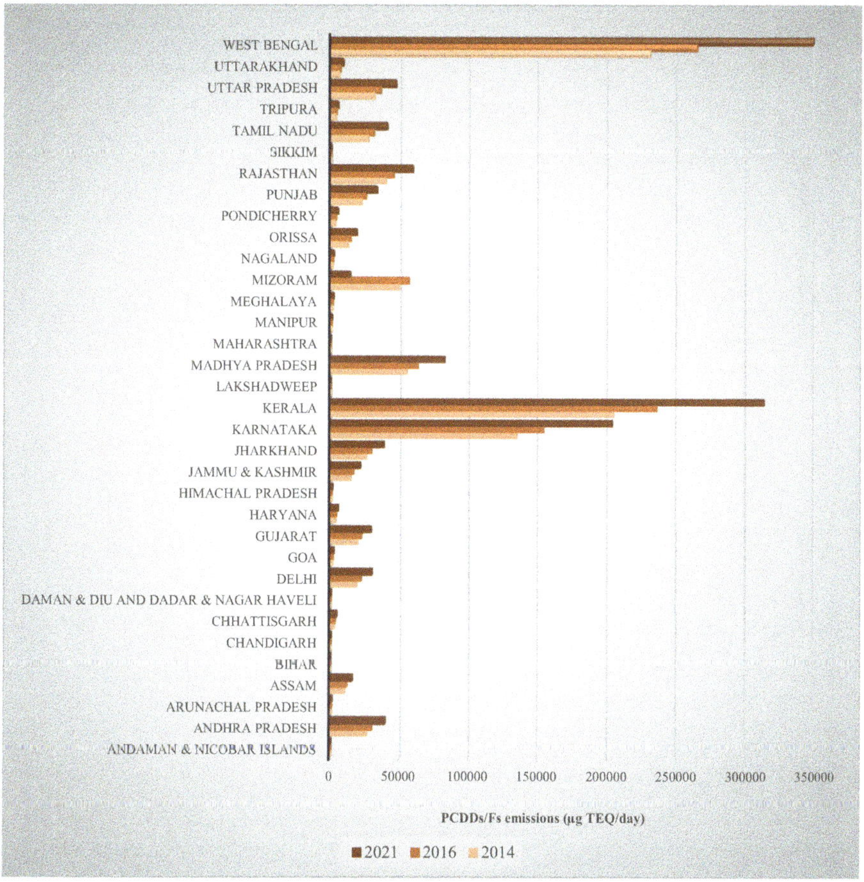

Fig. 3 Emission of PCDDs/Fs (µg TEQ/day) determined due to open burning of MSW based on the data of population and MSW generation for the year of 2014, 2016, and 2021 – a state/union territories wise list. (Kumari et al., 2019)

long-term passive air sampling studies at the waste processing centers to monitor the ambient air levels of dioxins and reported levels in the range of 2.2–27.1 pg/m^3. Ajay et al. (2022) assessed the open burning and accidental fire emissions of dioxins into ambient air at selected sites in Southern India and reported that the levels ranged from 2.7 to 41.4 pgTEQ/m^3. The study also monitored the burned residue levels at these as well and found that the levels ranged from 79.8 to 860 ngTEQ/kg. One critical observation of this study was that the exposure risk from street waste open burning emissions is much higher than the dump yard fire emissions particularly due to the higher frequency of action, low horizontal dispersion of pollutants, and higher ground level concentrations. The study conducted by Mayilsamy et al. (2022) reported the presence of congeners of dioxin and furan in bovine milk through chemically activated luciferase expression (CALUX) assay. POPs like dioxins and furans affect the transcription factor- aryl hydrocarbon receptor (AhR)

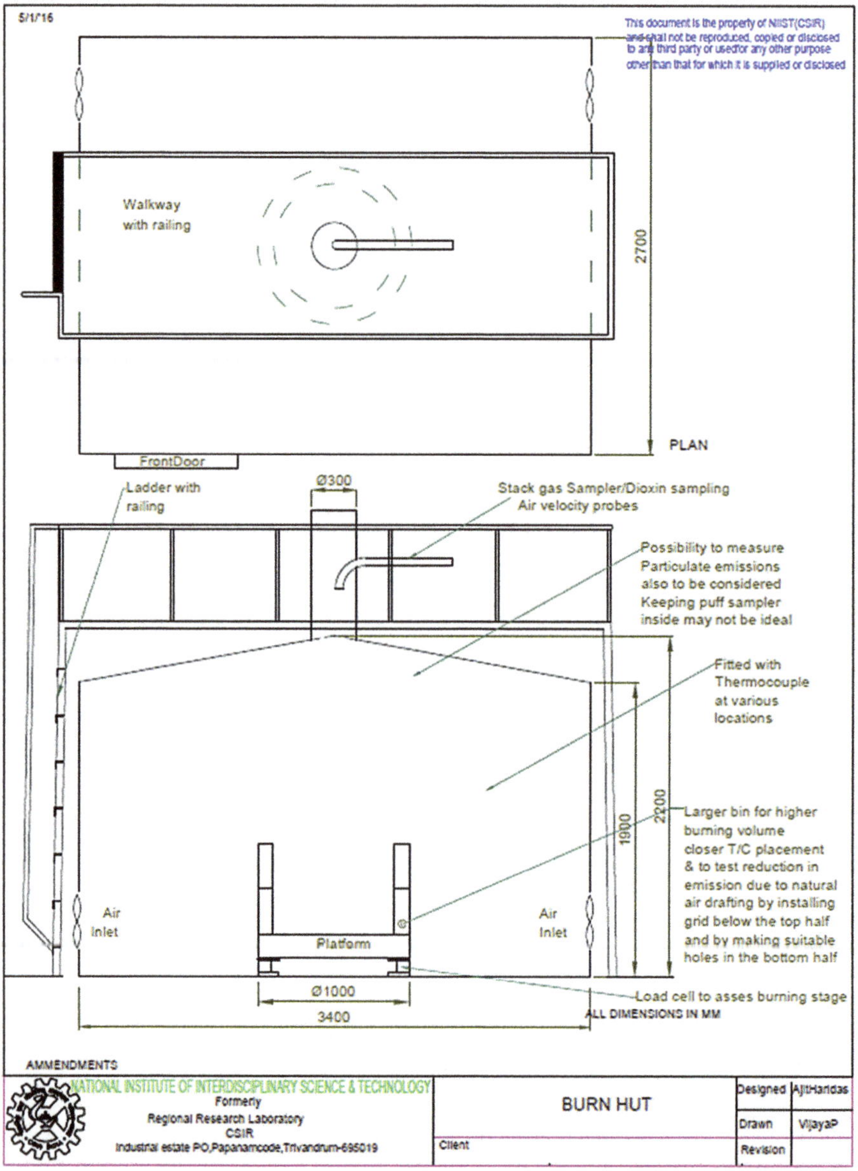

Fig. 4 Design of the burn hut

in cells. This adversely affects the cell growth and differentiation through modifications of gene expression. The reporter gene-based CALUX bioassay of cells measures the activation of the AhR thus evaluating the amount of PCDD/Fs accumulation in the sample. However, the samples were collected from mothers living near dumpsites and landfills and thus singlehandedly cannot conclude the level of exposure of people inhabiting in other parts of the country (Someya et al., 2010).

4 International Data

Being a subset of the Dirty Dozens, the concern to monitor dioxin and furan and its congener has been addressed quite seriously in countries such as the USA, Canada, Japan, European Union nations, and so forth. Soil response level for Dutch Intervention is 0,00018 mg/Kg. Soil screening level for the Protection of Environmental and Human Health for agricultural, residential/parkland, commercial, and industrial is 4 ngTEQ.Kg^{-1}. The US Food and Drug Administration estimates a lifetime excess cancer risk of one in 10,000 for eggs contaminated at 1 pg/g ITEQ. In the USA, the EPA has implemented a system known as the National Dioxin Air Monitoring Network (NDAMN) to monitor the amount of dioxin in the atmosphere where over 600 samples were collected and analyzed from the year of 1998 to 2004 (Lorber et al., 2013). Study conducted by Dwyer and Themelis (2015) estimated a 66.2% of release of dioxins into the atmosphere from electric and heat generating plants and thus were considered the highest contributors of PCDDs release in the USA followed by landfill fires (44%), forest fires (28%), and open burning sources (7.2%). The largest source of dioxin emissions in Ontario, Canada, are burning of household garbage whereas in Alberta a 200 million pounds of agricultural plastics are annually sold that is majorly managed through open burning. As per the Canadian Soil Quality guidelines, 75,000 kg of soil gets contaminated by burning 10,000 pounds of agricultural plastics due to exposure to dioxins (Sonnevera International Corporation, 2011). Being an integral part of the food web, soil behaves as a pathway for carrying this contaminant into the biota and subsequently to the higher organisms. Sharma et al. (2007) conducted a study to assess the human and ecological risk in southern Ontario and reported average soil contamination of TCDD to be 11.1 pg/g DW. In Europe, France along with Italy, and Spain has been identified to be among the biggest contributors of PCDDs/Fs emissions due to their lack of regulations for waste incinerators (European Commission, 2017). PCDDs/Fs in Turkey are disposed in the Black Sea which has subsequently affected the aquatic organisms of the waterbody (Çelik et al., 2010). Similarly, industrial facilities in and around the Saginaw Bay are a concern for PCDDs/Fs and its congeners contamination. Fishes from the Saginaw River and Tittabawassee River reported ΣTEQs values of 34 ± 28 pg/g ww and 13 ± 11 pg/g ww in channel catfish and northern pike species of fishes, respectively, while the highest ΣTEQs of PCDDs/Fs was observed in the carp fish (54 ± 59 pg/g ww) (Wan et al., 2010). Poultry samples in Canada reported PCDDs/Fs concentrations of 2.6 pg I-TEQ/g (Srogi, 2008). Comparative studies for France, Germany, and the Netherlands showed the PCDDs/Fs concentrations to be higher in home produced eggs than raw bovine milk which might be contributed due to the fat content (Kanan & Samara, 2018). Concentrations of dioxins (2.4 pg-TEQ/L/year) were detected in river water that exceeded the

Table 6 Residues of PCDDs/Fs in food sources

Food source	Location	Dioxin concentration	Furan concentration	References
Aquatic food, meat, eggs, dairy products, cereals, bean products, potatoes, vegetables	China	0.001–0.62 pg TEQ/g	0.001–0.62 pg TEQ/g	Zhang et al. (2013)
Fish, meat, dairy products, eggs	Sweden	0.007–0.391 pg TEQ/g	0.007–0.391 pg TEQ/g	Törnkvist et al. (2011)
Fish oil, meat and bone meal, chicken fat, pig fat	Japan	BDL-39 pg/g	BDL-2.9 pg/g	Guruge et al. (2005)

Table 7 Residual PCDDs/Fs found in various environmental compartments

Location	Environmental matrix	Dioxin concentration	Furan concentration	References
North America	Sediment	23,000–100,000 pg/g	1300–5500 pg/g	Marvin et al. (2002)
Japan	Soil	5031.55 pg/g	721.24 pg/g	Kadokami et al. (2002)
Croatia	Air	0.309–6.100 pg N/m^3	0.570–12.880 pg N/m^3	Krauthacker et al. (2006)
Taiwan	Ambient air	0.27–1.20 fg/m^3	0.27–1.20 fg/m^3	Huang and Liang (2011)
Hong Kong	Ambient air	0.018–0.429 pg/m^3	0.018–0.429 pg/m^3	Sin et al. (2002)
South Korea	Sediment	229–1589 pg/g	31–3095 pg/g	Hong et al. (2009)
France	Sediment	27–1451 pg/g	10–206 pg/g	Hong et al. (2009)
China	Air	670–10,380 fg/m^3	670–10,380 fg/m^3	Zhou et al. (2014)
Colombia	Air	39–883 fg/m^3	45–1107 fg/m^3	Cortés et al. (2014)
Korea	Air	169–882 fg/m^3	169–882 fg/m^3	Park and Kim (2002)
Portugal	Air	0.036–0.49 pg I-TEQ/m^3	0.036–0.49 pg I-TEQ/m^3	Ma and Yang (2007)
Spain	Air	0.158–0.264 pg I-TEQ/m^3	0.158–0.264 pg I-TEQ/m^3	Ma and Yang (2007)
UK	Air	0.02–0.51 pg I-TEQ/m^3	0.02–0.51 pg I-TEQ/m^3	Ma and Yang (2007)

environmental quality standards (EQS) for water in Japan. It was determined to be result of combustion that might have transported to the river via rainfall deposition (Minomo et al., 2018). Data presented in Tables 6, 7, 8 and 9 represents residual levels in different samples collected from various geographical locations.

Table 8 Residual PCDDs/Fs found in fauna

Sample	Location	Dioxin concentration	Furan concentration	References
Birds	USA	5 pg/g	5 pg/g	Custer et al. (2000)
Birds	Mexico	4.4–16 pg/g	3.3–4.7 pg/g	Jiménez et al. (2005)
Porpoise	China	1.07–4.47 pg/g	1.07–4.47 pg/g	Yang et al. (2008)
Mussels	North America	23–150 pg/g	3.9–14 pg/g	Marvin et al. (2002)
Frogs	Japan	8.49–108.92 pg/g	8.27–21.90 pg/g	Kadokami et al. (2002)
Birds	Spain	3.21–18.7 pg/g	402–23.3 pg/g	Gómara et al. (2008)
Polar bears, penguins, seals, fishes	USA	1.0–59 pg/g	2–122 pg/g	Kumar et al. (2002)
Birds	Japan	15–180,000 pg/g	18–310,000 pg/g	Senthilkumar et al. (2002)

Table 9 Residual PCDDs/Fs found in human

Study site	Sample type	Dioxin concentration	Furan concentration	References
Cambodia	Breast milk	49 pg/g	15 pg/g	Kunisue et al. (2004)
Vietnam	Breast milk	32 pg/g	20 pg/g	Kunisue et al. (2004)
Philippines	Breast milk	190 pg/g	21 pg/g	Kunisue et al. (2004)
Germany	Blood	1.43–268.43 pg/g	0.27–12.54 pg/g	Wittsiepe et al. (2007)
Spain	Breast milk	ND-48.5 pg/g	ND-48.5 pg/g	Schuhmacher et al. (2019)
New Zealand	Serum	BDL-961 ng/kg	BDL-21.7 ng/kg	Bates et al. (2004)
China	Breast milk	1.66–6.67 pg/g	1.66–6.67 pg/g	Zhou et al. (2011)
Northern Russia	Breast milk	3–7 pg/g	2–5 pg/g	Polder et al. (2008)
China	Adipose tissue	33.9–504 pg/g	33.9–504 pg/g	Shen et al. (2009)
Brazil	Breast milk	8.1 pg/g	8.1 pg/g	Paumgartten et al. (2000)

5 Adverse Effects of Dioxins and Furans

Numerous health effects in humans, including the disorders of immune system, chloracne, and the enzymatic disorders, have all been associated with dioxins (Fig. 5). These have also been classified as possible human carcinogens. For humans, food (particularly from animals) is the major source of exposure. Furans have also been detected in breast-fed infants. Furans are structurally similar to dioxins and thus share many of their toxic effects.

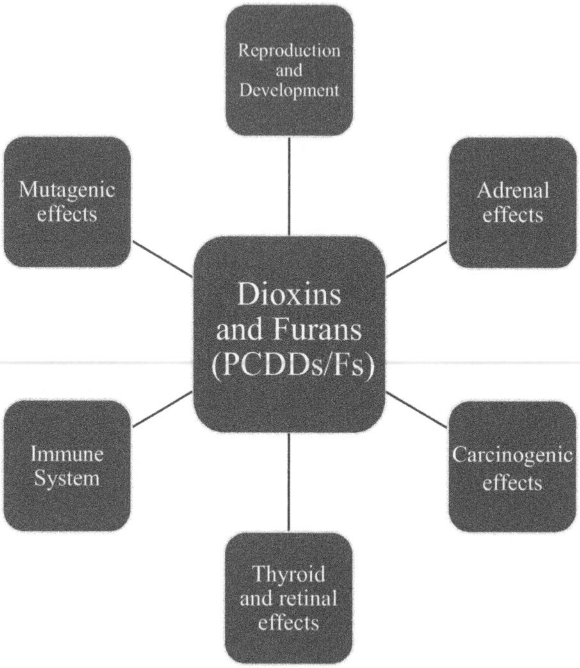

Fig. 5 Health effects of dioxins and furans

Toxicity due to PCDDs/Fs has been widely noted in animals where oral exposure of high dosage of 2,3,7,8-Cl_4DD has been majorly documented. PCDD lethality exposure in animals is wide ranged. Animals administered with dioxins in laboratories suffered a variety of negative effects, such an increase in birth defects and stillbirths. Fish eventually died shortly after when exposed to these substances. Animal systems can be permanently altered when exposed to chronic low-dosage of dioxin further exhibiting long-term health effects (Phol, 1998). Animals especially rodents have shown immune-toxicity and developmental alterations when exposed to congeners of dioxin and furan. These POPs have altered the level of testosterone, progesterne, and thyroid hormone in numerous species, while some have shown inhibition in estrogen activity. Primates, rats, and mice have shown reduction in uterine wastes and litter size, while some have shown a decrease in fertility (Schecter et al., 2006).

Evidences of alterations in microsomal enzyme induction, serum lipid levels, and gastrointestinal variations in humans have been reported due to high exposure to dioxin and furan (CACAR, 1997). Also, dioxin body burdens and exposure have documented metabolic and hormonal changes in humans by the US EPA (Phol, 1998). The mechanism of dioxin to affect the cells is majorly via a single cellular mechanism. When human or animal cells are exposed to dioxin, it binds with the cellular receptors for hormones, enters the nucleus, and manipulates the gene's functionality. Genes contain codes for proteins, hormones, and enzymes, aiding in their synthesis that collectively influences the development of tissues in an organism's body. This mechanism is thus inhibited and crippled by the inclusion of dioxins in the cells, thus eventually affecting the organism's body as a whole (CACAR, 1997). In utero and lactational exposures of PCDDs/Fs have been associated with thyroid, hypothalamic, and pituitary regulations in human infants, while occupational exposure of dioxins and furans has led to the development of types of cancer as PCDDs/Fs can affect the differentiation and growth of cancer-causing cells through corrupted gene regulation. The ability of dioxin to target cells at their developmental stages of growth can consequently affect developing fetuses if pregnant women are chronically exposed to low levels of dioxins in their diet (Schecter et al., 1990). Since breast milk and the placenta stores dioxin mostly, the fetuses are heavily exposed to these concentrations during their high susceptibility period (Koppe, 1995). Reduced level of testosterone has been recorded in men who are occupationally exposed to dioxins (Dhooge et al., 2006). Men exposed in the Vietnam war to dioxin-contaminated Agent Orange exhibited reduced testicular size (Van Luong et al., 2018). The ability of dioxin to suppress both humoral and cell-mediated responses in animals makes it a concerning toxin having a wide range of targets that eventually prevent normal functioning of body processes, including innate and acquired immunities (Vos et al., 1973; Warren et al., 2000).

Huff et al. (1994) studied that animal species chronically exposed to 2,3,7,8-TCDD even lower doses such as 1 part per trillion significantly showed carcinogenic responses. The US EPA declared TCDD in 1985 as the most potent synthetic carcinogen yet tested. In addition, the International Agency for Research on Cancer (IARC) in 1997 classified TCDD as a known human carcinogen, and the probability of associating all dioxins as human carcinogens is quite high. Studies on rats and monkeys exposed to 2,3,7,8-TCDD during gestation have shown prenatal mortality (Bjerke & Peterson, 1994; McNulty, 1984). Modifications in behavior and decrease in offspring survival has been noted in rhesus monkeys that are chronically exposed to 2,3,7,8-TCDD in their diet from pregnancy. Cognitive deficits as well as impaired learning performance and alterations in peer-group behavior were also observed in offspring (Schantz & Bowman, 1989). Lactating mice exposed to doses of 2,3,7,8-TCDD comparable to daily human intake showed thymic atrophy and immunosuppression in their offsprings's offspring (Blaylock et al., 1992). Male and female rats exposed in utero showed decreased fertility in both, while exposure including lactation resulted in reduction in testosterone level in male offsprings (Gray & Ostby, 1995; Mably et al., 1992). Female rats exposed to dioxin have reported altered hormone levels, reduced fertility, inhibition of estrogenic activity, estrous cycle, and ovulation (Li et al., 1995), whereas dioxin exposure on male rats showed reduced seminal vesicle weight and testosterone levels (Moore et al., 1985). Chronic doses of 2,3,7,8-TCDD in monkeys resulted in an increase in severity and incidence of endometriosis, reduced reproduction rates, and increased abortions (Rier et al., 1993). Moreover, a study on confined mink feeding on dioxin-contaminated Great Lakes fish showed reproductive loss and early mortality (Colborn et al., 1994).

6 Management and Remedial Measures

Since PCDDs and PCDFs are unintentional byproducts of incomplete combustion and related processes, specified chemical regulation and production prohibition cannot be the ultimate solution. Certain indirect measures such as banning of production and usage of chemicals associated with PCDDs/Fs contamination can be undertaken to reduce their emission into the environment. To contain the levels of dioxins and furans from the various wastes such as hazardous, biomedical, and municipal wastes, regulations have been introduced in the Indian subcontinent to manage the facilities of waste incineration. Equivalent to the stringent standards of EU the emissions of PCCDs/Fs are to be kept within 0.1 ng TEQ/Nm^3, as per the regulations in India.

From the National Implementation Plan (NIP) of India, it can be understood that 67.3% of the dl-POPs emissions in India is from municipal solid-waste management (MSWM) sector alone, which is twice the emissions from industrial sources. Also, the recent studies and reports suggest, open burning of MSW can become the single largest source for dl-POPs in developing nations, due to the inadequacies in MSWM sector and exacerbated MSW generation rate during the economic growth phase. Diffused sources of dl-POPs such as open burning of MSW are much more dangerous and have elevating effects on the intake doses than the point sources such as incineration or waste to energy plants. Therefore, solid waste management should be prioritized for dl-POPs emission curbing especially in the case of economies under transition. A plausible MSWM scheme, which can cut down 90–95% emissions from waste management sector in India, can reduce 63% of the national emission (NIP of India, 2011). Capturing or destruction of dl-POPs after being formed is much tedious and energy intensive process and hence the focus of remedial action plans should be eliminating favorable circumstances of dl-POPs formation. This is possible through an MSWM scheme, which is designed to accommodate rising MSW generation rate through low emitting management processes such as recycling, anaerobic digestion, composting, and waste to energy plants, when compared to the current unscientific practices such as open dumping/burning. Such a scheme can possibly account for maximum MSW collection and treatment before safe and sustainable disposal.

Nation-specific action level establishment and enactment of strict regulatory monitoring is another aspect which can improve the dl-POPs emission scenario in developing nations. Most of the developing nations like India are trying to enact stringent international standards of 0.1 ngTEQ/Nm^3 as maximum level for incinerator source irrespective of technology or capacity of the systems. These stringent standards are not taking into account of the huge requirements in the waste management sector the nation is facing. Small-scale systems are having economic advantage as low capital, operating cost is required, and hence community level waste management can be accelerated through these systems. In addition, many types of different capacity incinerator systems are in use across the country such as sanitary pad incinerators and single chamber MSW incinerators, but these systems are not properly monitored due to lack of targeted norms. It is quite difficult to accommodate all these systems under single emission norm, and it is high time to update the norms based on capacity of the facility. Four different maximum emission levels for dioxins are in place at the USA, Japan, South Korea, etc., depending on the capacity of the incinerator/waste to energy systems. Similarly, low-capacity systems require bit more relaxed norms as quantity of waste disposed through these systems will be comparatively much lower than the full-scale systems. When stringent norms are in place, the systems not conforming to it are not given the permission to operate, and this leads to the overflow of waste in the management centers causing open dumping and open burning. While with relaxed norms, more number of waste management systems will come in place and this can resist the waste overflowing in management centers ultimately curbing dioxins emissions.

Further sensitization of the common people to practice scientific waste management activities is also required for managing the increasing MSW generation rate. In most of the cases, people tend to plot for street littering and open burning without realizing the dangers of unintended emissions. The sensitization of the public is possible only through scientific data on its hazardous effects on the receiving community. Data generation and risk prediction could be achieved only through periodic or systematic national monitoring programs, which have led many of the developed nations to gradually reduce emissions through technological interventions. The scenario is quite complicated in developed countries, where often-meager financial support is provided for analytical or monitoring front, while investing reasonably well on technological solutions that ultimately lead to incompetent processes or technologies with unproven claims disguised by short-term low CAPEX/OPEX at the cost of long-term environmental and human health risk. On the contrary, a synchronous approach of capacity building in the analytical or monitoring sector along with translation of technologies could enable developing countries to effectively address the menace. More often, it could lead to technological advancements and create a healthy environment for the survival of sustainable technologies.

7 Conclusion

At present, there are very limited facilities for the analysis of dioxins and furans in environmental or biological matrices in India. Baseline data of dioxins in many of the sectors and medias are missing for the country, and this data gap needs to be closed as early as possible. Such state-of-the-art facilities and data are essential for the identification, implementation, and evaluation of Best Available Technologies/Best Environmental Practices (BAT/BEP). Nation-specific emission factor generation studies on open burning of MSW have been recently reported for the first time and this needs to be extended to some other sectors as well for developing ground specific national inventory. Scientific evidence creation on the harmful effects of dioxins and awareness creation based on those evidences are also necessary for the mindful cooperation of the people toward the sustainable remediation programs.

Setting up of facilities for the analysis of dioxins in various matrices needs to be undertaken first followed by implementation of BAT/BEP approaches. Further, solid waste management should be given prime importance as it is reported and projected as the major source of dl-POPs in developing nations. It is high time to introduce capacity wise stepped emission norms in India for incinerator and waste to energy systems which will be more acceptable for developing economies. Effectiveness evaluation of the policies developed or rules enacted is quite critical to ensure progress of any emission curbing programs which can win the support of general public as well.

References

Ajay, S. V., Kirankumar, P. S., Sanath, K., Prathish, K. P., & Haridas, A. (2022). An experimental simulation study of conventional waste burning practices in India for the assessment and inventorisation of PCDD/F/dl-PCB emissions. *Journal of Environmental Management, 303*, 114109. https://doi.org/10.1016/j.jenvman.2021.114109

Bates, M., Buckland, S., Garrett, N., Ellis, H., Needham, L., Patterson, D., Turner, W., & Russell, D. (2004). Persistent organochlorines in the serum of the non-occupationally exposed New Zealand population. *Chemosphere, 54*, 1431–1443. https://doi.org/10.1016/j.chemosphere.2003.09.040

Baughman, R., & Meselson, M. (1973). An analytical method for detecting TCDD (dioxin): Levels of TCDD in samples from Vietnam. *Environmental Health Perspectives, 5*, 27–35.

Bjerke, D. L., & Peterson, R. E. (1994). Reproductive toxicity of 2,3,7,8-tetrachlorodibenzo-p-dioxin in male rats: Different effects of in utero versus lactational exposure. *Toxicology and Applied Pharmacology, 127*(2), 241–249. https://doi.org/10.1006/taap.1994.1158

Blaylock, B. L., Holladay, S. D., Comment, C. E., Heindel, J. J., & Luster, M. I. (1992). Exposure to tetrachlorodibenzo-p-dioxin (TCDD) alters fetal thymocyte maturation. *Toxicology and Applied Pharmacology, 112*(2), 207–213. https://doi.org/10.1016/0041-008X(92)90189-Y

Bocio, A., & Domingo, J. L. (2005). Daily intake of polychlorinated dibenzo-p-dioxins/polychlorinated dibenzofurans (PCDD/PCDFs) in foodstuffs consumed in Tarragona, Spain: A review of recent studies (2001–2003) on human PCDD/PCDF exposure through the diet. *Environmental Research, 97*(1), 1–9. https://doi.org/10.1016/j.envres.2004.01.012

CACAR, I. (1997). Canadian Arctic Contaminants Assessment Report. In *Northern contaminants program*. Department of Indian and Northern Affairs Ottawa, Ontario, Canada.

Çelik, G., Kılıç, D., & Uçar, Y. (2010). Levels of polychlorinated dibenzo-p-dioxins, dibenzo-p-furans and polychlorinated biphenyls in farmed sea bass (Dicentrarchus labrax) and sea bream (Sparus aurata) from Turkey. *Food Control, 21*(9), 1245–1249. https://doi.org/10.1016/j.foodcont.2010.02.010

Chakraborty, P., Khuman, S. N., Selvaraj, S., Sampath, S., Devi, N. L., Bang, J. J., & Katsoyiannis, A. (2016). Polychlorinated biphenyls and organochlorine pesticides in River Brahmaputra from the outer Himalayan Range and River Hooghly emptying into the Bay of Bengal: Occurrence, sources and ecotoxicological risk assessment. *Environmental Pollution, 219*, 998–1006. https://doi.org/10.1016/j.envpol.2016.06.067

Chakraborty, P., Selvaraj, S., Nakamura, M., Prithiviraj, B., Cincinelli, A., & Bang, J. J. (2018). PCBs and PCDD/Fs in soil from informal e-waste recycling sites and open dumpsites in India:

Levels, congener profiles and health risk assessment. *Science of the Total Environment, 621*, 930–938. https://doi.org/10.1016/j.scitotenv.2017.11.083

Colborn, T., vom Saal, F. S., & Soto, A. M. (1994). Developmental effects of endocrine-disrupting chemicals in wildlife and humans. *Environmental Impact Assessment Review, 14*(5), 469–489. https://doi.org/10.1016/0195-9255(94)90014-0

Cortés, J., González, C. M., Morales, L., Abalos, M., Abad, E., & Aristizábal, B. H. (2014). PCDD/PCDF and dl-PCB in the ambient air of a tropical Andean city: Passive and active sampling measurements near industrial and vehicular pollution sources. *Science of the Total Environment, 491–492*, 67–74. https://doi.org/10.1016/j.scitotenv.2014.01.113

CPCB. (2004). *Annual report, 2004–2005*. Published by Central Pollution Control Board (CPCB), Delhi, India. http://www.cpcbenvis.nic.in/annual_report/AnnualReport_9_annualreport2004-2005.pdf

Custer, T. W., Custer, C. M., Hines, R. K., & Sparks, D. W. (2000). Trace elements, organochlorines, polycyclic aromatic hydrocarbons, dioxins, and furans in lesser scaup wintering on the Indiana Harbor Canal. *Environmental Pollution, 110*(3), 469–482. https://doi.org/10.1016/S0269-7491(99)00315-2

Dhooge, W., van Larebeke, N., Koppen, G., Nelen, V., Schoeters, G., Vlietinck, R., Kaufman, J.-M., & Comhaire, F. (2006). Serum dioxin-like activity is associated with reproductive parameters in young men from the general flemish population. *Environmental Health Perspectives, 114*(11), 1670–1676. https://doi.org/10.1289/ehp.9224

Dwyer, H., & Themelis, N. J. (2015). Inventory of U.S. 2012 dioxin emissions to atmosphere. *Waste Management, 46*, 242–246. https://doi.org/10.1016/j.wasman.2015.08.009

European Commission. (2017). *The role of waste-to-energy in the circular economy*. European Commission, Brussels. https://eur-lex.europa.eu/LexUriServ/LexUriServ.do?uri=COM:2017:0034:FIN:EN:PDF

Gómara, B., González, M. J., Baos, R., Hiraldo, F., Abad, E., Rivera, J., & Jiménez, B. (2008). Unexpected high PCB and total DDT levels in the breeding population of red kite (Milvus milvus) from Doñana National Park, south-western Spain. *Environment International, 34*(1), 73–78. https://doi.org/10.1016/j.envint.2007.07.002

Gray, L. E., & Ostby, J. S. (1995). In utero 2,3,7,8-tetrachlorodibenzo-p-dioxin (TCDD) alters reproductive morphology and function in female rat offspring. *Toxicology and Applied Pharmacology, 133*(2), 285–294. https://doi.org/10.1006/taap.1995.1153

Guruge, K. S., Seike, N., Yamanaka, N., & Miyazaki, S. (2005). Polychlorinated dibenzo-p-dioxins, –dibenzofurans, and biphenyls in domestic animal food stuff and their fat. *Chemosphere, 58*(7), 883–889. https://doi.org/10.1016/j.chemosphere.2004.09.049

Hong, S. H., Munschy, C., Kannan, N., Tixier, C., Tronczynski, J., Héas-Moisan, K., & Shim, W. J. (2009). PCDD/F, PBDE, and nonylphenol contamination in a semi-enclosed bay (Masan Bay, South Korea) and a Mediterranean lagoon (Thau, France). *Chemosphere, 77*(6), 854–862. https://doi.org/10.1016/j.chemosphere.2009.07.063

Huang, C.-K., & Liang, J.-J. (2011). Effects of basin topography and monsoon conditions on transport and occurrence of atmospheric PCDD/Fs in the Taichung Basin. *Environmental Science and Pollution Research, 18*(8), 1305–1315. https://doi.org/10.1007/s11356-011-0464-x

Huff, J., Lucier, G., & Tritscher, A. (1994). Carcinogenicity of TCDD: Experimental, mechanistic, and epidemiologic evidence. *Annual Review of Pharmacology and Toxicology, 34*, 343–372. https://doi.org/10.1146/annurev.pa.34.040194.002015

IARC Working Group on the Evaluation of Carcinogenic Risks to. (1997). *Polychlorinated dibenzo-para-dioxins and polychlorinated dibenzofurans*. International Agency for Research on Cancer.

Jiménez, B., Rodríguez-Estrella, R., Merino, R., Gómez, G., Rivera, L., José González, M., Abad, E., & Rivera, J. (2005). Results and evaluation of the first study of organochlorine contaminants (PCDDs, PCDFs, PCBs and DDTs), heavy metals and metalloids in birds from Baja California, México. *Environmental Pollution, 133*(1), 139–146. https://doi.org/10.1016/j.envpol.2004.05.014

Kadokami, K., Takeishi, M., Kuramoto, M., & Ono, Y. (2002). Congener-specific analysis of polychlorinated dibenzo- *p* -dioxins, dibenzofurans, and coplanar polychlorinated biphenyls in frogs and their habitats, Kitakyushu, Japan. *Environmental Toxicology and Chemistry, 21*(1), 129–137. https://doi.org/10.1002/etc.5620210119

Kanan, S., & Samara, F. (2018). Dioxins and furans: A review from chemical and environmental perspectives. *Trends in Environmental Analytical Chemistry, 17*, 1–13. https://doi.org/10.1016/j.teac.2017.12.001

Kashyap, R., Bhatnagar, V., Sadhu, H. G., Arora, B., Jhamb, N., & Karanjkar, R. (2008). Residues of dioxin in egg samples collected from West Zone of India. *Bulletin of Environmental Contamination and Toxicology, 80*(5), 428–430. https://doi.org/10.1007/s00128-008-9423-8

Koppe, J. G. (1995). Nutrition and breast-feeding. *European Journal of Obstetrics & Gynecology and Reproductive Biology, 61*(1), 73–78. https://doi.org/10.1016/0028-2243(95)02156-M

Krauthacker, B., Herceg Romanić, S., Wilken, M., & Milanović, Z. (2006). PCDD/Fs in ambient air collected in Zagreb, Croatia. *Chemosphere, 62*(11), 1829–1837. https://doi.org/10.1016/j.chemosphere.2005.07.042

Kumar, K. S., Kannan, K., Corsolini, S., Evans, T., Giesy, J. P., Nakanishi, J., & Masunaga, S. (2002). Polychlorinated dibenzo-p-dioxins, dibenzofurans and polychlorinated biphenyls in polar bear, penguin and south polar skua. *Environmental Pollution, 119*(2), 151–161. https://doi.org/10.1016/S0269-7491(01)00332-3

Kumar, B., Kumar, S., Mukherjee, D. P., & Sharma, C. S. (2012). Levels of polychlorinated dibenzo-p-dioxins (PCDDs) and polychlorinated dibenzo-p-furans (PCDFs) in the atmosphere (PM10) of a coastal area of West Bengal, India. *Human and Ecological Risk Assessment: An International Journal, 18*(6), 1307–1314. https://doi.org/10.1080/10807039.2012.702613

Kumari, K. (Ed.). (2021). *Persistent organic pollutants: Gaps in management and associated challenges* (1st ed.). CRC Press. https://doi.org/10.1201/9781003046806

Kumari, K., Kumar, S., Rajagopal, V., Khare, A., & Kumar, R. (2019). Emission from open burning of municipal solid waste in India. *Environmental Technology, 40*(17), 2201–2214. https://doi.org/10.1080/09593330.2017.1351489

Kunisue, T., Watanabe, M., Iwata, H., Subramanian, A., Monirith, I., Minh, T., Baburajendran, R., Tana, T., Viet, P., Prudente, M., & Tajima, Y. (2004). Dioxins and related compounds in human breast milk collected around open dumping sites in asian developing countries: Bovine milk as a potential source. *Archives of Environmental Contamination and Toxicology, 47*, 414–426. https://doi.org/10.1007/s00244-004-3172-4

Lemieux, P., Lutes, C., & Santoianni, D. (2004). Emissions of organic air toxics from open burning: A comprehensive review. *Progress in Energy and Combustion Science, 30*, 1–32. https://doi.org/10.1016/j.pecs.2003.08.001

Li, X. L., Johnson, D. C., & Rozman, K. K. (1995). Reproductive effects of 2,3,7,8-tetrachlorodibenzo-p-dioxin (TCDD) in female rats: Ovulation, hormonal regulation, and possible mechanism(s). *Toxicology and Applied Pharmacology, 133*(2), 321–327. https://doi.org/10.1006/taap.1995.1157

Lorber, M., Ferrario, J., & Byrne, C. (2013). EPA's National Dioxin Air Monitoring Network (NDAMN): Design, implementation, and final results. *Atmospheric Environment, 77*, 311–317. https://doi.org/10.1016/j.atmosenv.2013.05.008

Ma, S. W. Y., & Yang, R. R. (2007). Chapter 7: Persistent organic pollutants in Hong Kong. In *Developments in environmental science* (Vol. 7, pp. 313–373). Elsevier. https://doi.org/10.1016/S1474-8177(07)07007-6

Mably, T. A., Moore, R. W., & Peterson, R. E. (1992). In utero and lactational exposure of male rats to 2,3,7,8-tetrachlorodibenzo-p-dioxin: I. Effects on androgenic status. *Toxicology and Applied Pharmacology, 114*(1), 97–107. https://doi.org/10.1016/0041-008X(92)90101-W

Marvin, C. H., Howell, E. T., Kolic, T. M., & Reiner, E. J. (2002). Polychlorinated dibenzo-*p* -dioxins and dibenzofurans and dioxinlike polychlorinated biphenyls in sediments and mussels at three sites in the lower Great Lakes, North America. *Environmental Toxicology and Chemistry, 21*(9), 1908–1921. https://doi.org/10.1002/etc.5620210919

Mayilsamy, M., Sangeetha, S., Nakamura, M., Ko, S., Govarthanan, M., & Vimalkumar, K. (2022). Screening of PCDDs/Fs and DL-PCBs by AhR-CALUX bioassay in bovine milk and environment ash from India. *Toxicology and Environmental Health Sciences, 14*(3), 223–234. https://doi.org/10.1007/s13530-022-00133-w

McNulty, W. P. (1984). Fetotoxicity of 2,3,7,8-tetrachlorodibenzo-p-dioxin (TCDD) for rhesus macaques (Macaca mulatta). *American Journal of Primatology, 6*(1), 41–47. https://doi.org/10.1002/ajp.1350060105

Minh, N. H., Minh, T. B., Watanabe, M., Kunisue, T., Monirith, I., Tanabe, S., Sakai, S., Subramanian, A., Sasikumar, K., Viet, P. H., Tuyen, B. C., Tana, T. S., & Prudente, M. S. (2003). Open dumping site in asian developing countries: A potential source of polychlorinated dibenzo- *p* -dioxins and polychlorinated dibenzofurans. *Environmental Science & Technology, 37*(8), 1493–1502. https://doi.org/10.1021/es026078s

Minomo, K., Ohtsuka, N., Nojiri, K., & Matsumoto, R. (2018). Influence of combustion-originated dioxins in atmospheric deposition on water quality of an urban river in Japan. *Journal of Environmental Sciences, 64*, 245–251. https://doi.org/10.1016/j.jes.2017.06.027

Moore, R. W., Potter, C. L., Theobald, H. M., Robinson, J. A., & Peterson, R. E. (1985). Androgenic deficiency in male rats treated with 2,3,7,8-tetrachlorodibenzo-p-dioxin. *Toxicology and Applied Pharmacology, 79*(1), 99–111. https://doi.org/10.1016/0041-008X(85)90372-2

NIP. (2011). *Government of India National Implementation Plan, Stockholm Convention on Persistent Organic Pollutants*. Published by MoEF&CC and UNIDO, Delhi, India. http://www.indiaenvironmentportal.org.in/files/NIP%20final%20draft.pdf

Park, J.-S., & Kim, J.-G. (2002). Regional measurements of PCDD/PCDF concentrations in Korean atmosphere and comparison with gas–particle partitioning models. *Chemosphere, 49*(7), 755–764. https://doi.org/10.1016/S0045-6535(02)00375-2

Paumgartten, F. J. R., Cruz, C. M., Chahoud, I., Palavinskas, R., & Mathar, W. (2000). PCDDs, PCDFs, PCBs, and other organochlorine compounds in human milk from Rio de Janeiro, Brazil. *Environmental Research, 83*(3), 293–297. https://doi.org/10.1006/enrs.2000.4062

Petrlik, J., Digangi, J., Skalsky, M., Kuncova, H., Petrlikova, L., Link, T., & Association, A., & WG, I. (2005). *Contamination of chicken eggs near the Queen Mary's Hospital, Lucknow medical waste incinerator in Uttar Pradesh (India) by dioxins, PCBs and hexachlorobenzene*. https://doi.org/10.13140/RG.2.2.18999.21926

Phol, H. R. (1998). *Toxicological profile for chlorinated dibenzo-p-dioxins*. Agency for Toxic Substances and Disease Registry.

Polder, A., Gabrielsen, G. W., Odland, J. Ø., Savinova, T. N., Tkachev, A., Løken, K. B., & Skaare, J. U. (2008). Spatial and temporal changes of chlorinated pesticides, PCBs, dioxins (PCDDs/PCDFs) and brominated flame retardants in human breast milk from Northern Russia. *Science of the Total Environment, 391*(1), 41–54. https://doi.org/10.1016/j.scitotenv.2007.10.045

Rajan, S. (2021). Soil concentrations, compositional profiles, sources and bioavailability of polychlorinated dibenzo dioxins/furans, polychlorinated biphenyls and polycyclic aromatic hydrocarbons in open munic Mayilsamy ipal dumpsites of Chennai city, India. *Waste Management, 10*, 331.

Rier, S. E., Martin, D. C., Bowman, R. E., Dmowski, W. P., & Becker, J. L. (1993). Endometriosis in rhesus monkeys (Macaca mulatta) following chronic exposure to 2,3,7,8-tetrachlorodibenzo-p-dioxin. *Fundamental and Applied Toxicology, 21*(4), 433–441. https://doi.org/10.1006/faat.1993.1119

Santillo, D., Fernandes, A., Stringer, R., Alcock, R., Rose, M., White, S., Jones, K., & Johnston, P. (2003). Butter as an indicator of regional persistent organic pollutant contamination: Further development of the approach using polychlorinated dioxins and furans (PCDD/Fs), and dioxin-like polychlorinated biphenyls (PCBs). *Food Additives and Contaminants, 20*(3), 281–290. https://doi.org/10.1080/0265203021000057494

Schantz, S. L., & Bowman, R. E. (1989). Learning in monkeys exposed perinatally to 2,3,7,8-tetrachlorodibenzo-p-dioxin (TCDD). *Neurotoxicology and Teratology, 11*(1), 13–19. https://doi.org/10.1016/0892-0362(89)90080-9

Schecter, A., Fürst, P., Ryan, J. J., Fürst, C., Meemken, H.-A., Groebel, W., Constable, J., & Vu, D. (1989). Polychlorinated dioxin and dibenzofuran levels from human milk from several locations in the United States, Germany and Vietnam. *Chemosphere, 19*(1), 979–984. https://doi.org/10.1016/0045-6535(89)90444-X

Schecter, A., Päpke, O., & Ball, M. (1990). Evidence for transplacental transfer of dioxins from mother to fetus: Chlorinated dioxin and dibenzofuran levels in the livers of stillborn infants. *Chemosphere, 21*(8), 1017–1022. https://doi.org/10.1016/0045-6535(90)90124-C

Schecter, A., Birnbaum, L., Ryan, J. J., & Constable, J. D. (2006). Dioxins: An overview. *Environmental Research, 101*(3), 419–428. https://doi.org/10.1016/j.envres.2005.12.003

Schuhmacher, M., Mari, M., Nadal, M., & Domingo, J. L. (2019). Concentrations of dioxins and furans in breast milk of women living near a hazardous waste incinerator in Catalonia, Spain. *Environment International, 125*, 334–341. https://doi.org/10.1016/j.envint.2019.01.074

Senthil Kumar, K., Kannan, K., Paramasivan, O. N., Shanmuga Sundaram, V. P., Nakanishi, J., & Masunaga, S. (2001). Polychlorinated dibenzo- *p* -dioxins, dibenzofurans, and polychlorinated biphenyls in human tissues, meat, fish, and wildlife samples from India. *Environmental Science & Technology, 35*(17), 3448–3455. https://doi.org/10.1021/es010555+

Senthilkumar, N., Iseki, S., & Hayama, K. (2002). Polychlorinated dibenzo- *p* -dioxins, dibenzofurans, and dioxin-like polychlorinated biphenyls in livers of birds from Japan. *Archives of Environmental Contamination and Toxicology, 42*(2), 244–255. https://doi.org/10.1007/s00244-001-0030-5

Sharma, M., McBean, E., & Gowing, A. (2007). Bioconcentration of dioxins and furans in vegetation. *Water, Air, and Soil Pollution, 179*(1), 117–124. https://doi.org/10.1007/s11270-006-9218-7

Shen, H., Han, J., Tie, X., Xu, W., Ren, Y., & Ye, C. (2009). Polychlorinated dibenzo-p-dioxins/furans and polychlorinated biphenyls in human adipose tissue from Zhejiang Province, China. *Chemosphere, 74*(3), 384–388. https://doi.org/10.1016/j.chemosphere.2008.09.094

Someya, M., Ohtake, M., Kunisue, T., Subramanian, A., Takahashi, S., Chakraborty, P., Ramachandran, R., & Tanabe, S. (2010). Persistent organic pollutants in breast milk of mothers residing around an open dumping site in Kolkata, India: Specific dioxin-like PCB levels and fish as a potential source. *Environment International, 9*, 27.

Sonnevera Internationagricultural plasticsal Corporation. (2011, July). *Ontario Agricultural Waste Study: Environmental impacts of open burning*. Published by Clean Farms Inc. http://cleanfarms.ca/wp-content/uploads/2017/07/AgPlasticEnviroImpactOpenBurning_FINAL_201107.pdf

Srogi, K. (2008). Levels and congener distributions of PCDDs, PCDFs and dioxin-like PCBs in environmental and human samples: A review. *Environmental Chemistry Letters, 6*(1), 1–28. https://doi.org/10.1007/s10311 007 0105 2

Stockholm Convention. (2001). *Listing of POPs in Stockholm Convention*. Retrieved September 7, 2022, from http://chm.pops.int/TheConvention/ThePOPs/ListingofPOPs/tabid/2509/Default.aspx

Thacker, N. P., Nitnaware, V. C., Das, S. K., & Devotta, S. (2007). Dioxin formation in pulp and paper mills of India. *Environmental Science and Pollution Research International, 14*(4), 225–226. https://doi.org/10.1065/espr2007.02.386

Thacker, N., Kashyap, S., Sheikh, J., Trivedi, J., Thokchom, B., & Agnihotri, A. (2010). Dioxin releases in waste incinerations and thermal processes. *Bulletin of Environmental Contamination and Toxicology, 85*(6), 624–627. https://doi.org/10.1007/s00128-010-0137-3

Thacker, N., Sheikh, J., Tamane, S. M., Bhanarkar, A., Majumdar, D., Singh, K., Chavhan, C., & Trivedi, J. (2013). Emissions of polychlorinated dibenzo-p-dioxins (PCDDs), dibenzofurans (PCDFs), and dioxin-like polychlorinated biphenyls (PCBs) to air from waste incinerators and high thermal processes in India. *Environmental Monitoring and Assessment, 185*(1), 425–429. https://doi.org/10.1007/s10661-012-2564-6

Törnkvist, A., Glynn, A., Aune, M., Darnerud, P. O., & Ankarberg, E. H. (2011). PCDD/F, PCB, PBDE, HBCD and chlorinated pesticides in a Swedish market basket from 2005 – Levels and dietary intake estimations. *Chemosphere, 83*(2), 193–199. https://doi.org/10.1016/j.chemosphere.2010.12.042

UNEP. (1997). *Governing Council of the UNEP, Volume 6.* https://wedocs.unep.org/bitstream/handle/20.500.11822/17274/97_GC19_proceedings.pdf?sequence=23&isAllowed=y

van Leeuwen, F. X., Feeley, M., Schrenk, D., Larsen, J. C., Farland, W., & Younes, M. (2000). Dioxins: WHO's tolerable daily intake (TDI) revisited. *Chemosphere, 40*(9–11), 1095–1101. https://doi.org/10.1016/s0045-6535(99)00358-6

Van Luong, H., Tai, P. T., Nishijo, M., Trung, D. M., Thao, P. N., Van Son, P., Van Long, N., Linh, N. T., & Nishijo, H. (2018). Association of dioxin exposure and reproductive hormone levels in men living near the Bien Hoa airbase, Vietnam. *Science of the Total Environment, 628–629*, 484–489. https://doi.org/10.1016/j.scitotenv.2018.02.094

Vos, J. G., Moore, J. A., & Zinkl, J. G. (1973). Effect of 2,3,7,8-tetrachlorodibenzo-p-dioxin on the immune system of laboratory animals. *Environmental Health Perspectives, 5*, 149–162. https://doi.org/10.1289/ehp.7305149

Wai-mei Sin, D., Ya-yin Choi, J., & Kwok-keung Louie, P. (2002). A study of polychlorinated dibenzo-p-dioxins and dibenzofurans in the atmosphere of Hong Kong. *Chemosphere, 47*(6), 647–653. https://doi.org/10.1016/S0045-6535(01)00316-2

Wan, Y., Jones, P. D., Holem, R. R., Khim, J. S., Chang, H., Kay, D. P., Roark, S. A., Newsted, J. L., Patterson, W. P., & Giesy, J. P. (2010). Bioaccumulation of polychlorinated dibenzo-p-dioxins, dibenzofurans, and dioxin-like polychlorinated biphenyls in fishes from the Tittabawassee and Saginaw Rivers, Michigan, USA. *Science of the Total Environment, 408*(11), 2394–2401. https://doi.org/10.1016/j.scitotenv.2010.02.003

Warren, T. K., Mitchell, K. A., & Lawrence, B. P. (2000). Exposure to 2,3,7,8-tetrachlorodibenzo-p-dioxin (TCDD) suppresses the humoral and cell-mediated immune responses to influenza A virus without affecting cytolytic activity in the lung. *Toxicological Sciences, 56*(1), 114–123. https://doi.org/10.1093/toxsci/56.1.114

Watanabe, M. X., Iwata, H., Watanabe, M., Tanabe, S., Subramanian, A., Yoneda, K., & Hashimoto, T. (2005). Bioaccumulation of organochlorines in crows from an Indian open waste dumping site: Evidence for direct transfer of dioxin-like congeners from the contaminated soil. *Environmental Science and Technology, 39*, 4421–4430. https://doi.org/10.1021/es050057r

Watanabe, M. X., Kunisue, T., Tao, L., Kannan, K., Subramanian, A., Tanabe, S., & Iwata, H. (2010). Dioxin-like and perfluorinated compounds in pigs in an Indian open waste dumping site: Toxicokinetics and effects on hepatic cytochrome P450 and blood plasma hormones. *Environmental Toxicology and Chemistry, 29*(7), 1551–1560. https://doi.org/10.1002/etc.189

Wittsiepe, J., Fürst, P., Schrey, P., Lemm, F., Kraft, M., Eberwein, G., Winneke, G., & Wilhelm, M. (2007). PCDD/F and dioxin-like PCB in human blood and milk from German mothers. *Chemosphere, 67*(9), S286–S294. https://doi.org/10.1016/j.chemosphere.2006.05.118

Yang, F., Zhang, Q., Xu, Y., Jiang, G., Wang, Y., & Wang, D. (2008). Preliminary hazard assessment of polychlorinated biphenyls, polybrominated diphenyl ethers, and polychlorinated dibenzo-p-dioxins and dibenzofurans to yangtze finless porpoise in Dongting Lake, China. *Environmental Toxicology and Chemistry, 27*(4), 991. https://doi.org/10.1897/07-381.1

Zhang, L., Li, J., Liu, X., Zhao, Y., Li, X., Wen, S., & Wu, Y. (2013). Dietary intake of PCDD/Fs and dioxin-like PCBs from the Chinese total diet study in 2007. *Chemosphere, 90*(5), 1625–1630. https://doi.org/10.1016/j.chemosphere.2012.08.040

Zhou, P., Wu, Y., Yin, S., Li, J., Zhao, Y., Zhang, L., Chen, H., Liu, Y., Yang, X., & Li, X. (2011). National survey of the levels of persistent organochlorine pesticides in the breast milk of mothers in China. *Environmental Pollution, 159*(2), 524–531. https://doi.org/10.1016/j.envpol.2010.10.014

Zhou, Z.-G., Zhao, B., Qi, L., Xu, P.-J., Ren, Y., Li, N., Shen, L., Zheng, S., Zhao, H., Fan, S., Zhang, T., Liu, A.-M., & Huang, Y.-R. (2014). Distribution of polychlorinated dibenzo-p-dioxins and dibenzofurans in the atmosphere of Beijing, China. *Aerosol and Air Quality Research, 14*(4), 1269–1278. https://doi.org/10.4209/aaqr.2013.01.0032

Hexachlorocyclohexane (α, β-HCH) and Lindane (γ-HCH)

Kanchan Kumari, Md. Shahbaz, and Karthik Raghunathan

1 Introduction

Hexachlorocyclohexane (HCH) appears as a colorless to white crystalline substance with a distinctive odor, falling under the category of persistent organic pollutants (POPs) known for its significant toxicity to both humans and animals. HCH is a combination of five isomers, sharing the same chemical composition but differing in the arrangement of chlorine atoms. Among these isomers, alpha-HCH, beta-HCH, and gamma-HCH are the most prevalent. Notably, gamma-HCH (γ-HCH) is the most common and is widely known as Lindane. It is crucial to distinguish hexachlorocyclohexane as the accurate term for this compound and to avoid using "Benzene Hexachloride" to refer to 1, 2, 3, 4, 5, 6-hexachlorocyclohexane. Moreover, it is crucial to differentiate it from hexachlorobenzene. The technical formulation of HCH comprises a blend of isomers and associated compounds, which can exhibit variations in their respective concentrations (Hayes & Laws, 1991).

Their chemical structures and toxicological effects distinguish the three most prevalent isomers of HCH, namely, alpha-HCH, beta-HCH, and gamma-HCH (Lindane). Alpha-HCH, used primarily in agriculture as a pesticide, exhibits central nervous system stimulant properties, resulting in violent epileptic convulsions. In contrast, beta-HCH acts as a depressant on the central nervous system and was phased out because of environmental and health concerns. Gamma-HCH (Lindane) is widely recognized for its insecticidal properties and is employed in controlling

K. Kumari (✉) · M. Shahbaz
CSIR-National Environmental Engineering Research Institute (NEERI), Kolkata Zonal Center, Kolkata, West Bengal, India
e-mail: k_kumari@neeri.res.in

K. Raghunathan
CSIR-National Environmental Engineering Research Institute (NEERI), Nagpur, India

K. Kumari, *Pollutants of Global Concern*, Emerging Contaminants and Associated Treatment Technologies, https://doi.org/10.1007/978-3-031-50996-4_12

pests in various agricultural and veterinary applications. However, gamma-HCH is associated with health and environmental issues, displaying both central nervous system stimulant and depressant effects. While gamma-HCH (Lindane) is the most commonly known isomer, technical grade HCH typically contains a mixture of isomers, including alpha, beta, in addition, gamma forms. The presence of different isomers in the mixture contributes to variations in toxicological properties, making it essential to consider the relative concentrations of each isomer. Because of the persistence, bioaccumulation, and long-range transport capabilities of HCH and its isomers, various international and national regulatory agencies have classified them as hazardous substances. Consequently, many countries have imposed restrictions or bans on their use to mitigate adverse effects on human health and the environment. It is evident that the management and control of HCH are crucial in safeguarding the well-being of ecosystems and human populations. These HCH compounds were listed in Annex A of Stockholm Convention in 2009 (Kumari, 2021).

2 Technical Details of Hexacholorocyclohexane

Hexachlorocyclohexane (HCH) is a chemical compound derived from benzene through photo chlorination and is composed of six chlorine atoms attached to a cyclohexane ring. HCH exists in various isomeric forms, with the three most prevalent isomers being alpha-HCH, beta-HCH, and gamma-HCH (Lindane).

1. Alpha (α)-HCH: Also known as hexachlorocyclohexane α-isomer, its chemical formula is $C_6H_6Cl_6$.
2. Beta (β)-HCH: Also known as hexachlorocyclohexane β-isomer, its chemical formula is $C_6H_6Cl_6$.
3. Gamma (γ)-HCH (Lindane): Also known as hexachlorocyclohexane γ-isomer, it is the most widely recognized and extensively used isomer. Its chemical formula is $C_6H_6Cl_6$.

Historically, HCH and its isomers were employed as pesticides due to their insecticidal properties. However, their environmental persistence, ability to accumulate in living organisms, and potential to cause adverse effects on human health and the ecosystem have led to their regulation and restriction in many countries.

Each isomer of HCH exhibits distinct toxicological properties, and the composition of technical grade HCH typically comprises a mixture of these isomers (Fig. 1). Notably, the relative concentrations of alpha, beta, and gamma isomers in the mixture can significantly influence its overall toxicological effects.

It is imperative to monitor and regulate the use and disposal of HCH and its isomers to safeguard both human health and the environment. The recognition of their distinct toxicological profiles emphasizes the importance of employing safer alternatives and implementing responsible handling practices to mitigate the potential risks associated with these compounds (Table 1).

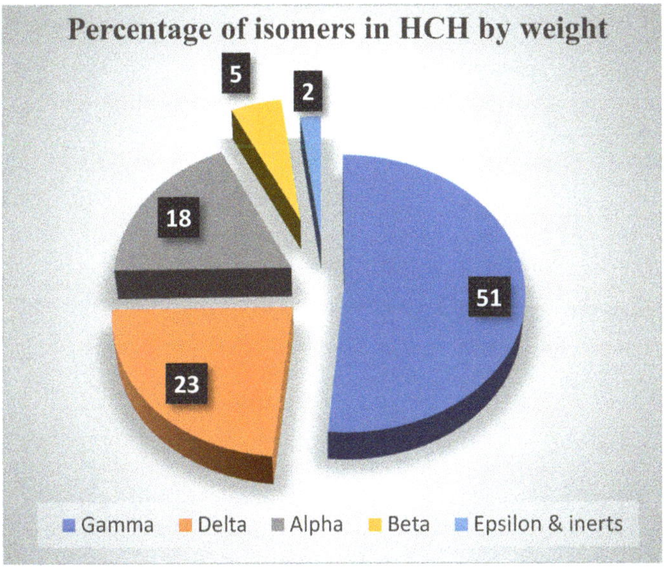

Fig. 1 Percentage of isomers in HCH by weight

2.1 *Other Characteristics*

When solid HCH is heated or its solution burns, dangerous combustion occurs, leading to the production of toxic gases (CHRIS, 2000). Upon decomposition, it emits smoke containing chlorides, hydrogen chloride, and phosgene (Lewis, 2000). Upon exposure to heat, light, air, carbon dioxide, and strong acids (except the β-isomer), HCH is generally stable but at ordinary temperatures, it is prone to dehydrochlorination by alkali (Spencer, 1982).

Regarding reactivity, HCH is not reactive with common substances or water, making it stable during transport (CHRIS, 2000). However, it can have potentially violent reactions with dimethylformamide in the presence of iron and carbon tetrachloride (Lewis et al., 1999). The gamma-isomer is particularly sharp to essence (Lewis, 2000). During decomposition or when heated, HCH produces hazardous dust and vapors containing poisonous compounds like hydrogen chloride, chloride, and phosgene (Lewis, 2000). Information on ignition and explosive limits in air is not available, as is data on its electrical and thermal conductivity.

In terms of environmental threats, HCH is known to cause water pollution and requires caution during secure disposal. It has been found to be simply adsorbed to river bottom sediments, indicating its persistence and potential to concentrate in the food chain (Clayton & Clayton, 1994; CHRIS, 2000). This bioaccumulation can pose risks to aquatic life and ultimately affect human health if consumed through contaminated food sources. Therefore, proper management and regulation are necessary to prevent environmental contamination and adverse health effects associated with HCH exposure.

Table 1 Chemical characteristics of alpha-, beta-, and gamma-HCH

Technical details	Beta-hexachlorocyclohexane (β-HCH)	Alpha-hexachlorocyclohexane (α-HCH)	Gamma-hexachlorocyclohexane
Chemical structure (Buser and Mueller 1995)	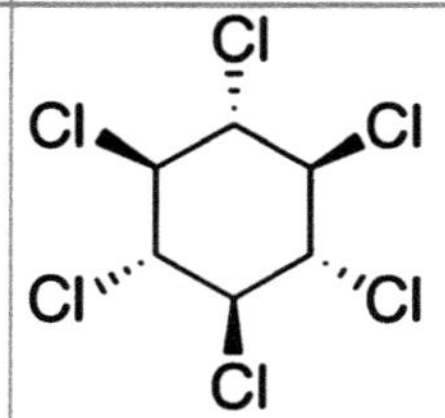	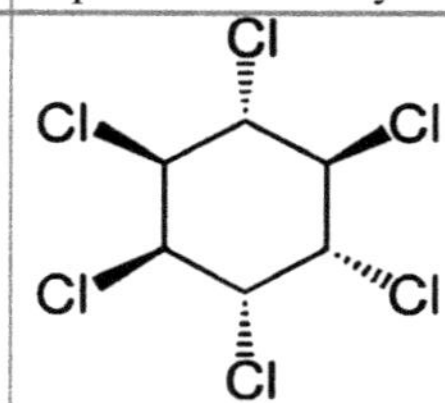	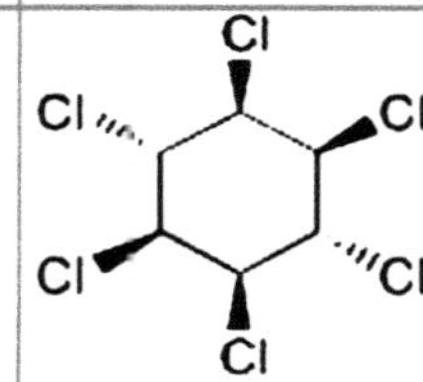
IUPAC name:	(1r,2r,3r,4r,5r,6r)-1,2,3,4,5,6-Hexachlorocyclohexane	(1R,2R,3S,4S,5S,6S)-1,2,3,4,5,6-Hexachlorocyclohexane	1,2,3,4,5,6-hexachlorocyclohexane
Chemical formula:	$C_6H_6Cl_6$	$C_6H_6Cl_6$	$C_6H_6Cl_6$
CAS number	319-85-7	319-84-6	58-89-9
Molecular weight:	290.83 g/mol	290.83 g/mol	290.8 g/mol
Color/form	No data	Brownish to white	White to yellow, crystalline powder
Odor	No data	Phosgene-like odor	Slight musty odor
Boiling point (°C)	60 °C at 0.5 mm of Hg	288 °C at 70 mm Hg	323.4 °C at 760 mm Hg
Melting point (°C)	314–315	156–161	112.5
Solubility: Water	10 ppm	5 ppm	17 ppm
Organic solvents (g/100gm)	Solvable in alcohol; ethyl alcohol-1.8; ether-6.2	Ethyl alcohol-1.1; ether-1.8; benzene- 1.9	Ethyl alcohol- 6.4; ether-20.8; benzene-28.9
Density (g/cm^3)	1.89 at 19 °C	1.87 at 20 °C	1.87 at 19 °C
Vapor pressure	4.5 × 10–5 mm Hg at 25 °C	3.6 × 10–7 mm Hg at 20 °C	4.2 × 10–5 mm Hg at 20 °C
Octanol/water partition coefficient Log Kow (25 °C)	3.9	3.9	3.72

Octanol-air partition coefficient Log Koa (25 °C)	8.7	7.5	8.29
Henry's law constant (Pa m^3 mol^{-1})	6.86×10^{-6b}	4.5×10^{-7b}	3.5×10^{-6b}
Decomposition	–	Upon heating, there is a possibility of decomposition leading to the emission of corrosive and potentially toxic fumes	During fire incidents, hazardous decomposition products such as carbon oxides and hydrogen chloride gas can form
Safety and hazards	–	Toxic when swallowed, harmful when in contact with skin, suspected of being carcinogenic, extremely toxic to aquatic life with chronic effects	Highly toxic carcinogen. May cause irritation on contact
Toxicity	Toxic	Toxic	Toxic

3 State of Use of HCH in India and the World

The primary and widely established usage of hexachlorocyclohexane (HCH) is as an insecticide, especially for controlling cotton insects (Gosselin et al., 1984). However, over the years, concerns regarding its environmental and health impacts have led to regulatory actions in various countries. In the United States, all HCH registrations have been changed administratively to Lindane registrations. HCH is now not produced in the USA, and its domestic use is prohibited by EPA regulations, as well as in many other countries (FCH, 1984). Some countries still allow the limited use of HCH for specific purposes, such as controlling leafhoppers and stem borers in lowland rice, as seed treatment to reduce wireworm damage in cereals, and for controlling pests of sugar beets, cereals, and oilseed rape (FCH, 1984).

India implemented a ban on agricultural use of technical HCH in 1990 while still permitting its use for public health purposes (AMAP, 2004). The Central Insecticides Board and Registration Committee of India from March 25, 2011, banned the manufacture, import, and formulation of Lindane, a type of HCH. Subsequently, its agricultural use was also banned from March 25, 2013 (Gazette Notification No S.O. 637(E) dated 25 March 2011).

As of 2023, specific information is not available on the stockpiles of α-hexachlorocyclohexane, β-hexachlorocyclohexane, or Lindane (National Center for Biotechnology Information). Regarding regulations, India lacks specific policies for alpha-hexachlorocyclohexane. While there is no ban on the production, manufacturing, or use, there is limited export of β -HCH from the country.

Import data from 2014 to 2018 are not available for 1,2,3,4,5,6-hexachlorocyclo hexane (HCH), and the available data only covers information on all chemical compounds rather than each chemical within 1,2,3,4,5,6-hexachlorocyclohexane (HCH)

The provided figure (Fig. 2) illustrates the average concentrations of hexachlorocyclohexane (HCH) in the air of the Sundarban region. The data reveal a noteworthy distinction in contamination levels between urban (U) and wetland (W) areas. Specifically, urban regions exhibit higher HCH contamination, highlighting the impact of human activities on air quality. In contrast, wetland areas show relatively lower contamination levels, emphasizing the potential resilience of these ecosystems. This underscores the importance of targeted pollution control efforts, particularly in urban environments, to ensure environmental health and balance.

India is recognized as a significant location with a high prevalence of HCH pollution. The residues observed in all compartments examined in this study (Fig. 3) were frequently above the thresholds set by several international regulatory bodies. This phenomenon can be attributed to the increased use of pesticidal persistent organic pollutants (POPs) in agricultural practices in past, which persisted until recently. The meta-analysis of temporal trends did not provide any evidence indicating a general decrease in HCH levels in the Indian environment, except for HCHs found in surface water (Fig. 4). Table 2 represents the concentrations recorded for HCH in different Indian rivers. The data obtained from human biomonitoring provide evidence of increased levels of environmental exposure. In essence, it was

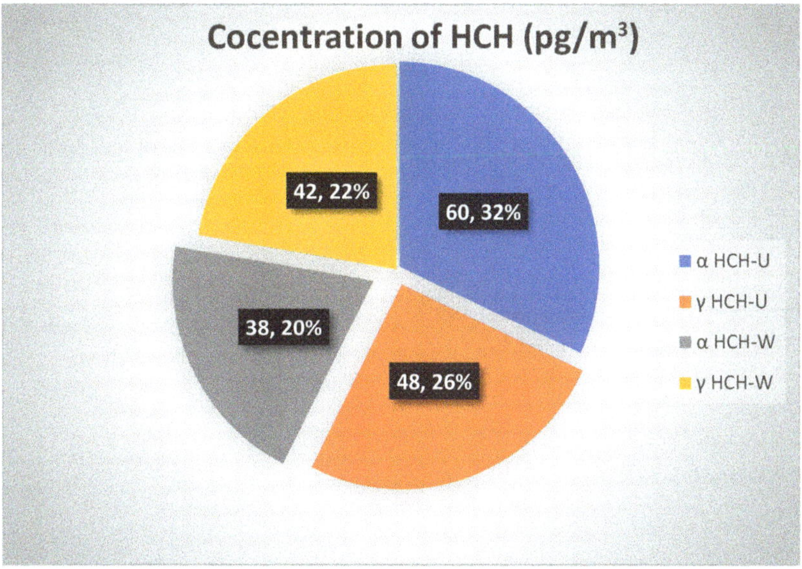

Fig. 2 Average concentration of HCH (pg/m^3) in the air of Urban and Wetlands areas of Sunderban. (Pozo et al., 2016)

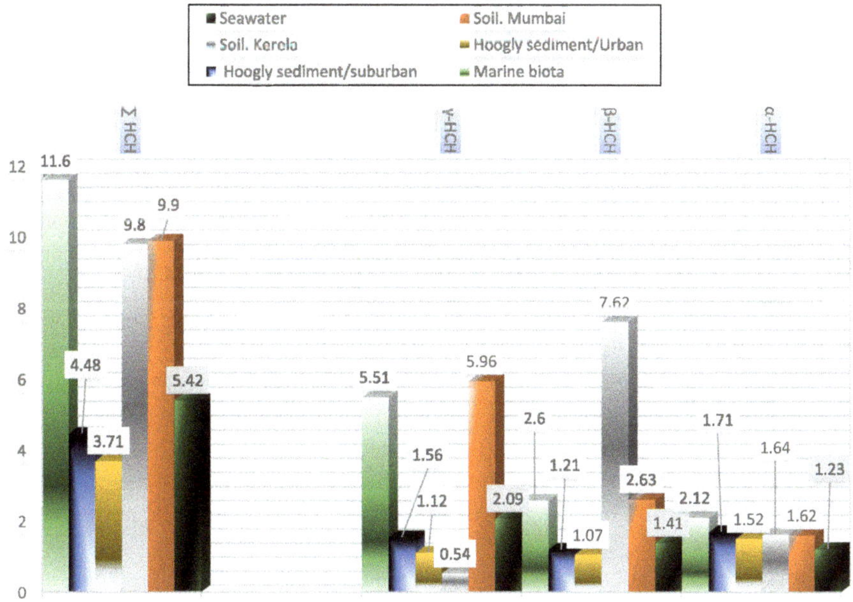

Fig. 3 Average concentration of HCH (ng/g) in seawater, sediment, and marine biota of Indian regions. (Source: Sruthi et al., 2016; Khuman et al., 2020; Pandit et al., 2006)

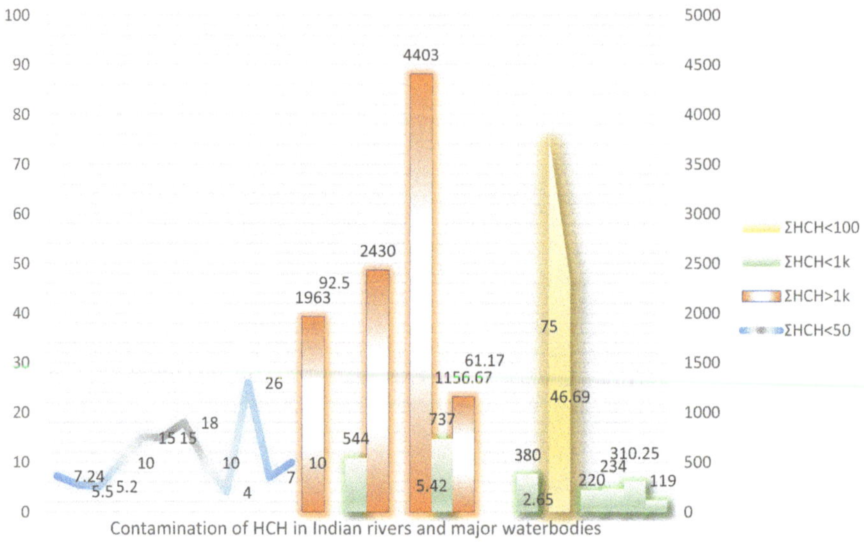

Fig. 4 Contamination of HCH in India. (Sharma et al., 2014)

determined that all accessible data of HCHs in human milk and blood surpassed established safety thresholds. Furthermore, no discernible patterns of decrease were observed after the implementation of regulations of persistent organic pollutants (POPs). The levels of environmental and human contamination caused by HCHs in India may be compared with those in China. However, it is worth noting that China has shown a discernible trend of decreasing environmental pollution, which may be attributed to the adoption of early substitution and more stringent measures for pesticidal POPs.

3.1 Environmental Standards and Regulations Related to Lindane (Gamma-HCH)

The Clean Air Act identifies Lindane as a hazardous air pollutant (HAP) with known or suspected adverse health effects. As per the 1990 amendment to the Act, the Environmental Protection Agency (EPA) is responsible for establishing standards that require significant reductions in toxic pollutant emissions from major sources. Lindane is specifically listed among the pollutants subject to specific performance-based standards.

The Acceptable Daily Intake (ADI) for Lindane is set at 0–0.005 mg/kg/bw (milligrams per kilogram of body weight). This value represents the safe amount of Lindane that can be consumed daily over a lifetime without causing significant health risks. Figure 5 represents the permissible limits of Lindane in different food products.

Table 2 Contamination of HCH in Indian rivers

Rivers and major waterbodies, India	ΣHCH < 1 k	ΣHCH > 1 k	ΣHCH < 50	ΣHCH < 100
Ganges, Devprayag (2010)			7.24	
Ganges, Rishikesh (2010)			5.5	
Ganges, Haridwar (2010)			5.2	
Sharda (1999)			10	
Saryu (1999)			15	
Gori (1999)			15	
Ram Ganga (1999)			18	
Dhuali (1999)			10	
Kosi (1999)			4	
Gomti (1999) Uttrakhand			26	
Ladhiya (1999)			7	
Lohwati (1999)			10	
Vellar, Tamil Nadu (1986)		1963		
Cauvery and Coreloon, South India (NA)				92.5
Kolleru lake, Andhra Pradesh (NA)	544			
Cauvery, Karnataka and Andhra Pradesh (NA)		2430		
Seawater, Mumbai (NA)			5.42	
Brahmaputra, Dibrugarh, Assam		4403		
[a]Ganges, Kolkata (2005)	737			
Tank water, Greater Kolkata (NA)		1156.67		
Ganges, Bhagalpur (2008)				61.17
Ganges, Patna (2011)			2.65	
[a]Reetha, Lucknow (NA)	380			
Ganges, Kanpur (NA)				75
Gomti, Uttar Pradesh (NA)				46.69
Sharda, Uttar Pradesh (NA)	220			
[a]Ramgarh water reservoir, Jaipur (NA)	234			
[a]Yamuna, Haryana and Delhi (1999)	310.25			
Ghaggar, Haryana (2000)	119			

[a]To represent more comprehensive and illustrated data, divided the data by thousand

In accordance with the Comprehensive Environmental Response, Compensation, and Liability Act (CERCLA), individuals in charge of vessels or facilities must immediately notify the National Response Centre (NRC) if there is a release of Lindane or any other designated hazardous substance, and the released amount equals or exceeds the reportable quantity of 1 pound (approximately 0.454 kg).

The Revised Human Health Water Quality Criteria for Lindane establish acceptable consumption levels at 4.2 micrograms per liter (μg/L) for water + organism and 4.4 μg/L for organism only. These criteria define the maximum allowable concentration of Lindane in water to safeguard human health when the water is directly consumed or ingested by aquatic organisms.

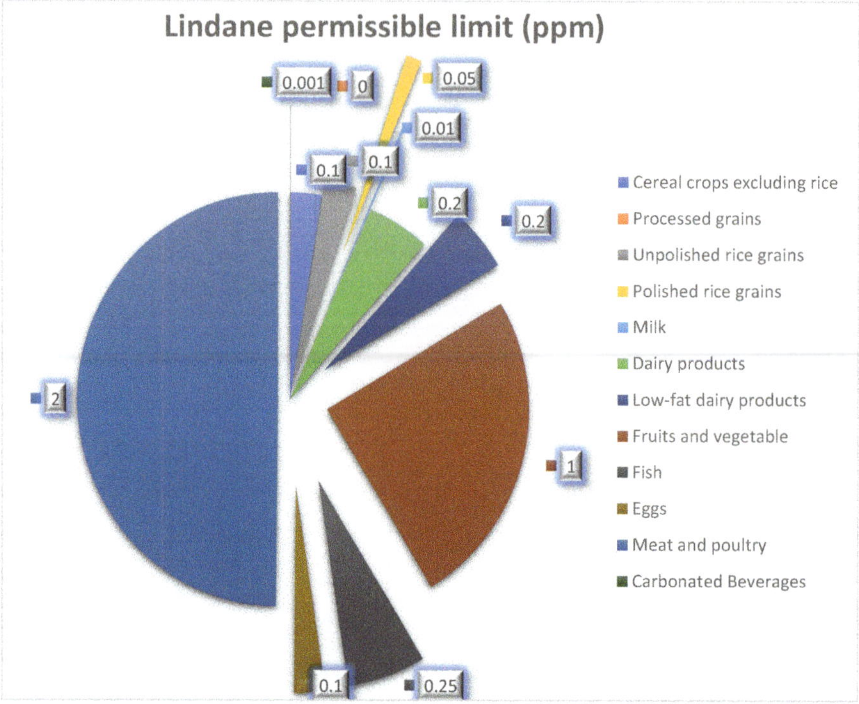

Fig. 5 Permissible limits of Lindane in food products. (WHO, 1984)

To ensure the safety of drinking water, Lindane is subject to specific maximum contaminant level (MCL) regulations. The MCL for Lindane is set at 0.0002 mg/L (milligrams per liter), which is the maximum permissible concentration of Lindane in drinking water.

Furthermore, the EPA provides water quality guidance by recommending an acceptable concentration level of 0.2 μg/L (micrograms per liter) for Lindane in water. This guidance assists in evaluating and managing Lindane concentrations in water bodies to protect both human health and the environment.

4 Adverse Effects of Hexacholorocyclohexane

Hexachlorocyclohexane (HCH) is associated with a range of adverse effects resulting from exposure through different routes. Oral ingestion of large amounts of HCH can lead to acute toxicity, while the chances of long-term toxicity development are through the consumption of contaminated edibles or drinks (Macnamara, 1970) on a daily basis. However, for HCH, intake is not the primary route of entry, except in cases of babies dependent on breastfeeding. Chronic occupational exposure to

HCH-II was reported to be associated with increased health risks, including respiratory, liver, and neurological issues (Brassow et al., 1981).

HCH inhalation has been observed among occupationally exposed individuals, although there have been no reports of acute poisoning through this route (Nigam et al., 1986; Chattopadhyay et al., 1988; Chandra et al., 1992). The chemical can be absorbed by the lungs as dust or liquid sprays, contributing to potential health risks (Clayton & Clayton, 1994). Dermal exposure to HCH poses significant risks, particularly in cases of occupational contact (Chattopadhyay et al., 1988). Serious accidental poisonings have been reported in a common population following skin exposure to HCH (Clayton & Clayton, 1994). The IPCS (1989) report highlights beta-HCH's potential to cause neurological damage, including tremors and convulsions, raising concerns about occupational and environmental exposure. Tomczak et al. (1981) in his study described occupational exposure to HCH was associated with elevated levels of liver enzymes and neurological symptoms like tremors and dizziness, suggesting potential hepatotoxic and neurotoxic effects. Beta-HCH exhibited persistent accumulation in rat tissues, highlighting its potential for long-term health risks (Richter et al., 1981). HCH exposure triggered significant oxidative stress and tissue damage in rats, indicating its potent deleterious effects (Joseph et al., 1992). Beta-HCH emerged as the most potent immunosuppressant, significantly impacting both thymus and spleen weight, suggesting heightened vulnerability to this isomer (Das et al., 1990).

HCH isomers exhibit distinct toxicological effects, with the **α** and γ isomers acting as stimulants for the central nervous system (CNS), inducing aggressive epileptiform convulsions, while the β and δ isomers primarily function as depressants (Gosselin et al., 1984; Lewis et al., 1999). HCH and its isomers alter nerve cell membrane properties, affecting ion flow kinetics (Na+ and K+) and potentially disturbing calcium transport or Ca2+-ATPase activity and phosphokinase activities (Hayes & Laws, 1991). The CNS is the primary target organ for HCH, with interactions at the synapse being a major site of action (Hayes & Laws, 1991). Histopathological examination revealed liver lesions, including fatty changes and necrosis, in rats exposed to high doses of beta-HCH (Van Velsen et al., 1986).

The exact mechanism of HCH's impact on the nervous system is complex and not entirely understood, but it is believed to involve interactions with the GABA receptor-linked chloride channel (Hayes & Laws, 1991). At the picrotoxin-binding site, the γ-HCH (Lindane) interacts with the GABA-A receptor-chloride channel complex and is supposed to interfere with the GABA system, leading to seizures that can be antagonized by GABA-A mimetic. Other evidence suggests magnified activity at synapses, changed GABA function, and hampering of activity of Na+-K+-ATPase contributing to neurological effects (Ratra & Casida, 2001; ATSDR, 1994). Jaeger et al. (1984) documented rapid onset of severe and multi-system toxicity, including seizures, respiratory depression, and pulmonary edema, following acute lindane-solvent poisoning, emphasizing the potent and immediate threat posed by this exposure.

Gamma-HCH also affects liver, as it interferes with hepatic oxidative capacity and glutathione metabolism, potentially leading to enhanced lipid metabolism and diminished activity of Mg2+ATPase in liver tissue of rat (ATSDR, 1994). Occupational exposure to HCH has been linked to neurological and immunological problems (NIOSH, 2000). Among the three BHC isomers (alpha, beta, gamma), alpha-BHC exhibited the highest accumulation in adipose tissues, while beta-Benzene hexachloride showed preferential distribution in the liver (Nakajima et al., 1970). Liver exhibited prominent histopathological changes like centrolobular necrosis and fatty infiltration upon BHC feeding, indicating significant hepatotoxic effects (Philip et al., 1989). Tryphonas & Iverson (1983) in their study demonstrated the time-dependent induction of hepatic megalocytosis and adenoma formation in mice exposed to α-HCH, supporting its potential carcinogenicity.

Acute poisoning cases involving HCH have been reported, with symptoms including hyperirritability, central nervous system excitation, vomiting, tremors, ataxia, muscle spasms, and convulsions (Gosselin et al., 1984). In severe cases, coma and respiratory failure may lead to fatalities, particularly in infants and children who may experience hyperpyrexia following convulsions (Gosselin et al., 1984). Gupta et al. (1982) demonstrated a rapid and significant increase in serum HCH residues following short-term occupational exposure in malaria spraymen, raising significant concerns about potential bioaccumulation, long-term health consequences.

Cardiovascular effects from technical grade HCH exposure can lead to abnormalities in the electrocardiogram (ECG), and chronic exposure may result in hepatic effects, leading to changes in enzyme levels (Ryan & Terry, 1997).

Skin exposure can lead to dermatitis and urticaria, whereas long-term inhalation or dermal exposure may result in systemic intoxication (Clayton & Clayton, 1994).

Occupationally exposed individuals, such as production workers in the manufacture of HCH, have reported symptoms such as headache, vomiting, paresthesia, apprehension, giddiness, and insomnia (Chattopadhyay et al., 1988).

Various case reports from different regions have documented HCH poisoning incidents resulting from accidental ingestion, contaminated food, and occupational exposure (Danopoulos et al., 1985; Heiberg & Wright, 1955; Macnamara, 1970; Nag et al., 1977; Clayton & Clayton, 1994).

5 Management

In case of ingestion, it is crucial not to induce vomiting and avoid giving milk, fat, or oils by mouth (ATSDR, 2020). Instead, maintain at least one intravenous route for the patient. Convulsions should be controlled using appropriate drug regimens, and activated charcoal should be administered (WHO, 2019). In addition, a thorough assessment of the patient's airway, breathing, circulation, and neurological status should be conducted to ensure that life support procedures are provided promptly.

Symptomatic and specific treatments include providing oxygen, intravenous dextrose, and thiamine to the patient. If myocardial dysrhythmias occur, Lidocaine

(1 mg/kg bw) can be administered as an intravenous infusion, followed by a continuous infusion of 2–4 mg/min. It is essential to continuously monitor the patient's vital signs, including ECG changes, to assess their condition accurately.

In cases of skin contact, contaminated clothing should be removed and discarded. The affected skin must be washed with copious amounts of water (and soap if available) for several minutes to ensure thorough decontamination (EHC 215 1997). If Lindane has been ingested, emesis should not be induced. Gastric lavage can be performed instead, and activated charcoal (50–100 g) should be administered along with a cathartic that has been shown to reduce the absorption of HCH (EHC 215 1997).

These general principles, life-supportive procedures, and decontamination measures are essential in managing cases of Lindane exposure to minimize its harmful effects on the individual's health. Proper medical attention and careful implementation of these protocols are crucial in ensuring the best possible outcome for the patient.

Long-term monitoring of the effects of prolonged HCH exposure on human health is still required as the chemical is being detected in human blood, milk and adipose tissues along with PCB and other organochlorine insecticides (EPA, 2016; CDC, 2023).

6 Conclusion

Hexachlorocyclohexane (HCH) is a group of chemical isomers that have played a significant role in pest control and agriculture due to their potent insecticidal properties. The three most common isomers of HCH, namely, alpha-HCH, beta-HCH, and gamma-HCH (Lindane), have distinct chemical structures and toxicological effects, which make them, vary in their impact on human health and the environment.

Despite its initial widespread use, the persistence, bioaccumulation, and long-range transport of HCH and its isomers have raised serious concerns about their potential adverse effects on living organisms and ecosystems. Scientific research and numerous case reports have indicated the potential health risks associated with HCH exposure. These risks encompass a range of issues, including neurological disorders, liver toxicity, carcinogenicity, and teratogenicity, among others.

Further research is required to perform additional studies on the efficacy of HCH (hexachlorocyclohexane) and to investigate the potential health risks associated with the consumption of contaminated food, particularly in relation to carcinogenic effects. The vegetables that are grown in areas with contamination may be contaminated themselves, and hence it is necessary to perform analysis on them to ensure the well-being of the population. Further research should be conducted to develop a comprehensive understanding of the effects of these activities on soil biota. It is imperative to investigate alternative options for use as pharmaceuticals in human health for management of head lice and scabies.

However, the management of HCH is not without challenges, particularly in regions where HCH has been widely used in the past, leading to soil and water contamination. Ensuring proper disposal of existing stocks of HCH and contaminated sites remains a priority, and international cooperation is crucial to tackling the global implications of HCH contamination.

References

AMAP. (2004). *AMAP Assessment 2002: Persistent Organic Pollutants in the Arctic. Arctic Monitoring and Assessment Programme (AMAP), Oslo, Norway. xvi +310 pp.*

ATSDR. (1994). *Profile for alpha-, beta-, gamma-, and delta-hexachlorocyclohexane (Update).* US Department of Health and Human Services, Agency for Toxic Substances and Disease Registry, Atlanta, GA.

ATSDR. (2020). *Agency for toxic substances and disease registry. Toxicological profile for lindane (Gamma-Hexachlorocyclohexane).* https://www.atsdr.cdc.gov/Toxprofiles/tp43.pdf

Brassow, H. L., Baumann, K., & Lehnert, G. (1981). Occupational exposure to hexachlorocyclohexane-II. Health conditions of chronically exposed workers. *The International Archives of Occupational and Environmental Health, 49*(1), 81–87.

Buser, H. R., & Mueller, M. D. (1995). Isomer and enantioselective degradation of hexachlorocyclohexane isomers in sewage sludge under anaerobic conditions. *Environmental Science & Technology, 29(3), 664–672.*

CDC. (2023). *Centers for Disease Control and Prevention. Second National Report on Human Exposure to Chemicals.* https://www.cdc.gov/exposurereport/index.html

Chandra, H., Pangtey, B. S., Modak, D. P., Singh, K. P., Gupta, B. N., Bharti, R. S., & Srivastava, S. P. (1992). Biological monitoring of chlorinated pesticides among exposed workers of mango orchards: A case study in tropical climate. *The Bulletin of Environmental Contamination and Toxicology, 48*(2), 295–301.

Chattopadhyay, P., Karnik, A. B., Thakore, K. N., Lakkad, B. C., Nigam, S. K., & Kashyap, S. K. (1988). Health effects among workers involved in the manufacture of hexachlorocyclohexane. *The Journal of the Society of Occupational Medicine, 38*(3), 77–81.

CHRIS. (2000). *CHRIS hazardous chemical data.* US Department of Transportation, US Coast Guard, Washington, DC. Micromedex, Inc, Englewood, CO (expires October 31, 2000).

Clayton, G. D., & Clayton, F. E. (Eds.). (1994). *Patty's industrial hygiene and toxicology* (Toxicology) (Vol. 2B, 4th ed.). Wiley.

Danopoulos, E., Mellisinos, K., & Katsas, G. (1985). Serious poisoning by hexachlorocyclohexane, clinical and laboratory observations on five cases. *Archives of Industrial Hygiene and Toxicology, 7*, 582–587.

Das, S. N., Paul, B. N., Saxena, A. K., & Ray, P. K. (1990). Effect of *in utero* exposure to hexachlorocyclohexane on the developing immune system of mice. *Immunopharmacology and Immunotoxicology, 12*(2), 293–310.

EHC 215. (1997). *World Health Organization. Lindane. Environmental Health Criteria 215.* https://www.epa.gov/sites/default/files/2016-09/documents/lindane.pdf

EPA. (2016). *Environmental Protection Agency. Lindane (Gamma-Hexachlorocyclohexane).* https://19january2021snapshot.epa.gov/pesticide-contacts/contacts-office-pesticide-programs-office-director

FCH. (1984). *Farm chemicals handbook.* Meister Publishing Co.

Gosselin, R. E., Smith, R. P., & Hodge, H. C. (1984). *Cinical toxicology of commercial products* (5th ed.). Williams and Wilkins.

Gupta, S. K., Parikh, J. R., Shah, M. P., Chatterjee, S. K., & Kashyap, S. K. (1982). Changes in serum hexachlorocyclohexane (HCH) residues in malaria spraymen after short-term occupational exposure. *Archives of Environmental Health, 37*(1), 41–44.

Hayes, W. J., Jr., & Laws, E. R., Jr. (1991). *Handbook of pesticide toxicology (3 Vol). Academic.*

Heiberg, O. M., & Wright, H. N. (1955). Benzene hexachloride poisoning. *A.M.A. Archives of Industrial Health, 11*, 457–458.

IPCS. (1989). *Environmental health criteria for beta-hexachlorocyclohexane*. WHO.

Jaeger, U., Podczeck, A., Haubenstock, A., et al. (1984). Acute oral poisoning with lindane-solvent mixtures. *Veterinary and Human Toxicology, 26*, 11–14.

Joseph, P., Shivanandappa, T., Narasinhamurthy, K., & Krishnakumari, M. K. (1992). Effect of Vitamin A on hexachlorocyclohane (HCH) toxicity in the rat. *General Pharmacology, 23*(6), 1159–1164.

Kumari, K. (Ed.). (2021). *Persistent organic pollutants: Gaps in management and associated challenges* (1st ed.). CRC Press. https://doi.org/10.1201/9781003046806

Khuman, S. N., Bharat, G., & Chakraborty, P. (2020). Spatial distribution and sources of pesticidal persistent organic pollutants in the Hooghly riverine sediment. *Environmental Science and Pollution Research, 27, 4137–4147.*

Lewis, R. J. (2000). *Sax's dangerous properties of industrial materials (10th ed.). Van Nostrand Reinhold Company.*

Lewis, D. L., Garrison, A. W., Wommack, K. E., Whittemore, A., Steudler, P., & Melillo, J. (1999). Influence of environmental changes on degradation of chiral pollutants in soils. *Nature, 401(6756), 898–901.*

LOLI(R). *The regulated chemicals list of lists*. Compiled by ChemADVISOR, Inc, Pittsburgh, PA (CD-ROM Version), Micromedex, Inc, Englewood, CO (edition expires 10/31/00).

Macnamara, B. G. P. (1970). benzene hexachloride poisoning. *British Medical Journal, 3*(5722), 585.

Nag, D., Singh, G. C., & Senon, S. (1977). Epilepsy epidemic due to benzahexachlorine. *Tropical and Geographical Medicine, 29*, 229–232.

Nakajima, E., Shindo, H., & Kurihara, N. (1970). Distribution of alpha-, beta-, and gamma BHC-^{14}C in whole body autoradiography in mice. *Radioisotopes, 19*(11), 532–538.

National Institute of Occupational Safety and Health (NIOSH). (2000). *Benzene hexachloride*. Registry of Toxic Effects of Chemical Substances (RTECS).

Nigam, S. K., Karnik, A. B., Majumder, S. K., Viswerwariah, K., Suryanarayana Raju, G., Muktha Bai, K., Lakkad, B. C., Thakore, K. N., & Chatterjee, B. B. (1986). Serum hexachlorocyclohexane residues in workers engaged at a HCH manufacturing plant. *International Archives of Occupational and Environmental Health, 57*(4), 315–320.

Pandit, G. G., Sahu, S. K., Sharma, S., & Puranik, V. D. (2006). Distribution and fate of persistent organochlorine pesticides in coastal marine environment of Mumbai. *Environment International, 32(2), 240–243. ISSN 0160–4120.* https://doi.org/10.1016/j.envint.2005.08.018

Philip, G. H., Sriraman, P. K., & Ramamurthi, R. (1989). Histopathological changes in liver and kidney of Mus Booduga following oral benzene hexachloride (BHC) feeding. *The Bulletin of Environmental Contamination and Toxicology, 42*(4), 499–502.

Pozo, K., Palmeri, M., Palmeri, V., Estellano, V. H., Mulder, M. D., Efstathiou, C. I., ... & Focardi, S. (2016). Assessing persistent organic pollutants (POPs) in the Sicily Island atmosphere, Mediterranean, using PUF disk passive air samplers. *Environmental Science and Pollution Research, 23, 20796–20804.*

Ratra, G. S., & Casida, J. E. (2001). GABA receptor subunit composition relative to insecticide potency and selectivity. Toxicology letters, 122(3), 215–222.

Richter, E., Luger, W., Klein, W., Korte, F., & Weger, N. (1981). Excretion of beta hexachlorocyclohexane by the rat as influenced by oral paraffin, squalene and Lutrol E 400. *Ecotoxicology and Environmental Safety, 38*(1), 117–124.

Ryan, R. P., & Terry, C. E. (Eds.) (1997). *Toxicology desk reference* (Vols. 1–3, 4th ed., 1455p.). Taylor & Francis.

Sharma, B. M., Bharat, G., Tayal, S., Nizzetto, L., & Larssen, T. A. (2014). The legal framework to manage chemical pollution in India and the lesson from Persistent Organic Pollutants (Pops). *The Science of the Total Environment, 490C*, 733–747. https://doi.org/10.1016/j.scitotenv,2014.05.043

Spencer, E. Y. (1982). *Guide to the chemicals used in crop protection* (Publication 1093) (7th ed.). Research Institute, Agriculture Canada.

Sruthi, S. N., Shyleshchandran, M. S., Mathew, S. P., & Ramasamy, E. V. (2016). Contamination from organochlorine pesticides (OCPs) in agricultural soils of Kuttanad agro-ecosystem in India and related potential health risk. *Environmental Science and Pollution Research, 24, 969–978.*

Tomczak, S., Baumann, K., & Lehnert, G. (1981). Occupational exposure to hexachlorocyclohexane. *The International Archives of Occupational and Environmental Health, 48*(3), 283–287.

Tryphonas, L., & Iverson, F. (1983). Sequential histopathologic analysis of alpha-hexachlorocyclohexane-induced hepatic megalocytosis and adenoma formation in the HPB mouse. *Journal of the National Cancer Institute, 71*(6), 1307–1318.

Van Velsen, F. L., Danse, L. H., Van Leeuwen, F. X., Dormans, J. A., & Van Logten, M. J. (1986). The subchronic oral toxicity of the beta-isomer of hexachlorocyclohexane in rats. *Fundamental and Applied Toxicology, 6*, 697–712.

WHO. (1984). *Guidelines for drinking water quality* (Vol. 2). Health Criteria and other supporting Information.

WHO. (2019). *World Health Organization. Lindane in drinking-water.* https://www.who.int/teams/environment-climate-change-and-health/water-sanitation-and-health/chemical-hazards-in-drinking-water/lindane

Chlordecone

Kanchan Kumari, Deepak Marathe, and Karthik Raghunathan

1 Introduction

Chlordecone, sometimes known as Kepone in commercial settings, is an organic pollutant with notable persistence that has attracted substantial attention due to its detrimental effects on human health and the environment. Chlordecone is synthesized through a process involving the condensation of two moles of hexachlorocyclopentadiene in the presence of sulfur trioxide, followed by subsequent hydrolysis (Sittig, 1980). The aforementioned molecule possesses characteristics that warrant its classification as an organochlorine pesticide, denoting its synthetic nature. Chlordecone was widely utilized as an insecticide, primarily for controlling banana weevils and other pests in banana plantations, in addition to its use in several other agricultural settings (Létondor et al., 2015). It has been shown that chlordecone has a high level of environmental persistence, which indicates that it is resistant to decomposition over the course of time. Persistent organic compounds like chlordecone are primarily defined by their enduring physical and chemical characteristics, which enable them to remain in the environment for extended durations, leading to harmful effects (Kumari et al., 2014). In addition, it has been demonstrated that chlordecone is capable of being transported across extended distances through the examination of its physicochemical features as well as the data obtained from its modeling. This material has been identified as a possible carcinogen for humans and possesses a high level of toxicity when it comes to aquatic organisms. Chlordecone

K. Kumari (✉)
CSIR-National Environmental Engineering Research Institute (NEERI),
Kolkata Zonal Center, Kolkata, West Bengal, India
e-mail: k_kumari@neeri.res.in

D. Marathe · K. Raghunathan
CSIR-National Environmental Engineering Research Institute (NEERI), Nagpur, India

K. Kumari, *Pollutants of Global Concern*, Emerging Contaminants and Associated Treatment Technologies, https://doi.org/10.1007/978-3-031-50996-4_13

is a synthetic chlorinated organic compound that was predominantly utilized as a pesticide in agricultural practices. The product had its initial manufacturing process in 1951 and afterwards made its debut in the market in 1958. Currently, there is a lack of recorded evidence on the utilization or production of the aforementioned chemical, since numerous governments have already enacted restrictions on its dissemination and utilization (UNEP-POPs). The chemical has been listed as persistent organic pollutants in Annex A in 2009 (Kumari, 2021) in Stockholm Convention, and India also banned its manufacture, uses, export, and import in 2018.

2 Technical Details of Chlordecone

Chlordecone, also known as Kepone, is denoted by its chemical name as 1,1a,3,3a,4,5,5,5a,5b,6-decachlorooctahydro-1,3,4-metheno-2H-cyclobuta[cd]pentalen-2-one. The compound's chemical makeup is distinguished by the existence of many chlorine atoms that are bound to a bicyclic carbon framework. Chlordecone is a crystalline compound that exhibits a colorless to white appearance and boasts a delicate and agreeable scent. The compound has a melting temperature of around 350 °C (662 °F) and exhibits restricted solubility in aqueous solutions but indicates solubility in organic solvents like hexane and chloroform (Table 1). Chlordecone demonstrates a notable level of environmental persistence due to its intrinsic resistance to degradation processes, such as photolysis and biodegradation. The persistent occurrence of this material has led to its accumulation in soil, aquatic ecosystems, and the ecological food chain. Chlordecone was utilized as an insecticide primarily for the aim of controlling banana weevils in banana plantations. Moreover, this material was utilized for many agricultural purposes, including the control of pests in crops such as coffee, cocoa, and sugarcane. There has been significant concern over the environmental persistence of chlordecone and its tendency to accumulate in the food chain. Human exposure to chlordecone can occur when individuals consume food and water that have been contaminated with this particular chemical component. Numerous health issues have been associated with it, including various types of cancer such as prostate and breast cancer, developmental problems in children, and disruptions to the endocrine system. Chlordecone has been subjected to regulatory measures, such as prohibition or limits, in several countries due to its adverse effects on both human health and the environment. The principal aim of the Stockholm Convention on Persistent Organic Pollutants is to establish regulations and ultimately eliminate the production and use of chemicals such as chlordecone. The cleanup of regions contaminated by chlordecone poses notable difficulties due to its enduring characteristics. The cleanup process involves several approaches, such as soil extraction, bioremediation (the use of microorganisms to dissolve pollutants), and phytoremediation (the use of plants to extract or break down contaminants). In summary, chlordecone is a persistent organic pollutant that was utilized as an insecticide, but its application has been restricted or banned due to its detrimental impacts on human health and the environment. The

Table 1 Technical details and physicochemical properties of chlordecone (NCBI, 2023)

Details	Description (chlordecone)
Structure	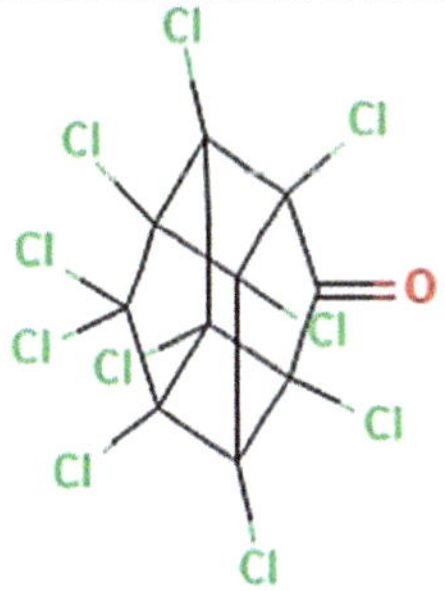
IUPAC name	1,2,3,4,6,7,8,9,10,10-Decachloropentacyclo [5.3.0.02,6.03,9.04,8] decan-5-one
Molecular formula	$C_{10}Cl_{10}O$
CAS No.	143-50-0
Molecular weight	490.6 g/mol
Color/form	Tan to white, crystalline solid
Odor	Odorless
Boiling point	662 °F
Melting point	662 °F
Flash point	>100 °C
Solubility	<1 mg/mL at 73 °F
Density	1.64 g/cu cm at 25 °C
Vapor pressure	< 0.0000003 mm hg at 77 °F
Octanol/water partition coefficient	5.41 (LogP)

importance of addressing contaminants of this kind resides in their capacity to persist in the environment and possibly bioaccumulate in organisms, underscoring the imperative to safeguard both ecosystems and human well-being.

3 Status of Usage Within the Country

Production and stockpiles: At present, there is a dearth of relevant data pertaining to the manufacture or stockpiling of the aforementioned chemical (IPEN, POPs, 2018). The production, selling, use, import, and export of chlordecone were restricted as a result of the notification issued by the Ministry of Environment, Forest and Climate Change on March 5, 2018, under G.S.R. 207 (E). The aforementioned action was executed in compliance with sections 3 and 6 of the Environment (Protection) Act, 1986 (29 of 1986), in combination with rule 13 of the Environment (Protection) Rules, 1986. The order indicated above was enacted on March 5, 2018.

3.1 Export and Import of Chlordecone

Two nations namely Bangladesh and Switzerland received a combined amount of 1.56 thousand kilograms of chlordecone from India during the financial year 2017–2018. During the time frame that is being considered, the United States shipped a shipment of chlordecone to India that had a combined weight of 0.09 thousand kilograms (IPEN, POPs, 2018).

3.2 Exemption (Use of Chemicals for Scientific Purpose)

In spite of the provisions that are outlined in these regulations, the chemicals that are referred to in regulation 2 may be used, traded, or imported in the necessary quantities for the purpose of conducting research and development endeavors within Central Universities, Council of Scientific and Industrial Research Laboratories, government institutions, or other research institutions accredited laboratories in the government or private sector. This provision applies regardless of the other provisions that are outlined in these regulations. However, the Government of India's Ministry of Environment, Forests, and Climate Change must provide its express permission before such endeavors may be undertaken. Table 2 outlines the regulations, and Table 3 summarizes the producers and trade names along with its uses and adverse effects.

3.3 Exposure Control and Personal Protection

Recommended Exposure Limit (REL): Ca TWA 0.001 mg/m^3 (NIOSH).
Permissible Exposure Limit (PEL): none (NIOSH).
Immediately Dangerous to Life or Health (IDLH): a potential occupational carcinogen (NIOSH, 2016).
Threshold Limit Values (TLV): skin absorption (H); carcinogen category: 2 (ILO).
Inhalation Risk: Evaporation at 20 °C is negligible; a harmful concentration of airborne particles can, however, be reached quickly when dispersed (ILO).

Table 2 Uses and sources of POPs and regulation in India (Kumari, 2021)

Identified POP	Uses and sources	Annex	Year of listing by SC	Regulation in India
Chlordecone	Pesticide	A	2009	Banned for manufacture, use, import, export, etc., since 5th March, 2018

Table 3 Producers, trade names, uses, and effects of chlordecone (candidate POPs criteria, processes, and existing status, factsheet number 34/October 2008)

Substance	Producers	Trade names	Uses	Effects
Chlordecone	Allied Chemical (USA), Life Sciences Products (USA), Hooker Chemical (USA), Nease Chemical (USA), De Laguarique (France), unnamed French and Brazilian companies	Kepone, GC-1189, Merex, ENT 16391, Curlone	Pesticide formerly used on banana root borer, fly larvicide, apple scab, powdery mildew, Colorado potato beetle, rust mite, wireworm, and household ant and roach traps	The pesticide is both acutely and chronically toxic, producing neurotoxicity, immunotoxicity, reproductive, musculoskeletal, and liver toxicity. Chlordecone is very toxic to aquatic organisms, with the most sensitive group being the invertebrates

4 Global Status of Use of Chlordecone

Clean Water Act Requirements
According to subsection 311(b)(2)(A) of the Federal Water Pollution Control Act, chlordecone is considered to be a hazardous substance. Because of this classification, it is subject to the extra requirements that are detailed in the Clean Water Act Amendments of 1977 and 1978. These constraints apply to the discharge of the substance that was described earlier. This classification's ambit extends to include not just isomers and hydrates, but also any and all combinations and solutions that comprise the substance in question (40 CFR 302.4 (USEPA), 2017).

CERCLA Reportable Quantities
Individuals responsible for the operation and management of boats or facilities are obligated to promptly inform the National Response Centre (NRC) in the event of a discharge of the specified hazardous chemical, reaching or exceeding the designated reportable quantity of 1 lb or 0.454 kg (40 CFR 302.4 (USEPA), 2017).

5 Adverse Effects

5.1 Biological Half-Life

The amount of time that passes before the concentration of a chemical in a biological system drops by one-half as a result of natural processes is referred to as the biological half-life. The retention time of chlordecone in the human body is significantly extended due to its high rate of absorption and distribution across the body's various tissues. The half-life of chlordecone in the blood is 165 days, while its half-life in adipose tissue is 125 days. This indicates that the clearance of chlordecone from the blood and fat of industrial workers occurs at a rather sluggish rate. It has been shown that the injection of cholestyramine results in a blood half-life of approximately 80 +/− 4 days, whereas the half-life in adipose tissue is around 64 days (DHsS/ATSDR, 1995).

5.2 Environmental Fate

Terrestrial Fate

The experimentally derived Koc values of 3330, 7440, and 15,850 suggest that chlordecone is likely to have restricted mobility within soil (Fernandez-Bayo et al., 2013; Schuurmann et al., 2006). This conclusion may be reached by applying a categorization approach to the data, as stated in the previous sentence. It is not anticipated that the ability of chlordecone to volatilize from the wet surfaces of the soil would play a significant role in determining its final destination. The basis for this conclusion is the computed value of the Henry's Law constant, which is 5.4 × 10–8 atm-cu m/mole (SRC). This value is obtained from the substance's vapor pressure, which is 2.25 × 10–7 mm Hg, and its water solubility, which is 2.7 mg/L. According to the findings of the investigation into its vapor pressure, it is not anticipated that chlordecone would undergo volatilization from dry soil surfaces. However, it is important to highlight that research conducted in laboratories has successfully quantified the emission of chlordecone from the ground because of the process of volatilization, albeit in very low concentrations. The process of chlordecone biodegradation has been seen and documented, despite the fact that it moves at a glacial rate in natural settings. A sediment inoculum taken from an estuary was used in the investigation, and the results showed that the half-life was greater than 312 days. Under aerobic conditions, the process of microbial mineralization of 14C-chlordecone was seen taking place in two tropical soils. Despite this, it was found that the rate of 14-CO_2 emission was low, with values ranging from 3.2% to 4.9% over the course of the 215-day incubation period. When soil surfaces are exposed to sun radiation, there is a possibility that chlordecone will show signs of being susceptible to direct photolysis. This might happen if chlordecone is present. It has been established, based on the monitoring data that is currently available, that the presence of chlordecone residues in soil can persist for an extended period following the application of pesticides. Even after it was taken off the market, the organochlorine pesticide known as chlordecone continued to have a significant negative influence on the ecosystem (ATSDR, 2016; Sanborn et al., 1977; Fernandez-Bayo et al., 2013).

Aquatic Fate

Both chlordecone and mirex are extremely resistant to degradation in aquatic environments, making them two of the most durable pesticides. There was no indication that these two compounds had degraded after being exposed to aerobic and anaerobic environments for a total of 56 days throughout the incubation process (Murty, 1986).

5.3 *Persistence/Toxicity*

Chlordecone exhibits a notably protracted degradation rate within the environment, leading to its potential persistence in both soil and water over extended periods of time. Furthermore, both the surface water and the surface soil lack the capability to make a substantial contribution to the process of evaporation. Chlordecone, a chemical compound, exhibits limited solubility in water, instead exhibiting a preference for selective binding to soil and silt particles. Chlordecone exhibits limited mobility within soil matrices, rendering it exceedingly unlikely to undergo significant dispersion through soil or infiltration into groundwater. Chlordecone, similar to Mirex and other persistent organic pollutants (POPs), has the capacity to undergo bioaccumulation in aquatic organisms and other fauna when exposed to contaminated food or water sources, or when inhabiting polluted settings.

The mechanisms by which toxic chemicals adversely affect human health remain incompletely understood. Conversely, those who were subjected to prolonged and significant exposure to chlordecone, beyond a duration of 1 year, manifested indications of detrimental impacts on many bodily systems, including the neurological system, skin, liver, and male reproductive system. The individuals in question were presumably exposed to chlordecone mostly through direct dermal contact. However, it is plausible that they may have also incurred exposure by inhalation or ingestion (NCBI, 2023).

5.4 *Exposure Routes*

The chemical has the potential to permeate the human body via dermal absorption, oral intake, and inhalation (Guzelian, 1982). The potential routes of exposure include inhalation, skin absorption, ingestion, as well as skin and/or ocular contact (NIOSH) (Fig. 1). The organs that are primarily affected by a certain stimulus or substance are referred to as target organs. In the case of this particular stimulus or substance, the target organs are the hepatic organ, which pertains to the liver and the renal organ, which pertains to the urinary system or kidneys (ATSDR).

The human body encompasses several vital organs and systems, including the eyes, skin, respiratory system, central nervous system, liver, kidneys, and reproductive system (NIOSH).

Effects of Short-Term Exposure

There exists a potential for the chemical to exert an impact on both the liver and the central nervous system. This phenomenon has the potential to result in impaired functioning and hepatic injury. There is a potential for a delay in the manifestation of the effects (ILO).

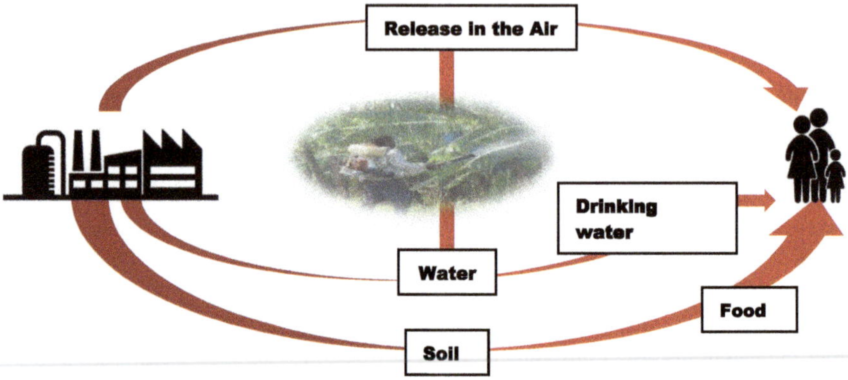

Fig. 1 Potential routes of exposure of chlordecone in humans

Effects of Long-Term Exposure

Dermatitis can manifest because of recurrent or extended exposure to the skin. The chemical has the ability to elicit effects on the endocrine system. There exists a potential for this substance to induce carcinogenic consequences in the human population. According to the referenced source, it has been observed that the drug under consideration can have adverse impacts on human reproduction or development, resulting in toxicity (ILO).

The possible carcinogenic properties of chlordecone have been detected in a study conducted by the US Department of Health and Human Services (DHHS). As of now, there exists a dearth of readily available scientific inquiry pertaining to the possible carcinogenic properties of chlordecone in the human population. However, prior studies done on rodents have shown that the ingestion of chlordecone may lead to the formation of tumors in the liver, adrenal gland, and kidneys (NCBI, 2023).

6 In-House (Countrywide) Scientific Work Done on the Chemical

Chlordecone is an organochlorine pesticide that exhibits estrogenic capabilities (Hodges et al., 2000). These features have been observed to impair the reproductive ability of the *Pseudetroplus maculatus* species of cichlid fish. This work was carried out by Asifa and colleagues and entailed the evaluation of chlordecone. During the course of this research project, fish were subjected to two different sub lethal dosages of chlordecone, precisely 3.5 and 7 micrograms per liter, for a period of 4, 7, 15, and 30 days, respectively. The vitellogenin levels in the fish that were exposed to the test substance were measured throughout the experiment, and control groups were maintained so that results could be compared. In order to determine the levels of vitellogenin in both male and female fishes, indirect markers such as alkali-labile phosphoprotein (ALP), total protein, and calcium concentrations in the blood plasma were measured. These quantities were taken from the fishes. Evidence that chlordecone has estrogenic characteristics may be seen in the considerable rise in vitellogenin levels that was detected in male fish. After being exposed to pesticides, the gonadosomatic index showed a significant drop, which may be related to the unique toxicity of chlordecone on the gonads. The fact that there was a significant decrease in the parameters of sperm, which was detected, implies that chlordecone has a negative influence on the healthy functioning of male gametes. According to the findings, chlordecone had a negative effect on fish reproduction. This was demonstrated by the elevated levels of vitellogenin in males and the alterations in sperm functional properties. These modifications have the potential to serve as a biomarker for determining whether or not chlordecone has an estrogenic impact in cichlid fish (Asifa & Chitra, 2019).

7 Relevance of the International Scientific Data as per Indian Conditions

An organochlorine pesticide known as chlordecone has been used extensively in the French West Indies for the goal of controlling the population of the banana root borer. This was done with the intention of regulating the pest's population. The current study made use of information gleaned from the Timoun Mother-Child Cohort Study, which was carried out in Guadeloupe between the years 2004 and 2007 (Rouget et al., 2020). The major purpose of this work was to examine the possible associations between chlordecone concentrations in maternal plasma and two critical pregnancy outcomes, namely, the duration of gestation and the occurrence of preterm delivery. In addition, secondary objectives included determining whether chlordecone concentrations in maternal plasma were associated with any adverse birth outcomes. The sample for the research included 818 different pregnant women. In a nutshell, the findings that were presented by Philippe Kadhel et al. (2014) provide certain indications that chlordecone, a chemical known for its estrogenic and progestogenic characteristics, could potentially influence the length of gestation (resulting in a minimum of 3 days shorter gestation period for 40% of women with blood chlordecone levels that exceeded 0.52 g/L), consequently elevating the likelihood of preterm birth. The aforementioned observations were made within a group that was distinguished by a noticeably increased risk of premature delivery at the outset. In spite of the fact that the levels of chlordecone in the blood were on the order of magnitude of 1 part per billion, it was still possible to make out the relationship between the two. Additional precautions are required to protect pregnant women from the possibility of being exposed to chlordecone, with the major emphasis being placed on reducing the risk posed by the chemical's principal sources in the diet (Kadhel et al., 2014). Cardiovascular, endocrine, and developmental effects of chlordane have been outlined in Table 4.

Table 4 Studies conducted with the aim of determining the associations existing between chlordecone in blood or breast milk and health outcomes in humans (Toxicological Profile for Mirex and Chlordecone Draft for Public Comment May 2019)

Reference	Exposure	Outcomes
Cardiovascular effects		
Saunders et al. (2014)	Serum chlordecone level (LOD 0.06 μg/L)	Negatively associated with hypertensive disorders during pregnancy
Saunders et al. (2014)	Serum chlordecone level (LOD 0.06 μg/L)	Not associated with diabetes mellitus during pregnancy
Endocrine effects		
Emeville et al. (2013)	Serum chlordecone level (LOD 0.06 μg/L)	No correlation found between the levels of steroid hormones in the blood and serum chlordecone
Developmental effects		
Boucher et al. (2013)	Chlordecone cord blood level (LOD 0.06 μg/L)	Negative correlation between the concentration of chlordecone in cord blood, findings suggest that prenatal exposure to chlordecone may have a detrimental impact on the fine motor development of male individuals
Cordier et al. (2015)	Chlordecone cord blood level (LOD 0.06 μg/L)	Positively associated with increased TSH level in male infants at 3 months: There is relationship between prenatal exposure to chlordecone and the development of fine motor abilities.
Costet et al. (2015)	Chlordecone cord blood level (LOD 0.06 μg/L)	Positive correlation between a higher body mass index at 3 months in male infants and an elevated concentration of chlordecone in their cord blood
Hervé et al. (2016)	Chlordecone cord blood level (LOD 0.02 μg/L):	No association between cord blood chlordecone and gestational weight
Kadhel et al. (2014)	Chlordecone maternal blood level (LOD 0.06 μg/L)	Increased concentration was associated with a decreased length of gestation
Prostate cancer		
Multigner et al. (2010)	Plasma chlordecone level (LOD 0.25 μg/L)	Positively associated with risk of prostate cancer, risk of prostate cancer among subjects with family history of prostate cancer, and risk of prostate cancer among subjects with past residence in western countries

8 Remedial Action Taken for Management of POPs

Chlordecone has greater resistance to decomposition compared to lindane. According to prevailing knowledge, under natural circumstances, soil degradation is mostly attributed to the process of leaching. However, the process of sequestration, wherein Kepone molecules can become trapped within mesopores found in allophane clays, particularly when compost is introduced, may reduce the bioavailability of Kepone to flora. Regrettably, owing to its inherent structure, Kepone exhibits a high degree of resistance to degradation, and the available research about its potential for bio decomposition is limited. Nevertheless, new research suggests that chlordecone undergoes limited degradation by bacteria present in tropical soil, but the specific microbes responsible for this process have not yet been identified. However, those who may come into contact with this particular persistent organic pollutant (POP) should priorities limited eating of possibly contaminated foods and the use of water filtration systems employing active carbon filters. Rana et al. (2017) developed a very effective technique for the removal of Kepone, utilizing an adsorbent composed of iron oxyhydroxide/iron oxide nanoparticles and activated carbon. This adsorbent demonstrated a remarkable adsorption capacity of up to 79.8 μg/mg of Kepone (Madaj, 2018).

9 Recommendations

As of now, there exists a dearth of available data pertaining to the manufacture and stockpiling of the aforementioned chemical. Furthermore, there is a conspicuous absence of evidence indicating any sort of contamination, whether it is environmental or human, inside the geographical boundaries of India. Therefore, a long term environmental monitoring is required to understand the levels of chlordecone prevailing in the various environmental compartments along with a health study of workers affected by chlordecone. Careful surveillance would improve future hazard identifications of chlordecone.

References

40 CFR 302.4 (USEPA); U.S. National Archives and Records Administration's Electronic Code of Federal Regulations. February 13, 2017. http://www.ecfr.gov

Asifa, K. P., & Chitra, K. C. (2019). Effect of Chlordecone on the reproductive potential of the cichlid fish, Pseudetroplus maculatus (Bloch, 1795). *Journal of Aquatic Research and Marine Sciences, 2019*, 163–177.

ATSDR. (1995). *Toxicological profile for mirex and chlordecone*. Agency for Toxic Substances and Disease Registry, US Public Health Service. Available from, as of December 30, 2016 http://www.atsdr.cdc.gov/toxprofiles/index.asp

ATSDR. (2016). Toxicological Profile for Mirex and Chlordecone. Atlanta, GA: Agency for Toxic Substances and Disease Registry, US Public Health Service (1995). Available from, as of Dec 30, 2016: http://www.atsdr.cdc.gov/toxprofiles/index.asp

ATSDR. https://www.atsdr.cdc.gov/substances/toxsubstance.asp?toxid=118

Boucher, O., Simard, M. N., Muckle, G., Rouget, F., Kadhel, P., Bataille, H., et al. (2013). Exposure to an organochlorine pesticide (chlordecone) and development of 18-month-old infants. *Neurotoxicology, 35*, 162–168.

Cordier, S., Bouquet, E., Warembourg, C., Massart, C., Rouget, F., Kadhel, P., et al. (2015). Perinatal exposure to chlordecone, thyroid hormone status and neurodevelopment in infants: The Timoun cohort study in Guadeloupe (French West Indies). *Environmental Research, 138*, 271–278.

Costet, N., Pelé, F., Comets, E., Rouget, F., Monfort, C., Bodeau-Livinec, F., et al. (2015). Perinatal exposure to chlordecone and infant growth. *Environmental Research, 142*, 123–134.

Emeville, E., Giton, F., Giusti, A., Oliva, A., Fiet, J., Thomé, J. P., et al. (2013). Persistent organochlorine pollutants with endocrine activity and blood steroid hormone levels in middle-aged men. *PLoS One, 8*(6), e66460.

Fernández-Bayo, J. D., Saison, C., Voltz, M., Disko, U., Hofmann, D., & Berns, A. E. (2013). Chlordecone fate and mineralisation in a tropical soil (andosol) microcosm under aerobic conditions. *Science of the Total Environment, 463*, 395–403.

Guzelian, P. S. (1982). Comparative toxicology of chlordecone (Kepone) in humans and experimental animals. *Annual Review of Pharmacology and Toxicology, 22*(1), 89–113.

Hervé, D., Costet, N., Kadhel, P., Rouget, F., Monfort, C., Thomé, J. P., et al. (2016). Prenatal exposure to chlordecone, gestational weight gain, and birth weight in a Guadeloupean birth cohort. *Environmental Research, 151*, 436–444.

Hodges, L. C., Bergerson, J. S., Hunter, D. S., & Walker, C. L. (2000). Estrogenic effects of organochlorine pesticides on uterine leiomyoma cells in vitro. *Toxicological Sciences, 54*(2), 355–364.

ILO International Labour Organization. http://www.ilo.org/dyn/icsc/showcard.display?p_version=2&p_card_id=1432

IPEN. (2018, June). *Persistent Organic Pollutants (POPs) in India: Country situation report by toxics link.*

Kadhel, P., Monfort, C., Costet, N., Rouget, F., Thomé, J. P., Multigner, L., & Cordier, S. (2014). Chlordecone exposure, length of gestation, and risk of preterm birth. *American Journal of Epidemiology, 179*(5), 536–544.

Kumari, K. (Ed.). (2021). *Persistent organic pollutants: Gaps in management and associated challenges* (1st ed.). CRC Press. https://doi.org/10.1201/9781003046806

Kumari, K., Sharma, J. K., Kanade, G. S., et al. (2014). Investigation of polybrominated diphenyl ethers in old consumer products in India. *Environmental Monitoring and Assessment, 186*, 3001–3009. https://doi.org/10.1007/s10661-013-3596-2

Létondor, C., Pascal-Lorber, S., & Laurent, F. (2015). Uptake and distribution of chlordecone in radish: Different contamination routes in edible roots. *Chemosphere, 118*, 20–28.

Madaj, R. (2018). Lindane, kepone and pentachlorobenzene: Chloropesticides banned by Stockholm convention. *International Journal of Environmental Science and Technology, 15*, 471–480.

Multigner, L., Ndong, J. R., Giusti, A., Romana, M., Delacroix-Maillard, H., Cordier, S., et al. (2010). Chlordecone exposure and risk of prostate cancer. *Journal of Clinical Oncology, 28*(21), 3457–3462.

Murty, A. S. (1986). *Toxicity of pesticides to fish* (Vols. I, II, p. V1 25). CRC Press.

National Center for Biotechnology Information. (2023). *PubChem compound summary for CID 299, Chlordecone*. Retrieved October 17, 2023, from https://pubchem.ncbi.nlm.nih.gov/compound/Chlordecone

NIOSH. (2016). National Institute of Occupational Safety and Health. NIOSH Pocket Guide to Chemical Hazards (full website version). https://www.cdc.gov/niosh/npg (accessed October, 2023).

Rana, V. K., Kissner, R., Jauregui-Haza, U., Gaspard, S., & Levalois-Grützmacher, J. (2017). Enhanced chlordecone (Kepone) removal by FeO-nanoparticles loaded on activated carbon. *Journal of Environmental Chemical Engineering, 5*(2), 1608–1617.

Rouget, F., Kadhel, P., Monfort, C., Viel, J. F., Thome, J. P., Cordier, S., & Multigner, L. (2020). Chlordecone exposure and risk of congenital anomalies: The Timoun mother-child cohort study in Guadeloupe (French West Indies). *Environmental Science and Pollution Research, 27*, 40992–40998.

Sanborn, J. R., et al. (1977). *The degradation of selected pesticides in soil: A review of the published literature*. NTIS PB-272 353.

Saunders, L., Kadhel, P., Costet, N., Rouget, F., Monfort, C., Thomé, J. P., et al. (2014). Hypertensive disorders of pregnancy and gestational diabetes mellitus among French Caribbean women chronically exposed to chlordecone. *Environment International, 68*, 171–176.

Schüürmann, G., Ebert, R. U., & Kühne, R. (2006). Prediction of the sorption of organic compounds into soil organic matter from molecular structure. *Environmental Science & Technology, 40*(22), 7005–7011.

Sittig, M. (1980). *Pesticide manufacturing and toxic materials control encyclopedia* (p. 171). Noyes Data Corporation.

The National Institute for Occupational Safety and Health (NIOSH). https://www.cdc.gov/niosh/npg/npgd0365.html

Toxicological Profile for Mirex and Chlordecone Draft for Public Comment May 2019, Agency for toxic substances and disease registry. https://www.atsdr.cdc.gov/toxprofiles/tp66.pdf

UNEP-POPS-PUB-Brochure-16NewPOPs 201706.English.pdf. http://www.pops.int/TheConvention/ThePOPs/TheNewPOPs/tabid/2511/Default.aspx

Hexabromobiphenyl (HBB)

Kanchan Kumari and T. Arfin

1 Introduction

Belonging to the group of polybrominated biphenyls, hexabromobiphenyl (HBB) is highly bioaccumulative, chronic, and highly durable in environmental transport. It is a carcinogen and has some persistent toxic effects. Based on their characteristics, they have been listed in the category of persistent organic pollutants (POPs). Persistent organic compounds are mainly characterized by their physical-chemical properties, which allow them to persist for long periods in the environment, causing toxicity (Kumari et al., 2014).

Hexabromobiphenyl has been used as a flameproof in the industrial field, mainly in the 1970s. The manufacture and application of HBB ended in most countries under the restrictions of national and international regulations (UNEP, 2017).

2 Technical Details of Hexabromobiphenyl

Hexabromobiphenyl is characterized as a solid substance with a white, chalky appearance. The molecular formula for the compound is $C_{12}H_4Br_6$ with a molecular weight of 627.6 g/mol. The alternative nomenclature for the compound is 2,2′,4,4′,6,6′-Hexabromo-1,1′-biphenyl which can be identified by its IUPAC

K. Kumari (✉)
CSIR-National Environmental Engineering Research Institute (NEERI), Kolkata Zonal Center, Kolkata, West Bengal, India
e-mail: k_kumari@neeri.res.in

T. Arfin
CSIR-National Environmental Engineering Research Institute (NEERI), Nagpur, India

K. Kumari, *Pollutants of Global Concern*, Emerging Contaminants and Associated Treatment Technologies, https://doi.org/10.1007/978-3-031-50996-4_14

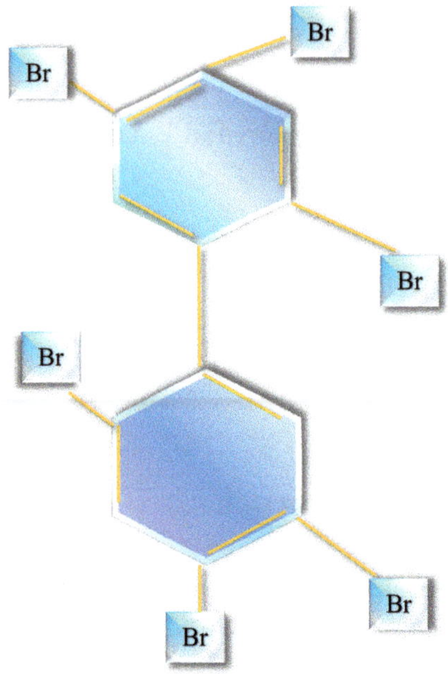

Fig. 1 Structure of hexabromobiphenyl

designation, 1,2,4-tribromo-5-(2,4,5-tribromophenyl) benzene (Fig. 1). The substance's boiling point is measured at 72 °C, while its melting point ranges between 118.889 and 121.111 °C. The solubility of hexabromobiphenyl in water at a temperature of 64 °F is less than 1 mg/ml (Fig. 2). At a temperature of 194 °F, the hexabromobiphenyl compound exhibits a vapor pressure of 7.6e-05 mm Hg. Additionally, it exhibits solubility in acetone and benzene.

3 Status of Usage Within the Country

It is conceivable that certain nations are still producing HBB. In the most common commercial items, including motor housing, lacquers, and vehicle upholstery, an HBB substance was utilized as a fire retardant. The uses and its source are listed in Table 1. The trade names, producers, and its effect are also mentioned in Table 2.

Hexabromobiphenyl is an industrial compound mainly employed as a flame retardant in plastics and wires throughout the 1970s. Currently, the majority of nations need to engage in limiting the production or utilization of hexabromobiphenyl. Hexabromobiphenyl has been included in the year 2009 by Stockholm Convention in Annex A, which means that the listed substance is prohibited from being manufactured, placed on the market, and used.

Between 1970 and 1974, 11 million pounds of hexabromobiphenyl were manufactured and distributed in the market, primarily under the commercial designations of Fire Master BP-6 and Fire Master FF-1. Between the years 1970 and 1979, a

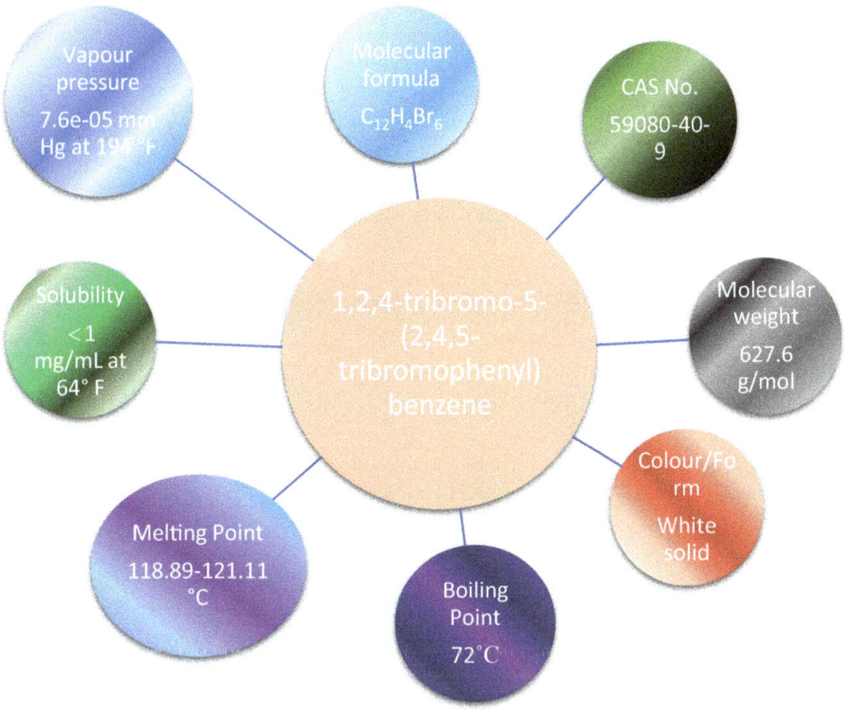

Fig. 2 Technical details and physicochemical properties of hexabromobiphenyl. (NCBI, 2023)

Table 1 Uses and sources of POPs and regulation in India (Singh & Kumari, 2022)

Identified POP	Uses and sources	Annex	Year of listing by SC	Regulation in India
Hexabromobiphenyl	Industrial chemical	A	2009	Banned for manufacture, use, import, export etc (notified in 2018).

particular American company was engaged in the production of octabromobiphenyl and decabromobiphenyl. Additionally, another American company was involved in the production of decabromobiphenyl from 1973 until the date preceding 1977. The estimated output of polybrominated biphenyls (PBBs) in the United States was 170,000 pounds in 1975. This production increased to 807,000 pounds in 1976, which was subsequently exported. In 2009, the production of decabromobiphenyl was undertaken by a single business in China and Europe. This chemical compound was made accessible through five vendors, three of which were based in the United States. There needed to be more global identification of providers for hexabromobiphenyl and octabromobiphenyl.

Table 2 Producers, trade names, uses, and effects of hexabromobiphenyl

Substance	Producers	Trade names	Uses	Effects
HBB	Atochem (France), Berk Corp (UK), Chemische Fabrik Kalk (Germany). Michigan Chemical Corp (the United States), White Chemical Corp (the United States). Hexcel Corp (the United States)	Fire master BP-6 Fire master FF-1	HBB is employed in the form of fire retardant within the chemical of acrylonitrile-butadiene-styrene (ABS) thermo- based plastics for various purposes such as setting up businesses and electric invention	Thyroid, hepatotoxicity, and disturbance in the endocrine gland affect the reproductive ability in various animals such as monkeys, rats, and mink. Epidemiological indications on employees were affected by hypothyroidism as they were mainly uncovered to PBBs, and even evidence of breast cancer is observed in women

3.1 *Production and Stockpiles*

According to the current data, information regarding the manufacture and stockpile of hexabromobiphenyl in the country has yet to be made available (IPEN, 2018).

Regulation

The notification number is G.S.R. 207 (E) dated 5th March, 2018 under Section 3 and also with Section 6 of the Environment (Protection) Act, 1986 (29 of 1986) and in compliance with Rule 13 of the Environment (Protection) Rules, 1986, to prohibit the manufacture, use, trade, export, and import of hexabromobiphenyl (IPEN, 2018).

Regulations and Guidelines (IARC, 2016)

The USA FDA has lowered the permissible restrictions to 0.3 mg/kg on a fat-weight basis for PBBs in the animal milk content, and fat of milk has been observed to be 0.05 ppm in the range of the egg, and earlier in 1974, it was observed to be 1.0 mg/kg and 0.1 ppm, respectively (ATSDR, 2004).

The European Union in 1983 restricted PBB uses in textile industries that are in direct interaction with the skin (EFSA, 2010).

The European Parliament declared not to contain PBBs at concentrations higher than 0.1% in electronic and electrical apparatus tools in the European Union 2002. Six substances are included in this category of such a degree of restriction, and the remaining five are PBDEs, hexavalent chromium, lead, mercury, and cadmium. Plastics that contain brominated flameproof must be separated and treated differently from e-wastes (Hadi et al., 2015).

The treaty on Long-period Transboundary Air Pollution from 1998 and the Stockholm agreement on POP from 2009 have included the commercial product "hexabromobiphenyl". Affairs related to such conventions are undergoing to yield actions accordingly to abolish the manufacture and usage of such contaminants (EFSA, 2010).

PBBs were not listed as harmful chemicals in animal feed, or as pollutants in food, so there were no federal or international laws or recommendations for them.

EPA Guidelines (EPA, 2017a)
Concerning the PBBs in EPA, no evidence exists of long-term treatment reference doses (RfDs). The testing thresholds for water from the tap, household soil, and industrial soil are all listed in the EPA (EPA, 2017b):

According to guidelines, tap water has been found to contain 0.0026 μg/L. In the case of residential chemicals, the soil value has been observed to be 0.018 mg/kg, but from an industrial point of view, the soil value is 0.077 mg/kg. For PBBs, the EPA has said that residential air screening stages should not exceed 3.3 × 10–4 mg/m^3, and industrial air screening stages should not exceed 1.4×10^{-3} mg/m^3 (EPA, 2017b).

According to the Agency for Toxic Substances and Disease Registry, a minimal risk limit (MRL) of 0.01 mg/kg/day for severe indicates that oral consumption of PBBs for 14 days or less is acceptable (ATSDR, 2016). The California Environmental Protection Agency has proposed a no-major-risk threshold of 0.02 μg per day for PBBs (Cal/EPA, 2017).

Exemption (use of chemicals for scientific purpose): Permission to use, sell, or import the substances listed in Rule 2 in the quantities required for research and development activities at CSIR-Laboratories, government establishments, central universities, or other research centers or accredited labs in the public or private sector following the laws and policies of the MoEF CC, GOI (IPEN, 2018).

Probable Routes of Human Exposure
Fire master FF-1 and Fire master BP-6, used in thermoplastics, indulged in electronic devices like televisions and computers, had a significant constitution of hexabromobiphenyl which is gradually landfilled or incinerated (ATSDR, 2004).

3.2 *Environmental Fate*

Terrestrial Fate The categorization method states that HBB exhibits little to no motility in soil, with Koc values between 2138 and 7413. With a predictable Henry's Law constant of 4.3 × 10–6 atm-cu m/mole consequent from a vapor pressure of 5.2 × 10–8 mm Hg and a water solubility of 0.011 mg/L, HBB has the critical property of volatilizing from moist soil surfaces and is projected to be an essential fate process. However, adsorption is predictable to decrease the rate of volatilization. Based on the vapor pressure, HBB does not volatilize from dry soil surfaces. Under aerobic conditions, HBB highly persists and biodegrades slowly under anaerobic conditions. HBB was preserved by incubating in a Spinks loamy sand (81.1% sand, 14.5% silt, 4.4% clay, 1.1% organic carbon, pH 5.8) and Brookston Loam (58% sand, 27.9% silt, 14.1% clay, 3.14% organic carbon, pH 7) maintained under aerobic conditions for 24 weeks (ATSDR, 2004; IARC, 1978).

Aquatic Fate A categorization method reveals that HBB may bind to suspended particles and silt because its Koc value ranges from 2138 to 7413. The approximate Henry's Law constant of 4.3×10^{-6} atm-cu m/mole has been calculated using HBB's vapor pressure of 5.2×10^{-8} mm Hg and water solubility of 0.011 mg/L. However, adsorption leads to a decrease in volatilization in suspended solids and sediments. An estimation method was conducted using this Henry's Law constant and found that volatilization half-lives for a classical river are 14 days and model lake are 165 days if adsorption is ignored. And the estimated volatilization half-life is 160 days for a model pond when adsorption is considered. Under a classification scheme, a BCF of 18,100 measured in fathead minnows proposes high bioconcentration in marine life. HBB is degrade-proof under aerobic conditions but may slowly undergo debromination at the meta and para positions below anaerobic circumstances. Due to the absence of hydrolysable functional sets in HBB, it is not predictable to experience hydrolysis in the atmosphere. HBB can degrade over 60% by undergoing 12 h of irradiation in cyclohexane solution, which paves a scope of direct photolysis in sunlit surface waters (ATSDR, 2004; IARC, 1978; Frank et al., 1994).

Atmospheric Fate Given that HBB has a vapor pressure of 5.2×10^{-8} mm Hg at 25°, the gas or particle fractionation concept of semi volatile organic compounds in the environment predicts that it will reside in the particulate stage in the ambient environment. The particulate set of HBB in the air can vanish through a dry and wet deposition method. In the solution phase, HBB shows direct photolysis, giving less brominated biphenyls as the photo degradation by products. At the same time, there is no evidence for immediate photolysis occurrence of particulate phase HBB in the air (ATSDR, 2004; IARC, 1978; Robertson et al., 1983).

Persistence

According to the reports, PBBs are chronic under field conditions. PBBs were found in soil samples taken from a site where they used to be made, years after they were accidentally dumped there. But the composition varies, indicating partial degradation of PBBs in the soil. Even after banning PBB production, EHC Reviews and surveys over 3 years do not show a significant deduction of PBB levels in river sediments. A blend of PBBs looks pretty resistant to microbial destruction in laboratory testing (Jacobs et al, 1976).

Bioaccumulation

PBBs may bioconcentrate in the food supply chain because they are very lipophilic, and the observations of wildlife inspections hold it up. For instance, it is observed that fathead minnows (*Pimephales promelas*) confined in a river for 2 weeks showed 10,000-fold contaminants in their body, but PBB level constantly remained at less than 0.1 μg/liter (Hesse & Powers, 1978).

Toxicity

According to the IARC, HBB may cause human cancer (IARC group 2B). The PBBs are grouped as endocrine-troublesome substances, and the belongings are observed in the multiplicative capacities of monkeys, minks, and rats. There are epidemiological proofs for hypothyroidism in people visible to PBBs and an augmented number of people living with breast cancer in uncovered women.

The US ATSDR (2004) mentions that the present introduction of PBBs to humans is minute since PBBs are not produced and used. Therefore, the probability of PBBs exposure to the general population will be only from earlier discharge.

4 In-House (Countrywide) Scientific Work Done on the Chemical

Considering the present levels reported for PBBs in surface and groundwater, aquifers, sediment, or soil, there is scientific need to compile all information or data which should be easily available to all stakeholders. As on today, much data are not available on background level of PBBs in different environmental matrices.

5 Relevance of the International Scientific Data as per Indian Conditions

China started processing imported electronic waste ("e-waste") like obsolete electric capacitors, scrap metals, household appliances, cable wires, and electric generators in the early 1990s. The United States, Japan, the Russian Federation, and Western European nations import 90% of all e-waste. The recycling methods include burning in the open air, physical dismantling using a hammer, screwdriver, chisel, bare hands, and acid leaching. Thyrotropin (thyroid-stimulating hormone), thyroid hormone, and brominated flame retardant blood levels among Chinese employees in the e-waste recycling industry were noted in the 2008–2009 study (Wang et al., 2021). This study includes all people who have direct exposure to brominated flame retardant occupationally throughout dismantlement and recycling procedures and non-occupationally unprotected individuals. The mean concentration of PBBs in the sera from these in the workplace subjected employees was found to be 3.02 ng/mL plasma (n = 239), which was higher than that of the regulates ($\sum$PBB = 1.43 ng/mL plasma, n = 116) but lower than that of farmers in the regions encompassing the e-waste site ($\sum$PBB = 4.34 ng/mL plasma, n = 39) (IARC, 2016). The toxicological behavior has been listed in Table 3.

Despite hexabromobiphenyl's relatively low acute toxicity, some chronic toxic effects, such as hepatotoxicity, have been reported in laboratory animals when exposed to doses of approximately one mg/kg bw/day over an extended period. Additionally, adverse impacts on the thyroid of rats have been reported at dosages as small as 0.05 mg/kg bw/day. The acute toxicity of HBB was higher than that of

Table 3 Crucial toxicological investigations on the dangers of hexabromobiphenyl (US ATSDR, 2004)

Species	Type of work	Consequence	LOAEL/ NOAEL
Rat	Short-term/acute toxicity 10-day repeat dose gavage research on acute/ short-term toxicity	Lower thyroid serum T4 hormones levels	LOAEL:3 mg/kg bw/day NOAEL:1 mg/kg bw/day
Rat, Sprague Dawley	Diet feeding trial lasting 30 days	Thyroid follicles have risen in number and shrunk in size	LOAEL:0.05 mg/ kg bw/day
Mice B6C3F1	Gd 0-ppd 56 exposure during pregnancy and after delivery	Children's hepatocellular adenoma and carcinoma	LOAEL:1.5 mg/kg bw/day NOAEL: 0.15 mg/ kg bw/day
Rhesus Monkey	25–50-week dietary feeding experiment	The proliferation of mucosal cells, chronic inflammation, severe ulcerative colitis, alopecia, keratinization of hair follicles and sebaceous glands, clinical chemistry and hepatic abnormalities, and a weight loss of 34% in adult males and a gain of 0% in juveniles are all associated with this condition	LOAEL:0.73 mg/ kg bw/day
Rat, Sprague Dawley	7-month nutritional feeding experiment	T3 and T4 levels in thyroid serum are lower	LOAEL:0.45 mg/ kg bw/day
Monkey, Rhesus		Implantation hemorrhage in 2/7; lengthened menstrual cycles in 4/7. Abortion affected 1/7 pregnancies, stillbirth—1/7 fetuses, birth weight fell by 12%, and postpartum weight growth decreased by 22% in 4/7 survivors	LOAEL:0.012 mg/ kg bw/day

Source: https://www.informea.org/en/risk-profile-hexabromobiphenyl

OBB when absorbed via the skin in rabbits. Additionally, HBB produced larger livers at lower dosages compared to OBB.

6 Remedial Action Taken for Management of POPs

Hexabromobiphenyl is said to be quite persistent in the natural world. There needs to be more soil, water, and sediment degradation study in the lab and the field. To address the environmental and health related threats posed by widespread POPs contamination, remediation technologies continue to be developed to treat such deadly pollutants.

Replacement
Replaceable hexabromobiphenyl is expected, thus forbidding its manufacture and use is feasible and inexpensive.

7 Recommendations

Even though the manufacture and usage of hexabromobiphenyl are ceased in most countries, there are probabilities of reintroduction, which may lead to extensive release into the environment. Therefore, strict and effective evaluations should be followed up. A more integrated approach is required involving all the stakeholders in order to develop sustainable approach to mitigate the risks posed by hexabromobiphenyl and other listed chemicals in POPs category.

References

ATSDR. (2004). *Toxicological profile for polybrominated biphenyls and polybrominated diphenyl ethers*. Department of Health Human Services, Agency Toxic Sub Disease Reg, Atlanta, GA.
ATSDR. (2016). *Agency for Toxic Substances and Disease Registry's priority list of hazardous substances*. Retrieved from http://www.atsdr.cdc.gov/SPL/index.html
Cal, E. P. A. (2017). *Chemicals known to the state to cause cancer or reproductive toxicity*. State of California, Environmental Protection Agency, Office of Environmental Health Hazard Assessment. https://oehha.ca.gov/media/downloads/proposition-65/p65122917.pdf
EFSA Panel on Plant Health (PLH). (2010). Guidance on a harmonised framework for pest risk assessment and the identification and evaluation of pest risk management options by EFSA. *EFSA Journal, 8*(2), 1495.
EPA. (2017a). *Renewable fuel standard program, renewable fuel standard program: Standards for 2017 and biomass-based diesel volume for 2018*. United States Environmental Protection Agency. Retrieved from https://www.epa.gov/renewable-fuel-standard-program/final-renewablefuel-standards-2017-and-biomass-based-diesel-volume
EPA. (2017b, November). *Technical fact sheet – Polybrominated biphenyls (PBBs)*. https://www.epa.gov/sites/production/files/2017-12/documents/ffrro_factsheet_pbb_11-16-17_508.pdf
Franke, C., et al. (1994). *Chemosphere, 29*, 1501–1514.
Hadi, P., Xu, M., Lin, C. S., Hui, C. W., & McKay, G. (2015). Waste printed circuit board recycling techniques and product utilization. *Journal of Hazardous Materials, 283*, 234–243.
Hesse, J. L. & Powers, R. A. (1978). Polybrominated biphenyl (PBB) contamination of the Pine River, Gratiot, and Midland Counties, Michigan. *Environmental Health Perspectives, 23, 19–25.*
http://www.popstoolkit.com/about/chemical/hbb.aspx
IARC. (1978). *International Agency For Research on Cancer polychlorinated and polybrominated biphenyls* (Vol. 18). Lyon, France.
IARC Working Group on the Evaluation of Carcinogenic Risk to Humans. (2016). *Polychlorinated biphenyls and polybrominated biphenyls* (IARC monographs on the evaluation of carcinogenic risks to humans, No. 107. Exposure data). International Agency for Research on Cancer. Available from https://www.ncbi.nlm.nih.gov/books/NBK361698/
IPEN. (2018). Persistent Organic Pollutants (POPs) in India: Country Situation Report by Toxics Link, June 2018

Jacobs, L. W., Chou, S. F., & Tiedje, J. M (1976). Fate of polybrominated biphenyls (PBB's) in soils. Persistence and plant uptake. *Journal of Agricultural and Food Chemistry, 24, 1198–1201.*

Kumari, K., Sharma, J. K., Kanade, G. S., et al. (2014). Investigation of polybrominated diphenyl ethers in old consumer products in India. *Environmental Monitoring and Assessment, 186*, 3001–3009. https://doi.org/10.1007/s10661-013-3596-2

National Center for Biotechnology Information. (2023). *PubChem compound summary for CID 42948, 2,2′,4,4′,5,5′-hexabromobiphenyl*. Retrieved October 11, 2023, from https://pubchem.ncbi.nlm.nih.gov/compound/42948

Robertson, L. W., et al. (1983). *Journal of Agricultural and Food Chemistry, 31*, 454–457.

Singh, A., & Kumari, K. (2022). Classical and emerging persistent organic pollutants in environment. In *Persistent organic pollutants: Gaps in management and associated challenges* (1st ed., pp. 33–55). CRC Press.

UNEP-POPS-PUB-Brochure-16NewPOPs-201706.English.pdf.

US ATSDR. (2004). *Toxicological profile for polybrominated biphenyls and polybrominated diphenyl ethers (PBBs and PBDEs)*. http://www.atsdr.cdc.gov/toxprofiles/tp68.html, http://www.pops.int/TheConvention/ThePOPs/TheNewPOPs/tabid/2511/Default.aspx

Wang, H., Zhang, Y., Liu, Q., Wang, F., Nie, J., Qian, Y. (2010). Examining the relationship between brominated flame retardants (BFR) exposure and changes of thyroid hormone levels around e-waste dismantling sites. *International Journal of Hygiene and Environmental Health, 213(5), 369–380. ISSN 1438-4639,* https://doi.org/10.1016/j.ijheh.2010.06.004

Wang, Y., Zhong, H., Luo, Y., Xian, H., Li, F., Gao, W., et al. (2021). Temporal trends of novel brominated flame retardants in mollusks from the Chinese Bohai Sea (2011–2018). *Science of the Total Environment, 777*, 146101.

Hexabromocyclododecane (HBCD)

Kanchan Kumari, Deepak Marathe, and Md. Shahbaz

1 Introduction

Hexabromocyclododecane (HBCD) is a brominated flame retardant (BFR) that is produced and utilized globally, primarily as an additive flame retardant in products such as polystyrene and textiles. Brominated flame retardants (BFRs) are employed in consumer goods to mitigate the risks of fire-related harm and damage to property (Birnbaum & Staskal, 2004). HBCD ranks as the third most widely used brominated flame retardant globally, following decabromodiphenyl ether (decaBDE) and tetrabromobisphenol A (TBBPA) (Zegers et al., 2005). HBCD finds primary application in exterior building insulation materials such as expanded polystyrene (EPS) and extruded polystyrene (XPS). It is also incorporated into electrical equipment, high-impact polystyrene (HIPS), and textile coatings (POPRC, 2011). HBCD was added to the Stockholm Convention on Persistent Organic Pollutants in 2013 (Kumari, 2021) with the objective of globally phasing out its utilization. Current risk assessments suggest that the primary HBCD risk is associated with occupational exposure, particularly during the manufacturing of HBCD and flame-retarded EPS and XPS (Environment Canada and Health Canada, 2011; Managaki et al., 2009). Persistent organic pollutants (POPs) are chemical substances that do not degrade easily and persist in the environment (Kumari & Khare, 2018). It is prevalent in the environment worldwide, and monitoring data generally indicate higher

K. Kumari (✉) · M. Shahbaz
CSIR-National Environmental Engineering Research Institute (NEERI), Kolkata Zonal Center, Kolkata, West Bengal, India
e-mail: k_kumari@neeri.res.in

D. Marathe
CSIR-National Environmental Engineering Research Institute (NEERI), Nagpur, India

K. Kumari, *Pollutants of Global Concern*, Emerging Contaminants and Associated Treatment Technologies, https://doi.org/10.1007/978-3-031-50996-4_15

levels in proximity to known sources; however, consistent temporal trends are not always observed (de Wit et al., 2010).

As substitutes for PBDEs, the worldwide market exhibited a substantial surge in demand for TBBPA and HBCD in recent years. Tetrabromobisphenol A (TBBPA), for instance, was estimated at approximately 120,000 tonnes in 2001, which further escalated to over 170,000 tonnes in 2004. Notably, a significant portion, roughly 80%, of these products found consumption in Asia, particularly in China (Liu et al., 2016). HBCD is notable for its propensity to accumulate in organisms and bio magnify through food chains, making it a persistent environmental concern. This persistence allows it to be transported over long distances, posing a significant threat to aquatic ecosystems. Additionally, HBCD has demonstrated neuroendocrine and developmental toxicity, raising concerns for potential adverse effects on human health (Aznar-Alemany et al., 2019).

Historically, HBCD was extensively used as a flame-retardant additive for polystyrene-based materials, in compliance with safety regulations for various products, including vehicles and buildings (UNEP, Stockholm Convention).

2 Technical Details of Hexabromocyclododecane (HBCD) (NCBI, 2023)

HBCD is a chemical compound widely employed to enhance the fire resistance of various materials in a range of applications. It possesses a cyclic molecular structure with covalently bonded bromine atoms. The chemical formula $C_{12}H_{18}Br_6$ signifies its composition, comprising 12 carbon atoms, 18 hydrogen atoms, and 6 bromine atoms (Fig. 2). Notably, HBCD consists of several isomers, including α-HBCD, β-HBCD, and γ-HBCD, each characterized by distinct spatial arrangements of bromine atoms. These isomers exhibit differing physical properties, such as melting points and solubility (Figs. 1 and 2).

Typically, HBCD presents as a white crystalline solid with limited solubility in water. Its fire-retardant properties make it suitable for use in products such as polystyrene foams for insulation and packaging, textiles, and plastics. This is achieved by generating bromine radicals when exposed to heat, effectively reducing the material's flammability.

However, concerns about the environmental and health impacts of HBCD have arisen due to its persistence in the environment, resulting in its accumulation in soil, water, and organisms. Certain types of HBCD have been categorized as persistent organic pollutants (POPs) under the Stockholm Convention, leading to regulatory actions in various countries to restrict or phase out specific HBCD variants. The potential adverse effects on ecosystems and human health have prompted increased scrutiny, resulting in regulatory measures and limitations on HBCD production and usage. In response to these concerns, various industries are actively researching alternative solutions to replace HBCD.

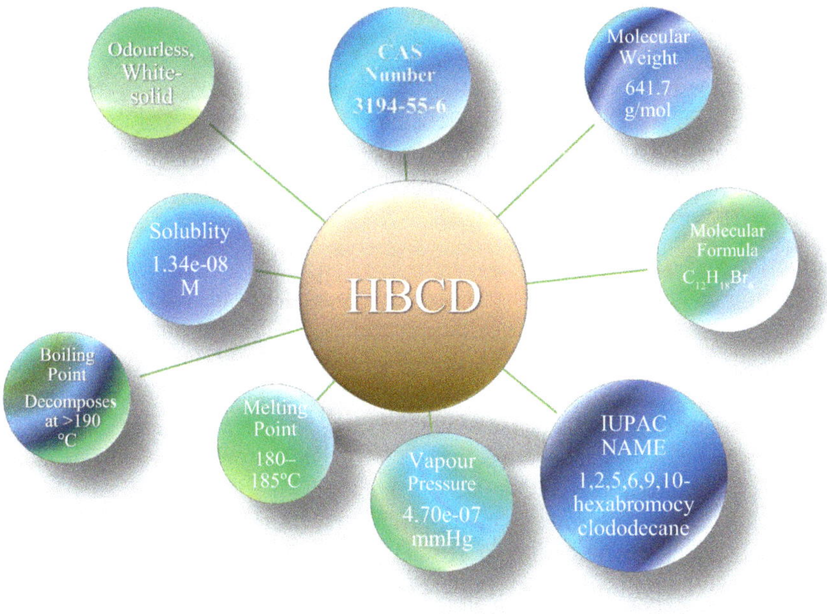

Fig. 1 Physicochemical properties of hexabromocyclododecane (HBCD) (NCBI, 2023)

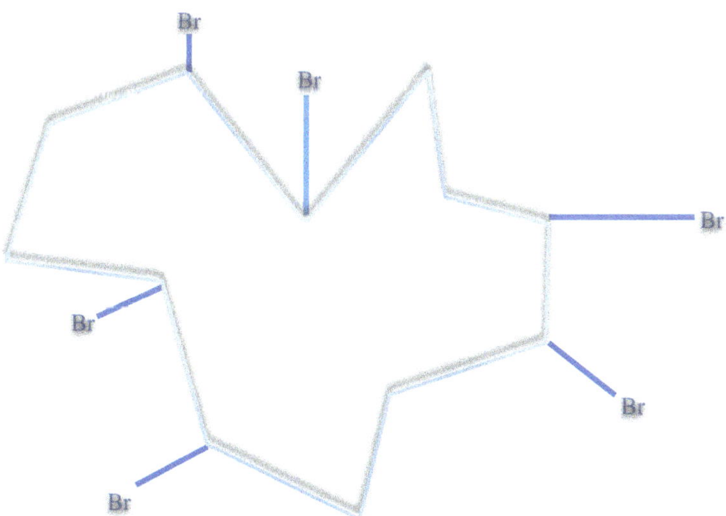

Fig. 2 Structure of Hexabromocyclodecane (HBCD)

3 Status of Usage Within the Country

In 2013, Stockholm Convention (SC) took the significant step of adding to Annex A of SC regulation. HBCD, an industrial chemical widely employed in the construction sector for expanded and extruded polystyrene. A comprehensive ban of HBCD has been enforced in India as of March 5, 2018, prohibiting the manufacturing, utilization, import, and export of HBCD.

Production and Stockpiles

HBCD is not manufactured within the territorial boundaries of India, whether in its raw form or as a finished product (Mohapatra et al., 2023). The absence of readily available stockpile data for HBCD is attributed to the lack of domestic production in India (IPEN, POPs, 2018). Figure 3 shows the import data of HBCD in India.

Regulation involves the exercise of control or governance over a specific entity, typically achieved by establishing rules and laws. On March 5, 2018, the Ministry of Environment, Forestry, and Climate Change issued a notice, specifically identified as G.S.R. 207 (E). This notification pertains to sections 3 and 6 of the aforementioned ministry. The prohibition of HBCD (IPEN POPs, 2018) was enacted in accordance with the Environment (Protection) Act of 1986 (29 of 1986) and rule 13 of the Environment (Protection) Rules of 1986, both promulgated in 1986. This directive effectively prohibits the manufacturing of the specified substance, as well as all associated activities, including its production, usage, import, and export.

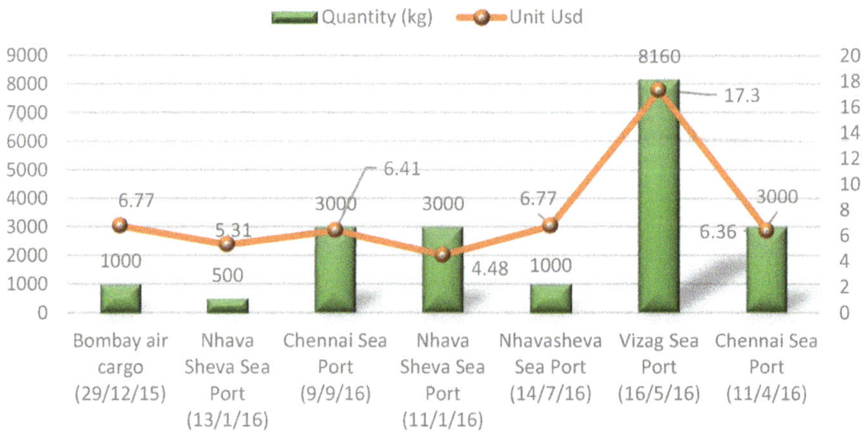

Fig. 3 Import of HBCD in India from China Port (SEAIR HBCD)

3.1 *Exemption for the Utilization of Chemical Substances in the Context of Scientific Investigation*

Notwithstanding the existing regulations, there exists a provision for the authorized use, trade, and import of chemicals as outlined in rule 2, but with strict limitations on the quantities involved. This authorization is exclusively granted for the purposes of conducting research and development activities within Central Universities, Council of Scientific and Industrial Research Laboratories, government institutions, or other duly recognized research organizations or laboratories, irrespective of their affiliation with the public or private sector.

It is of paramount importance to highlight that prior authorization from the Ministry of Environment, Forest, and Climate Change, an entity operating under the Government of India, must be sought and obtained before commencing the aforementioned procedures (IPEN, POPs, 2018).

***In-House (Countrywide) Scientific Work Done on the Chemical**

No relevant data found.

4 Relevance of the International Scientific Data as per Indian Conditions

Three brominated flame retardants (BFRs) that have witnessed widespread use over the years are hexabromocyclododecane (HBCD), tetrabromobisphenol A (TBBPA), and polybrominated diphenyl ether (PBDEs) (Birnbaum & Staskal, 2004). Jang et al. (2017) provided data indicating that the annual production volume of TBBPA is roughly 30,136 tons in Korea. Furthermore, Korea stands as one of the prominent consumers of HBCD in Asia. To illustrate, in 1993, this country consumed 530 tons of HBCD, and this annual consumption escalated to 1896 tons by 2010 (Al-Odaini et al., 2015). While these substances are categorized as legacy chemicals, China continues to manufacture and employ them. In a study conducted by Wang et al. (2019), it was unveiled that animal-based meals contained median levels of total HBCD ranging from 1.14 to 5.65 nanograms per gram of dry weight. It is noteworthy that α-HBCD was identified as the predominant isomer in these dietary items.

Furthermore, the research indicated that the median concentration of HBCD in vegetables was measured at 0.266 nanograms per gram of wet weight (ng/g ww), with γ-HBCD being the primary form in the samples. The study also revealed that the daily intake of HBCD among adults in Beijing amounted to 2.74 ng/kg of body weight per day. An analysis comparing estimated daily intakes (EDIs) with threshold reference values (TRVs) demonstrated that the ingestion of brominated flame retardants (BFRs) through food consumption does not pose substantial health risks. Similarly, margins of exposure (MOEs) were calculated using the European Food Safety Authority (EFSA) methodology, revealing values significantly higher than the established threshold. This data supports the notion that the estimated daily intakes (EDIs) of brominated flame retardants (BFRs) do not present significant risks to human health.

The graph shown in Fig. 4 depicts the concentrations of HBCD (cumulative sum of three isomers) within various distinct food categories. Notably, a substantial 73% of the samples exhibited HBCD concentrations surpassing the lowest detectable limit (LOD). Moreover, HBCD was detected in a significant 82% of the 62 animal-derived food items subjected to testing, with quantities ranging from below the limit of detection (bLOD) to 26.8 nano grams per gram of lipid weight (ng/g lw), contingent on the specific product.

When comparing the presence of TBBPA in animal-derived food products, fish and shellfish emerged with the highest median concentration of HBCD, measuring at 5.65 ng/g lw. Following closely were eggs (3.98 ng/g lw), beef (2.43 ng/g lw), and dairy products (1.14 ng/g lw). It is noteworthy that catfish, among all types of fish and seafood, exhibited the highest quantity of hexabromocyclododecane (HBCD), registering at 26.8 ng/g lw based on lipid weight (Fig. 4).

The study also revealed that α-HBCD was present in a significant majority of fish and seafood samples, specifically in 78% of cases. However, the coexistence of both β-HBCD and γ-HBCD was observed in only three samples, accounting for 17% of the total sample population. In the domain of meat and meat products, α-HBCD was detected in 71% of the meat samples. Nevertheless, the simultaneous presence of both β-HBCD and γ-HBCD was identified in only three samples, as these were the sole instances where both variants of HBCD were observed. Intriguingly, none of the tested samples of eggs or dairy products, encompassing one egg sample and two dairy product samples, displayed any traces of HBCD.

However, the analysis distinctly revealed that among all the HBCD isomers, only α-HBCD exceeded the limit of detection (LOD) in eggs. Conversely, dairy samples exhibited lower concentrations of β and γ-HBCD. These experimental findings consistently showcased α-HBCD as the dominant isomer in animal-derived meals, with

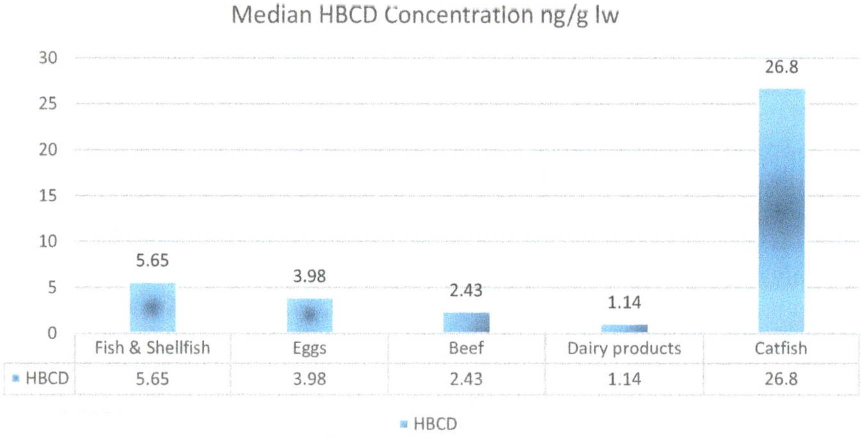

Fig. 4 Median HBCD concentration in food items in China (Wang et al., 2019)

the exception of a specific sample of duck flesh. According to the study's results, α-HBCD constituted 42% of the overall HBCD mixture, while γ-HBCD accounted for 49% of the total. Remarkably, the range of α-HBCD levels in animal-based diets exhibited variability, ranging from 69% to a full 100% (Wang et al., 2019).

The increased concentrations of BDE209 and HBCD observed in the downstream segment of the Hunhe River and the Dahuofang Reservoir can be attributed to local sources originating from the cities of Fushun and Shenyang. This is evident when comparing these concentrations with those found in the upstream segment of the river and reservoir. Interestingly, a higher concentration of hexabromocyclododecane (HBCD) was detected in the upstream section of the Hunhe River (Fig. 5). Further investigation is required to determine the potential correlation between the observed increase in α-HBCD levels and the presence of heavy metal contamination (Su et al., 2015).

Despite the widespread use of the brominated flame retardant HBCD across Europe, there is a notable lack of information regarding the levels of this chemical in the environment. An initial assessment of the ambient presence of HBCD has been carried out in Sweden, a country where the production of this compound is absent, and its industrial use is now limited. A comprehensive examination of their utilization within a life cycle context facilitated the identification of potential emission sources. Polymeric materials have the potential to act as sources of diffuse emissions, alongside the limited number of recognized point sources.

HBCD and TBBPA are released into the environment, and human exposure to these substances primarily occurs through dietary intake and the ingestion of dust (Barghi et al., 2017). It is conceivable to envision those comprehensive measurements have been conducted in the immediate vicinity of potential point sources, encompassing both urban and rural areas. These measurements covered a diverse array of environmental parameters, including atmospheric conditions, deposition

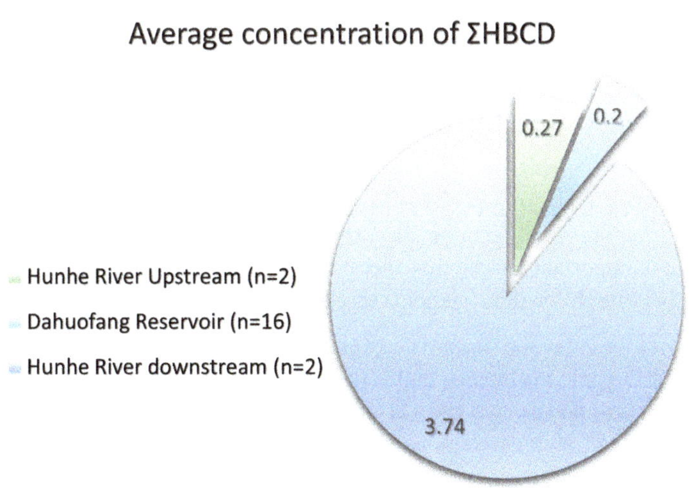

Fig. 5 Average concentrations (ng/g dry weight) of HBCD in the Hunhe River (Su et al., 2015)

Table 1 Concentration of HBCD in Laizhou Bay, China (Su et al., 2015)

Sample type	HBCD concentration (ng/g dw, lw)
Soil	0.88–6901
Sediment	2.93–1029
Plant specimens	8.88–160,241

rates, water quality, soil composition, sediment characteristics, sludge properties, biotic factors, and various food items, among other variables. The presence of HBCD was found in all environments and media examined.

A more extensive body of data substantiates the presence of diffuse emissions of HBCD in the urban environment. In a research study conducted in the Laizhou Bay region of East China, a comprehensive assortment of 55 samples was obtained from various environmental sources. These samples included soil, sediment, plant specimens (cypress, reed, and seepweed), and aquatic organisms. The analysis aimed to determine the concentrations of several environmental pollutants, facilitating a study on the occurrence and spatial distribution of HBCD near manufacturing facilities.

Bioaccumulation factors (BAFs) computed from the data indicate the potential for HBCD to accumulate in plants. Selective tissues in aquatic organisms were found to accumulate HBCD diastereoisomers via the process of bioaccumulation. For instance, in crabs, the gill exhibited the highest levels of HBCD, reaching 815 ng g^{-1} lw for females and 446 ng g^{-1} lw for males, with greater concentrations in females. α-HBCD had a greater tendency to accumulate in the spermary and ovary, while both β and γ-HBCD showed a higher propensity for accumulation in the fish's muscle tissue. A similar pattern was observed in the muscle and roe of goby fish as well (Su et al., 2015). Su et al., 2015 has also reported HBCD concentration in various environmental matrix (Table 1).

5 Remedial Action Taken for Management of POPs

In the context of managing persistent organic pollutants (POPs) in India, several corrective measures and actions have been undertaken to address this issue. However, it is important to note that specific details and data on these measures can vary, and there might be developments that are more recent.

- Regulatory Framework: India is a signatory to the Stockholm Convention on Persistent Organic Pollutants, and it has developed a comprehensive regulatory framework to manage and control the production, use, and disposal of POPs. This framework includes regulations and guidelines for specific POPs, including HBCD.

- Restrictions on Production and Use: India has implemented restrictions on the production and use of certain POPs, aligning with international agreements like the Stockholm Convention. These restrictions are aimed at reducing or eliminating the production and use of particularly hazardous POPs, such as HBCD.
- Environmental Monitoring: India conducts environmental monitoring programs to assess the presence and levels of POPs in various environmental compartments, such as air, water, soil, and biota. This monitoring helps in understanding the extent of contamination and tracking trends over time.
- Risk Assessment: Risk assessments are regularly conducted to evaluate the potential health and environmental risks associated with POPs like HBCD. These assessments inform decision-making processes related to risk management and regulatory actions.

Regarding Alternatives to HBCD
Chemical Substitutes: Safer chemical substitutes are available on the market for use in place of HBCD, particularly in applications like high-impact polystyrenes (HIPs) and textile back-coating. These alternatives are designed to provide flame-retardant properties while posing fewer environmental and health risks.

6 Recommendations

HBCD is not manufactured or produced in India. As it is not produced in India, there is no stockpile information available about hexabromocyclododecane. HBCD was imported from China. Effectiveness evaluation shall be carried out.

For future courses of action regarding HBCD, several recommendations can be made to address its environmental and health concerns:

- Regulatory Measures: Continue and strengthen regulatory actions to restrict or phase out the production, use, and release of HBCD, particularly its isomers with higher toxicity or persistence. Align regulations with international agreements like the Stockholm Convention on Persistent Organic Pollutants.
- Alternative Flame Retardants: Encourage the development and adoption of safer, more environmentally friendly flame retardants as alternatives to HBCD. Promote research into innovative flame retardant technologies.
- Environmental Monitoring: Establish and maintain robust monitoring programs to track HBCD levels in the environment, wildlife, and food chains. This helps identify emerging trends and potential hotspots of contamination.
- Risk Assessment: Conduct comprehensive risk assessments to evaluate the potential health and ecological risks associated with HBCD exposure. This includes studying its impact on human health, wildlife, and ecosystems.
- Education and Awareness: Raise awareness among industries, consumers, and policymakers about the risks associated with HBCD and the importance of proper disposal and handling of products containing it.

- Waste Management: Implement effective waste management strategies for products containing HBCD to prevent its release into the environment. Encourage recycling and safe disposal practices.
- Research and Development: Support research focusing on the fate, transport, and transformation of HBCD in the environment. Investigate its long-term effects on ecosystems and biodiversity.
- Global Cooperation: Collaborate with other countries and international organizations to share information, best practices, and solutions for managing HBCD on a global scale.
- Health Surveillance: Establish health surveillance programs to monitor and assess the health of individuals who may be exposed to HBCD in occupational or environmental settings.
- Policy Integration: Integrate HBCD management into broader policies related to chemical safety, environmental protection, and sustainable development.
- Public Engagement: Engage the public in discussions and decision-making processes related to HBCD, ensuring transparency and public input in regulatory actions.
- Innovation and Green Chemistry: Encourage the development and adoption of green chemistry principles to reduce the use of hazardous chemicals, including HBCD, in manufacturing and product design.

By implementing these recommendations, stakeholders can work together to mitigate the environmental and health risks associated with HBCD while promoting safer alternatives and sustainable practices.

References

Al-Odaini, N. A., Shim, W. J., Han, G. M., Jang, M., & Hong, S. H. (2015). Enrichment of hexabromocyclododecanes in coastal sediments near aquaculture areas and a wastewater treatment plant in a semi-enclosed bay in South Korea. *Science of the Total Environment, 505*, 290–298.

Aznar-Alemany, Ò., Sala, B., Plön, S., Bouwman, H., Barceló, D., & Eljarrat, E. (2019). Halogenated and organophosphorus flame retardants in cetaceans from the southwestern Indian Ocean. *Chemosphere, 226*, 791–799. https://doi.org/10.1016/j.chemosphere.2019.03.165. ISSN 0045-6535.

Barghi, M., Shin, E. S., Kim, J. C., Choi, S. D., & Chang, Y. S. (2017). Human exposure to HBCD and TBBPA via indoor dust in Korea: Estimation of external exposure and body burden. *Science of the Total Environment, 593*, 779–786.

Birnbaum, L. S., & Staskal, D. F. (2004). Brominated flame retardants: Cause for concern? *Environmental Health Perspectives, 112*(1), 9–17.

de Wit, C. A., Herzke, D., & Vorkamp, K. (2010). Brominated flame retardants in the Arctic environment—Trends and new candidates. *Science of the Total Environment, 408*(15), 2885–2918.

Environment Canada, Health Canada. (2011). *Screening Assessment Report on 359 Hexabromocyclododecane-Chemical Abstracts Service Registry Number 3194-55-6*. Canada, 360 November.

IPEN, POPs, Persistent Organic Pollutants (POPs) in India: Country situation report by Toxics Link, June 2018.

Jang, M., Shim, W. J., Han, G. M., Rani, M., Song, Y. K., & Hong, S. H. (2017). Widespread detection of a brominated flame retardant, hexabromocyclododecane, in expanded polystyrene marine debris and microplastics from South Korea and the Asia-Pacific coastal region. *Environmental Pollution, 231*, 785–794.

Kumari, K. (Ed.). (2021). *Persistent organic pollutants: Gaps in management and associated challenges* (1st ed.). CRC Press. https://doi.org/10.1201/9781003046806

Kumari, K., & Khare, A. (2018). Integration of biomarker approach in pollution monitoring programme of aquatic ecosystem. In S. Varjani, B. Parameswaran, S. Kumar, & S. Khare (Eds.), *Biosynthetic technology and environmental challenges. Energy, environment, and sustainability*. Springer. https://doi.org/10.1007/978-981-10-7434-9_18

Liu, K., Li, J., Yan, S., Zhang, W., Li, Y., & Han, D. (2016). A review of status of tetrabromobisphenol A (TBBPA) in China. *Chemosphere, 148*, 8–20. https://doi.org/10.1016/j.chemosphere.2016.01.023. ISSN 0045-6535.

Managaki, S., Miyake, Y., Yokoyama, Y., Hondo, H., Masunaga, S., Nakai, S., et al. (2009). Emission load of hexabromocyclododecane in Japan based on the substance flow analysis. *Organohalogen Compounds, 71*, 2471–2476.

Mohapatra, P., Nøklebye, E., Arora, T., & Basu, A. R. (2023). International regulatory frameworks and best practices in management of POPs. In *Managing persistent organic pollutants in India: Case studies on Vapi and Surat, Gujarat* (pp. 127–151). Springer Nature Switzerland.

National Center for Biotechnology Information. (2023). PubChem Compound Summary for CID 18529, 1,2,5,6,9,10-Hexabromocyclododecane. Retrieved 13 Oct 2023 from https://pubchem.ncbi.nlm.nih.gov/compound/1_2_5_6_9_10-Hexabromocyclododecane

Persistent Organic Pollutants Review Committee (POPRC). (2011). *Addendum to the risk management evaluation on hexabromocyclododecane*. UNEP/POPS/POPRC.7/19/Add.1.

Su, J., Lu, Y., Liu, Z., Gao, S., Zeng, X., Yu, Z., et al. (2015). Distribution of polybrominated diphenyl ethers and HBCD in sediments of the Hunhe River in Northeast China. *Environmental Science and Pollution Research, 22*(21), 16781–16790.

Wang, J., Zhao, X., Wang, Y., & Shi, Z. (2019). Tetrabromobisphenol A, hexabromocyclododecane isomers and polybrominated diphenyl ethers in foodstuffs from Beijing, China: Contamination levels, dietary exposure and risk assessment. *Science of the Total Environment, 666*, 812–820.

Zegers, B. N., Mets, A., Van Bommel, R., Minkenberg, C., Hamers, T., Kamstra, J. H., et al. (2005). Levels of hexabromocyclododecane in harbor porpoises and common dolphins from western European seas, with evidence for stereoisomer-specific biotransformation by cytochrome P450. *Environmental Science & Technology, 39*(7), 2095–2100.

Tetra-, Penta-, Hexa-, and Heptabromodiphenyl Ethers

Kanchan Kumari and Md. Shahbaz

1 Introduction

Tetra-, penta-, hexa-, and heptabromodiphenyl ethers listed in Annex A to Stockholm Convention in the year 2009 (Kumari, 2021) are a group of chemicals that belong to the class of polybrominated diphenyl ethers (PBDEs). These substances are artificial chemicals that have been used primarily in a variety of products as flame retardants such as foam-based products, clothing, and electrical equipment.

Hexabromodiphenyl ether and heptabromodiphenyl ether are primary constituents of marketed octabromodiphenyl ether. The commercial octaBDE composition is known for its significant persistence, potential for bioaccumulation, involvement in food web biomagnification, and long-range transport. The primary degradation mechanism involves debromination, resulting in the formation of various bromodiphenyl ethers (IPEN & POPs, 2006).

Tetra and pentabromodiphenyl ethers are the core constituents of commercial pentabromodiphenyl ether. Tetrabromodiphenyl ether (TBDPE) is not produced as an individual product. The 2, 2′, 4, 4′-TBDPE isomer is a major component of the Pentabromodiphenyl ether (PeBDPE) commercial product. These substances are categorized as "polybromodiphenyl ethers" (PBDEs). The commercial pentaBDE mixture demonstrates notable environmental persistence, bioaccumulation propensity, and the potential for extensive environmental dispersal over long distances, as evidenced by its detection in human populations spanning diverse geographic regions. Robust scientific evidence substantiates its adverse impacts on wildlife, particularly mammals (IPEN & POPs, 2018).

K. Kumari (✉) · M. Shahbaz
CSIR-National Environmental Engineering Research Institute (NEERI),
Kolkata Zonal Center, Kolkata, West Bengal, India
e-mail: k_kumari@neeri.res.in

K. Kumari, *Pollutants of Global Concern*, Emerging Contaminants and Associated Treatment Technologies, https://doi.org/10.1007/978-3-031-50996-4_16

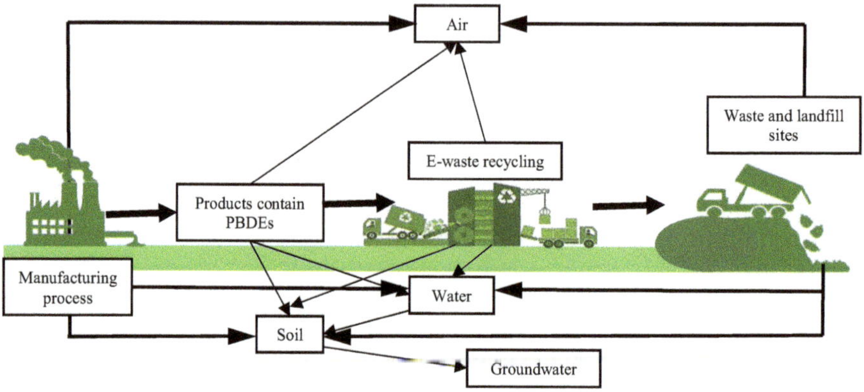

Fig. 1 Major routes of exposure of PBDEs in environment

Polybrominated diphenyl ethers (PBDEs) are environmental pollutants that can originate from various sources (Fig. 1). The sources of PBDEs in the environment include the following:

- Manufacturing and Production: A predominant origin of PBDEs in the ecosystem stems from the manufacturing and production of products containing PBDEs. These compounds were extensively employed as flame retardants in a diverse array of items, encompassing electronics, textiles, foam cushioning within furniture, and plastic enclosures for appliances (Van der Veen & de Boer, 2012). Throughout the manufacturing phase, the release of PBDEs into the surrounding environment can occur via emissions and waste management practices.
- Product Use: PBDEs were added to consumer products to enhance their fire resistance. Over time, PBDEs in these products can degrade or leach out into the environment. This occurs as products age, especially when they are exposed to heat and sunlight. For example, flame-retardant-treated furniture and electronics can release PBDEs into indoor dust (Stubbings & Harrad, 2014).
- Waste and Landfills: The disposal of PBDE-containing products in landfills or incineration facilities can lead to the release of PBDEs into the surrounding environment. When these products break down in landfills, PBDEs can leach into soil and groundwater. Similarly, incineration can release PBDEs into the air and contribute to atmospheric contamination (Li et al., 2018).
- Atmospheric Deposition: PBDEs can be transported over long distances through the atmosphere. They may enter the air from indoor sources, manufacturing facilities, or waste incineration. Once in the air, PBDEs can be carried by wind and eventually be deposited onto land and water surfaces (Corsolini et al., 2019). This deposition can contaminate soils, water bodies, and vegetation.
- Wastewater and Sewage Treatment: PBDEs can enter wastewater systems through domestic and industrial sources. Although wastewater treatment can remove some PBDEs, they may still be discharged into water bodies through

effluent (Wang et al., 2013). This can lead to the contamination of aquatic environments.

- Global Transport: PBDEs are persistent organic pollutants that can undergo long-range transport. They have been detected in remote regions far from their original sources. This global transport can result in PBDE contamination in areas where their use may be limited or banned (Ockenden et al., 2003).
- Legacy Contamination: Even though some PBDE formulations have been phased out or restricted, legacy contamination remains a source of PBDEs in the environment (Hoang et al., 2021). PBDEs released into the environment in the past continue to cycle through ecosystems and may still pose risks.

Polybromodiphenyl ethers including tetra-, penta-, hexa-, and hepta BDEs (Fig. 2) possess the ability to hinder or reduce combustion in organic materials. Consequently, they are employed as flame-retardant additives. In specific regions of the world, the production of tetra- and pentaBDEs has been discontinued, with no reported production of hexa- and hepta BDEs. Recycling of articles containing these chemicals has specific exemptions until 2030 as per Stockholm Convention. The main congeners found in products are (Fang et al., 2008):

- BDE 47 – Tetrabromodiphenyl ether
- BDE 99 – Pentabromodiphenyl ether
- BDE 153 – Hexabromodiphenyl ether
- BDE 183 – Heptabromodiphenyl ether

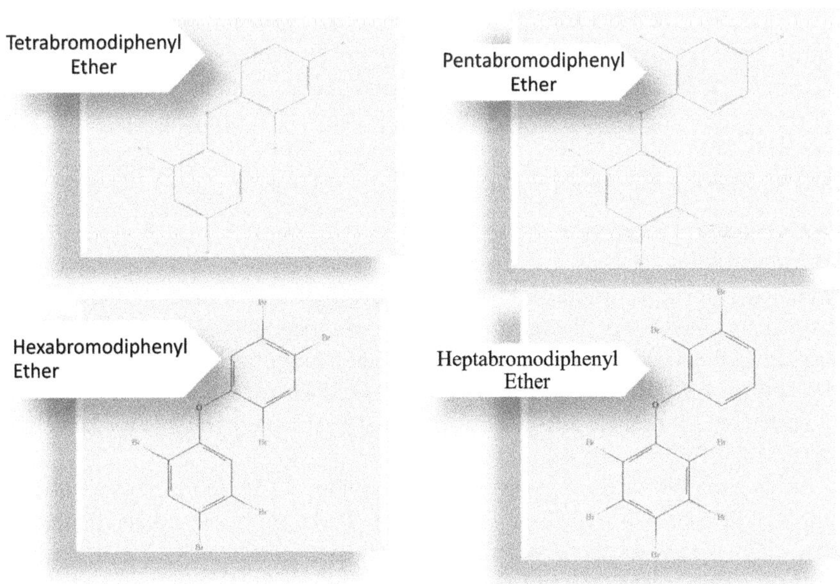

Fig. 2 Structures of tetra-, penta-, hexa-, and heptabromodiphenyl ethers (NCBI, 2023)

2 Technical Details of Tetra-, Penta-, Hexa-, and Heptabromodiphenyl Ether

The physicochemical properties of tetra-, penta-, hexa-, and hepta BDEs have been presented in detail in Table 1.

3 Status of Usage Within the Country

Production and Stockpiles
Presently, there is a lack of data pertaining to the production or stockpiles of tetra-, penta-, hexa-, and heptabromodiphenyl ether compounds in India (IPEN & POPs, 2018).

Regulation
On March 5, 2018, the Ministry of Environment, Forest, and Climate Change issued a notification. This notification, G.S.R. 207 (E), invoked sections 3 and 6 of the Environment (Protection) Act, 1986 (29 of 1986), along with rule 13 of the Environment (Protection) Rules, 1986. It effectively prohibited the manufacture, trade, use, import, and export of hexabromodiphenyl ether and heptabromodiphenyl ether (IPEN & POPs, 2018).

Exemption (Use of Chemicals for Scientific Purpose)
Without prejudice to any provisions within these regulations, the compounds delineated in regulation 2 can be utilized, distributed, or imported in quantities requisite for research and developmental pursuits at central universities, Council of Scientific and Industrial Research Laboratories, governmental institutions, or accredited laboratories within the public or private sphere. This is contingent upon obtaining approval from the Ministry of Environment, Forest, and Climate Change, Government of India (IPEN & POPs, 2018).

Production and Stockpiles
Research studies have raised concerns about the presence of Persistent Organic Pollutants (POPs) in various products, including those for children. Extensive measurements of PBDEs in diverse matrices have been documented in regions such as Sweden, Holland, Japan, the United States, and other global locations. However, the availability of PBDE measurements specific to India is notably limited, as indicated by Kumari et al. in their 2014 study. This dearth of data highlights a research gap concerning PBDEs in the Indian context.

As an example, a research investigation conducted by the International POPs Elimination Network (IPEN) in 2017, titled 'POPs recycling contaminates children's toys with toxic flame retardants', unveiled the presence of PBDEs (specifically octaBDE, decaBDE, and HCBD) within Indian toys, including well-known items like Rubik's cubes. This inquiry elucidated that the recycling of plastics

Table 1 Technical details of PBDEs (NCBI, 2023)

Physical properties and chemical Properties	Tetrabromodiphenyl ether	Pentabromodiphenyl ether	Hexabromodiphenyl ether	Heptabromodiphenyl ether
IUPAC name	2,4-dibromo-1-(2,4-dibromophenoxy) benzene	1,2,4-tribromo-5-(2,4-dibromophenoxy)benzene	1,2,4-tribromo-5-(2,4,5-tribromophenoxy)benzene	1,2,3,5-tetrabromo-4-(2,4,5-tribromophenoxy)benzene
Molecular formula	$C_{12}H_6Br_4O$	$C_{12}H_5Br_5O$	$C_{12}H_4Br_6O$	$C_{12}H_3Br_7O$
CAS no.	5436-43-1	60348-60-9	68631-49-2	207122-16-5
Molecular weight (g/mol)	485.79	564.7	643.6	722.5
Color/form	White to pale beige solid	Viscous white to amber-colored liquid	White to off-white solid	White to light yellow solid
Boiling point (°C at 760 mmHg)	398.1 ± 42.0	434.2 ± 45.0	N.A	491
Melting point (°C)	82–82.5	−5	157.6	70–150
Density (g/cm³)	2.2 ± 0.1	2.3 ± 0.1	2.502	2.643 ± 0.06
Vapor pressure mm Hg at 25 °C	2.20×10^{-6}	N.A	N.A	N.A

harboring detrimental flame-retardant compounds sourced from electronic waste resulted in the contamination of fresh plastic materials utilized in children's products.

Under the *Canadian Environmental Protection Act, 1999,* Federal Sediment Environmental Quality Guidelines (FSeQG) were developed to protect aquatic life from adverse effects of PBDEs (Table 2).

EPA Guidelines (EPA, 2017)

In accordance with the 2017 EPA guidelines, specific limitations have been established for the ingestion of main PBDE congeners via oral exposure, as illustrated in Fig. 3. Particularly, BDE 99 and BDE 47 are deemed highly concerning, with recommended daily ingestion limits not exceeding 1 milligram per kilogram of body weight. Furthermore, Fig. 4 underscores that the concentration of hexabromodiphenyl ether (PBDEs) in tap water is notably elevated, primarily attributed to their insoluble nature. This emphasizes the pivotal role of water as a key medium for PBDE exposure. However, it is essential to acknowledge that PBDE concentrations in soil also exert a significant influence on biota within ecosystems.

Concerning lower brominated PBDEs, the Agency for Toxic Substances and Disease Registry (ATSDR) has defined a minimal risk level (MRL) for inhalation exposure over an intermediate duration, set at 0.006 milligrams per cubic meter. Additionally, ATSDR has established MRLs for acute oral exposure at 6×10^{-5} milligrams per kilogram per day and for intermediate oral exposure at 3×10^{-6} milligrams per kilogram per day (ATSDR, 2016).

It is important to highlight that several states, including California, Illinois, Hawaii, Maine, Michigan, Maryland, New York, Minnesota, Oregon, Rhode Island, and Washington, have implemented bans on pentaBDE and octaBDE (EPA, 2014). Additionally, in 2006, the EPA introduced a Significant New Use Rule (SNUR) with the aim of phasing out pentaBDE and octaBDE gradually. This regulatory measure specified that no fresh production or importation of these two congeners would be permissible after January 1, 2005, unless a 90-day notice was submitted to the EPA for evaluation (EPA, 2014).

Table 2 Federal Sediment Environmental Quality Guidelines (FSeQG) for PBDEs (Tiwari et al., 2018)

Homologue		FSeQG (ng g^{-1} dw)
triBDE	Total[a]	44
tetraBDE	Total[b]	39
pentaBDE	Total[c]	0.4
pentaBDE	BDE-99	0.4
pentaBDE	BDE-100	0.4
hexaBDE	Total[d]	440
decaBDE	BDE 209	19

[a]triBDE includes only BDE-28

[b]tetraBDE includes BDE-47, BDE-66, BDE-71, BDE-75, and BDE-77

[c]pentaBDE includes BDE-85, BDE-99, and BDE-100

[d]hexaBDE includes BDE-138, BDE-153, and BDE-154

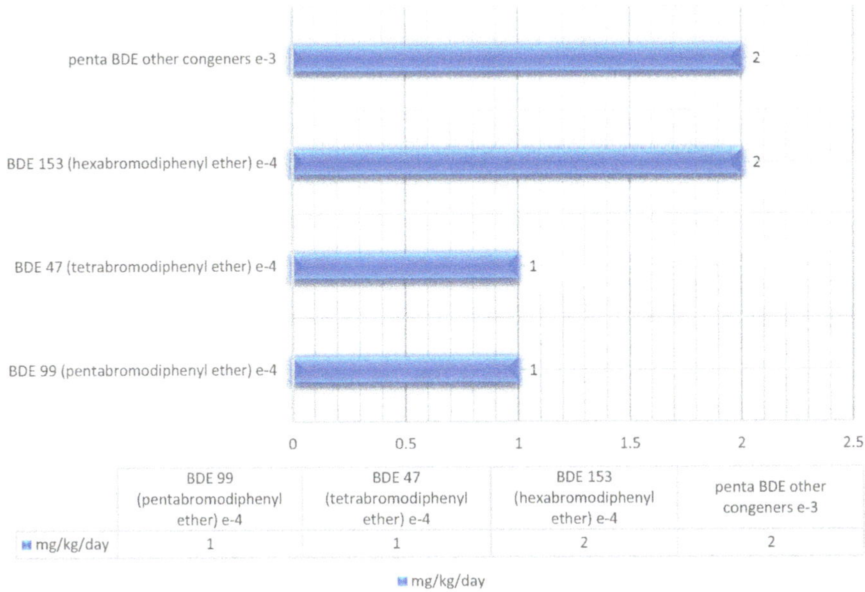

Fig. 3 Chronic oral reference doses (RfDs) as per EPA for PBDEs (ACT CW, EPA, 2017)

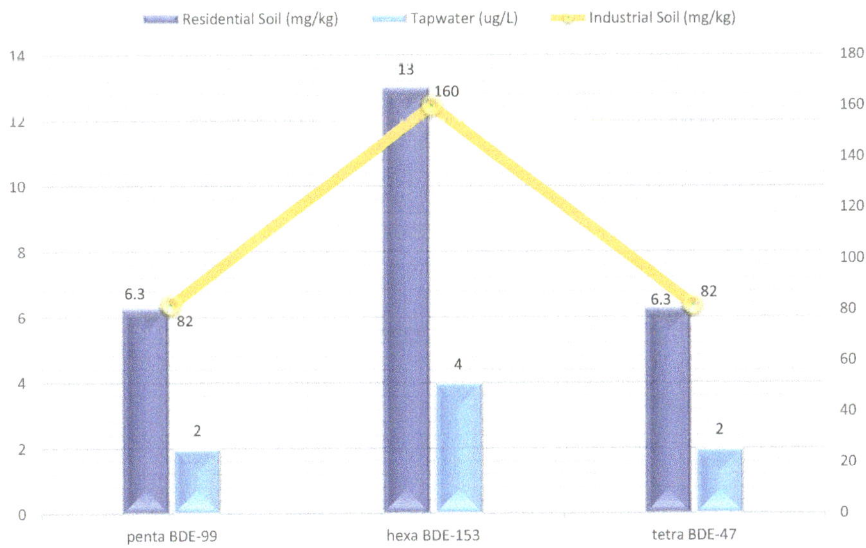

Fig. 4 Screening levels for residential soil, industrial soil, and tap water (ACT CW, EPA, 2017)

4 Adverse Effects

PBDEs are associated with a spectrum of health concerns, including disruptions to thyroid hormone function, enduring deficits in learning and memory, alterations in behavior, impairment in auditory capabilities, delays in the onset of puberty, fetal malformations, and potential carcinogenic effects (Kumari et al., 2014).

NIOSH, as part of the NOES Survey carried out during 1981 and 1983, revealed potential occupational exposure to pentabromodiphenyl ether among 975 workers in the United States, with a notable majority of 838 of these workers being female (UNEP, 2012). Occupational exposure to pentabromodiphenyl ether is primarily attributed to dermal contact in workplaces engaged in its production or utilization. Additionally, monitoring data suggest that the general populace may encounter pentabromodiphenyl ether through ambient air inhalation, consumption of food, and dermal contact with articles containing this compound (SRC). In Sweden, the primary source of human exposure appears to be fish-related food consumption (NIOSH, 2010; NCBI, 2023). It gives significant insights into the potential routes and sources of human exposure to pentabromodiphenyl ether, which are essential for understanding associated risks and implementing appropriate mitigation strategies.

Figure 5 depicts the extent of contamination of various polybrominated diphenyl ethers (PBDEs) congeners in human biological samples. Notably, BDE-47 emerged as the most prevalent congener across all tested human media, underscoring its

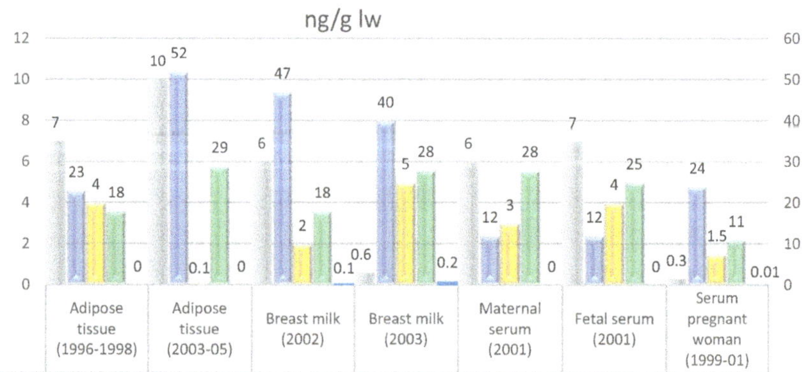

	Adipose tissue (1996-1998)	Adipose tissue (2003-05)	Breast milk (2002)	Breast milk (2003)	Maternal serum (2001)	Fetal serum (2001)	Serum pregnant woman (1999-01)
BDE-99	7	10	6	0.6	6	7	0.3
BDE-153	4	0.1	2	5	3	4	1.5
BDE-183	0	0	0.1	0.2	0	0	0.01
No. of donors	23	52	47	40	12	12	24
BDE-47	18	29	18	28	28	25	11

BDE-99 BDE-153 BDE-183 No. of donors BDE-47

Fig. 5 Median PBDE congener concentration in human media in the U.S. (Source: Mazdai et al., 2003; She et al., 2007; Johnson-Restrepo et al., 2005; Schecter et al., 2006)

significant contribution to the contamination profile. Conversely, BDE-183 exhibited the lowest presence among the PBDE congeners within human biological matrices, implying a comparatively lower impact in terms of contamination levels.

Additionally, pentaBDE is capable of being transferred from mothers to embryos and lactating infants (Hakk et al., 2002), which could have implications for early development and exposure risk (Koren et al., 2019). Once absorbed by the body, pentaBDE tends to accumulate primarily in body fat, where it can persist for extended periods, potentially lasting for years (ATSDR, 2017). In a study conducted in 2007, notably high bioaccumulation factors were identified for PBDE 47 (a tetraBDE) and PBDE 99 (a pentaBDE) in terrestrial carnivores and humans, with bioaccumulation factors reaching an impressive factor of 98 (Eriksson et al., 2001). This finding underscores the unique bioaccumulative properties of pentaBDE and its potential for long-term persistence within organisms.

Moreover, in a comprehensive investigation conducted by the World Wide Fund for Nature (WWF), it was ascertained that PBDE 153, a brominated flame-retardant compound and a constituent of penta- and octa-brominated diphenyl ether flame-retardant products, was detectable in the blood samples of all 14 ministers responsible for health and environmental affairs, representing 13 European Union member states. While the confirmed health impacts of pentaBDE in humans remain elusive, animal experimentation has raised apprehensions concerning potential deleterious effects on hepatic function, thyroid activity, and neurobehavioral development. These findings underscore the imperative for continued research and vigilant monitoring to gain deeper insights into potential health risks associated with human exposure to pentaBDE (Mir et al., 2020).

Figure 6 furnishes vital scientific data regarding the half-life durations, measured in days, of specific polybrominated diphenyl ether (PBDE) congeners within distinct environmental matrices. Two prominent congeners, BDE-47 and BDE-99, are the focal points of investigation across three diverse mediums: aerobic sediment, soil, and water. It is noteworthy that BDE-47 exhibits an extended half-life of

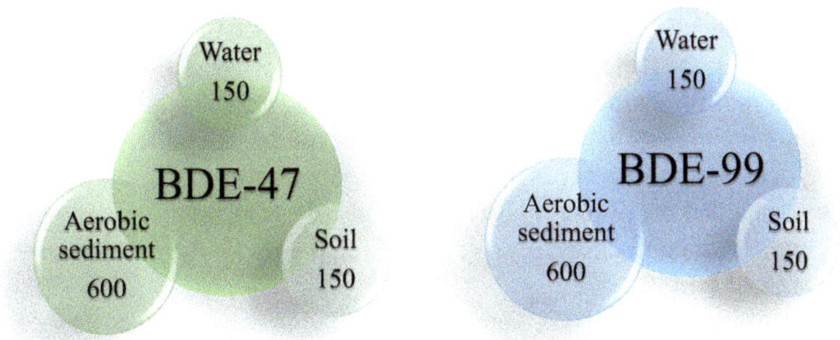

Fig. 6 Estimated total half-lives (in days) of pentaBDE in soil, sediment and water (Söderström et al., 2004)

600 days in aerobic sediment, signifying its remarkable persistence in this ecological context. Conversely, both soil and water environments demonstrate a notably shorter half-life of 150 days for BDE-47. In a similar vein, BDE-99 shares this temporal trend, featuring a 600-day half-life in aerobic sediment, with soil and water displaying equivalent 150-day half-lives. This evaluation was done by Palm (2001) with the help of Syracuse Corporation's EPIWIN program. These scientifically derived findings establish the degradation kinetics of these PBDE congeners across different environmental compartments, contributing to our comprehension of their environmental fate and potential risks.

5 In-House (Countrywide) Scientific Work Done on the Chemical

In a comprehensive investigation conducted within Thane Creek, a detailed analysis was carried out to assess the spatial and temporal distribution of 15 specific polybrominated diphenyl ether (PBDE) congeners in both grab sediment samples and sediment cores. Concentrations of total PBDE (referred to as RPBDE) in the grab sediment samples ranged from 15.98 to 132.72 Nano grams per gram (ng g^{-1}) of dry weight. Among these congeners, BDE-209 predominated, constituting a percentage ranging from 19% to 35% of the overall PBDE content.

The sediment depth analysis unveiled a complex pattern of PBDE concentration, featuring multiple modes. A conspicuous mode was particularly evident at a depth of 10 centimeters. Moreover, the examination of sediment cores furnished indications of notable PBDE usage in the vicinity over the preceding two decades. The Pb210 dating technique was applied to ascertain the sedimentation rate at Thane Creek, providing valuable insights into the sediment deposition dynamics. The composition of commercial pentaBDE (fP), octaBDE (fO), and decaBDE (fD) in the PBDE profile within Thane Creek sediments was found to be 24% ± 5%, 5% ± 1%, and 69% ± 7%, respectively, with statistically significant distinctions ($p < 0.001$).

OctaBDE, decaBDE, and HBCD are widely employed as noncandescent in the plastic enclosures of electronic devices. In cases where these flame retardants remain unremoved during recycling procedures, they exhibit persistence and can potentially contaminate newly manufactured products derived from recycled plastics. A global survey encompassing product analysis from 26 countries, including India, has unveiled significant findings. Specifically, 90% of the examined samples were found to contain octaBDE or decaBDE, and 43% of these samples exhibited the presence of HBCD.

Further examination of six samples from India corroborated these findings, revealing elevated concentrations of octaBDE and decaBDE in four of the samples. Notably, one sample displayed a notably high concentration of HBCD, arousing concerns due to the enduring environmental presence of these chemicals and their well-documented detrimental effects on reproductive health and endocrine disruption. These effects, in turn, raise apprehensions regarding potential impacts on cognitive functions such as intelligence, attention, learning, and memory (Malik et al., 2022).

6 Relevance of the International Scientific Data as per Indian Conditions

Various global studies (Fig. 7) have investigated the occurrence of PBDE congeners in diverse environmental contexts, revealing considerable variability in concentration ranges and dominant contributors.

In Russia, the work of Labunska et al. (2013) revealed PBDE concentrations ranging from 0.1 to 3.4 ng/g, with BDE 99 and 47 emerging as the primary congeners of interest. In the United Kingdom, Vane et al. (2010) reported a wider spectrum, spanning from 1 to 2645 ng/g, with BDE 209 taking precedence as the predominant congener. Moving to China, as studied by Mai et al. (2005), PBDE concentrations displayed a broad range from 0.04 to 7340 ng/g, with BDE 209 standing out as the dominant contributor.

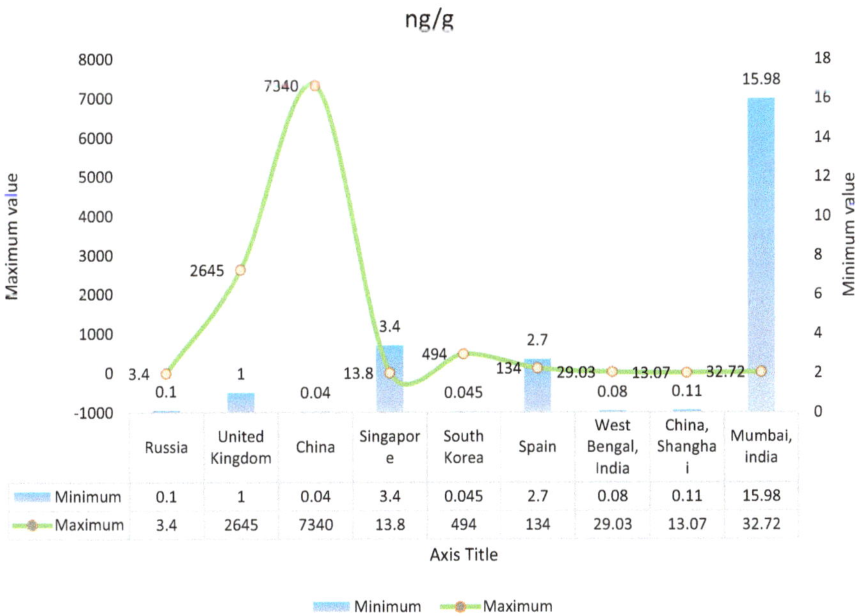

Fig. 7 Concentration of PBDEs in sediments of different regions of the world. (Source: Labunska et al., 2010; Vane et al., 2010; Mai et al., 2005; Wurl & Obbard, 2005; Moon et al., 2007; Eljarrat et al., 2005; Binelli et al., 2007; Wu et al., 2016)

Conversely, the scenario in Singapore, as analysed by Wurl et al. (2006), portrayed lower concentrations, ranging from 3.4 to 13.8 ng/g, primarily influenced by BDE 47. South Korea, subject to investigation by Moon et al. (2007), exhibited concentrations varying from 0.45 to 494 ng/g, with BDE 47, 49, 99, and 209 contributing notably. In Spain, Eljarrat et al. (2005) reported PBDE concentrations ranging from 2.7 to 134 ng/g, featuring BDE 47 and 209 as prominent congeners.

Turning to West Bengal, India (Binelli et al., 2007), PBDE concentrations ranged from 0.08 to 29.03 ng/g, with BDE 47, 185, 99, and 154 as significant contributors. Shanghai, China, studied by Wu et al. (2016), exhibited concentrations spanning between 0.11 and 13.07 ng/g, with BDE-209 emerging as the major congener of interest. Mumbai, India, displayed concentrations varying from 15.98 to 32.72 ng/g, with notable contributions from BDE 47, 99, and 209. Notably, the highest concentrations were observed in China and the United Kingdom, primarily influenced by BDE 209, while Singapore and West Bengal, India, exhibited lower concentrations primarily driven by BDE 47.

Polybrominated diphenyl ethers (PBDEs), a class of brominated flame retardants, have undergone significant regulatory phase-out initiatives, most notably the discontinuation of technical mixtures like pentaBDE and octaBDE in 2004. The utilization of decaBDE continued to a limited extent until 2013. Biomonitoring endeavors have unveiled the widespread presence of PBDEs within both the United States and global populations. Numerous biomonitoring studies have been conducted to assess human exposure to PBDEs and monitor temporal trends. While some investigations have suggested a decline in human serum concentrations of PBDEs, the persistence of this downward trajectory remains uncertain. In a research endeavor led by Emily Parry et al., a comprehensive exploration into temporal trends pertaining to pentaBDEs and their hydroxylated metabolites (OH-PBDEs) was undertaken, spanning the years 2008 to 2014. This investigation centered on a cohort of 111 ethnically diverse pregnant women residing in Northern California. The core methodology involved the collection of serum samples from the participants, which were subsequently subjected to meticulous analysis utilizing high-resolution mass spectrometry techniques. This analytical approach was employed to discern and quantify five distinct pentaBDE congeners, in addition to two OH-PBDE metabolites, offering valuable insights into the temporal dynamics of these compounds within this specific demographic.

7 Remedial Action Taken for Management of POPs

No relevant data has been found in Indian context.

The aerobic biodegradation of hexa- to nonaBDEs is generally deemed improbable, as indicated by BIOWIN estimates, suggesting their recalcitrance to biodegradation. Furthermore, a closed bottle test (OECD 301D) spanning 28 days yielded no degradation based on oxygen consumption under aerobic conditions.

Nonetheless, when subjected to anaerobic conditions, documented instances of nonaBDE 206 and 207 degradation to octaBDEs have been observed, facilitated by the use of sewage sludge inoculum, with support from multiple research studies. Predictive estimates generated via the AOPWIN model have anticipated half-life durations for reactions with atmospheric hydroxyl radicals, ranging from 30.4 to 161.0 days across the spectrum of hexa- to nonaBDE congeners. It is important to note that these projections, while informative, lack empirical validation and are thus provided as reference values.

Within the atmospheric domain, octaBDE is anticipated to exhibit substantial adsorption to suspended particles, undergoing subsequent removal through both wet and dry deposition mechanisms. Photodecomposition investigations have further illuminated this phenomenon, indicating that higher brominated DE congeners generally experience accelerated degradation in comparison to their lower brominated counterparts.

As the phased-out adoption of C-pentaBDE gains momentum in significant markets, manufacturers are actively engaged in the quest for alternative flame retardants. Notably, certain corporations, including IKEA, have initiated global cessation of C-pentaBDE utilization within their product offerings. The impetus behind this drive for alternative solutions is further reinforced by the presence of green procurement directives, instituted by both governmental entities and prominent corporate entities, which mandate the exclusion of PBDEs in electronic products. Corporations have submitted data regarding potential replacements for C-pentaBDE, as highlighted in a localized survey conducted in the United States (Washington State 2006). The cataloguing of these alternatives can be found within the UNEP/POPS/POPRC.3/INF/23 repository. It is of paramount importance to underscore that the authors have not undertaken a comprehensive inquiry into the potential human health and environmental ramifications of these substitute options.

As an illustrative example, hexabromocyclododecane, once proposed as a potential substitute for C-pentaBDE in coatings and adhesives, has elicited concerns in various geographical regions due to its inherent chemical properties. An earlier document, namely the RPA (2000) report, posited tetrabromobenzoate (TBBE) and chlorinated alkyl phosphate esters, with particular emphasis on tris(2-chloroisopropyl) phosphate (TCPP), as prime chemical candidates deserving consideration as alternatives to pentaBDE. It is incumbent upon us, however, to acknowledge that the landscape of flame-retardant alternatives may have evolved since then, with novel options possibly gaining commercial viability. Consequently, a judicious strategy involves rigorous scrutiny of manufacturing procedures, meticulous evaluation of synthetic materials, and the endorsement of alternatives that present minimal risks when contemplating flame resistance solutions.

Substitutes to C-pentaBDE in PUR Foam

The United States Environmental Protection Agency (EPA), under the auspices of its Design for the Environment initiative, conducted an extensive evaluation of potential substitutes for C-pentaBDE within polyurethane foam. This comprehensive assessment culminated in a published report in September 2005, as documented in the US EPA's 2005 publication. In a concerted effort to further investigate environmentally sustainable alternatives and bolster fire safety standards within the realm of furniture production, the EPA inaugurated the Furniture Flame Resistance Partnership, listing the active participation of a diverse array of stakeholder in this collaborative pursuit. In this collaborative initiative, prominent flame-retardant chemical manufacturers in the United States have discerned 14 chemical formulations that are regarded as feasible substitutes for C-pentaBDE in the large-scale production of low-density flexible polyurethane foam. These alternatives are characterized as drop-in replacement chemicals for C-pentaBDE, given their compatibility with the existing equipment employed in foam manufacturing facilities, thus rendering them economically prudent selections.

Beyond these 14 identified formulations, a multitude of other chemical candidates have been proposed as prospective alternatives to C-pentaBDE, catering to diverse foam types and specialized markets associated with low-density polyurethane foam. These compounds are employed with a specific purpose: to confer flame-retardant properties to high-density, flexible polyurethane foam.

The general cited alternatives to C-pentaBDE include the following:

- Melamine: Flame retardants formulated with melamine as the foundational compound are currently deployed across diverse applications, encompassing flexible polyurethane foams, intumescent coatings renowned for their expansion under thermal exposure to confer flame resistance, polyamides, and thermoplastic polyurethanes. Notably, their efficacy has been substantiated within the European context, particularly in the realm of high-density flexible polyurethane foams. It is worth noting that their effective utilization frequently necessitates a notable proportion, typically ranging between 30% to 40% by weight of the polyol, of melamine.
- Tris (1,3-dichloro-2-propyl) phosphate (TDCPP): It is a chlorinated phosphate ester and frequently features in polyurethane foam formulations. Its application extends to high-density foam scenarios, and it has found utility in low-density foam applications where the foremost concern is not discoloration (light scorching).
- Ammonium Polyphosphate (APP): This flame-retardant additive finds application in conferring flame resistance to both flexible and rigid polyurethane foams, as well as in intumescent laminations, molding resins, sealants, and adhesives. Nevertheless, chemical producers and trade associations within the foam-manufacturing sector do not generally regard it as a widespread substitute for C-pentaBDE.

These assessments and discussions highlight ongoing efforts to find environmentally safer alternatives to C-pentaBDE in the production of polyurethane foam, balancing the need for fire safety with environmental and health considerations.

Nonchemical Substitutes to C-pentaBDE in PUR Foam

The United States Environmental Protection Agency (EPA) has recognized nonchemical substitutes for flame resistance in 2005. These alternatives encompass a range of technologies for making furniture flame-resistant. Three noteworthy alternatives currently available include the following:

- Barrier Technologies: Barrier technologies have been employed to create flame-resistant barriers within furniture. Notably, these are widely used in the manufacturing of mattresses rather than residential upholstered furniture. However, they may have the potential for broader applications.
- Graphite Impregnated Foam: Graphite impregnated foam is another alternative, although its commercial use has been somewhat limited.
- Surface Treatment: Surface treatment methods have been explored to enhance the flame resistance of furniture materials. Like graphite-impregnated foam, these treatments have seen limited commercial adoption.

These non-chemical alternatives offer potential solutions for reducing or eliminating the reliance on traditional chemical flame retardants in furniture, addressing concerns related to the environmental and health impacts of such chemicals.

Substitutes to C-pentaBDE in EE-Appliances and Textiles

By mid-November 2005, several prominent manufacturers had commenced the phased removal of all PBDEs from their product lines. This proactive approach was driven by the imperative to align with the European Union's regulatory ban on hazardous substances within electrical and electronic appliances, notably including C-pentaBDE among the restricted compounds. In the context of global electrical and electronic appliance production, a substantial segment, comprising approximately 35%, foresaw a marginal price increment of less than 5% for their product lines. Concurrently, an additional 23% of these manufacturers projected a moderate price escalation spanning the range of 5–10%. A smaller subset, approximately 6% of producers, anticipated more substantial price escalations exceeding 10%, as reported in Environmental International Reporter in 2006.

Numerous instances of alternative flame-retardant methodologies were already in practice among these manufacturers. These alternatives encompassed the following:

- The adoption of bromine-free circuit boards by one leading company.
- The utilization of phosphorus-based flame retardants for printed circuit boards.
- The integration of flame-resistant plastics in their products.
- The incorporation of halogen-free materials and low-voltage internal wires, as exemplified by one company, as per findings from the Norwegian EPA in 2003.

The use of alternative flame retardants such as zinc borate, magnesium hydroxide, or expandable graphite could be implemented without causing significant problems, providing viable options to replace brominated flame retardants (Kausar et al., 2016).

There are alternatives to bromine-based flame retardants that are bromine-free and can be used in textiles. However, it is important to note that not all of these alternatives are environmentally friendly. For instance, substances like antimony trioxide and borax, which have been used as flame retardants, are known to raise environmental concerns.

Conversely, alternative sustainable and resilient flame-retardant materials, such as wool and polyester fibers, are readily accessible. These materials offer flame resistance without relying on the use of potentially harmful chemicals. Certain manufacturers contend that the enforcement of a prohibition on the utilization of C-pentaBDE in textiles might present challenges in preserving equivalent standards of quality and longevity in textile products. This perspective highlights the ongoing debate surrounding the trade-offs between flame resistance, product quality, and environmental impact in the textile industry.

8 Recommendations

At present, there is a lack of accessible information pertaining to the production or stockpiling of hexabromodiphenyl ether and heptabromodiphenyl ether within India. Additionally, there is no reported production or stockpile data for pentaBDE in the Indian context. Research studies have highlighted a concerning issue of cross-contamination of Persistent Organic Pollutants (POPs) in various products, including those intended for children's use. This cross-contamination involving banned chemicals raises significant concerns due to its potential impact on human health.

Moreover, it is essential to emphasize the need for further research to investigate the presence of polybrominated diphenyl ethers (PBDEs), including hexa- and heptabromodiphenyl ethers, in various matrices such as seawater, fish, and other biota. Such studies are valuable for gaining a deeper understanding of the fate and distribution of these chemicals in the environment.

References

Act, C. W. (2017). United States Environmental Protection Agency. *Appendix A to, 40.*

ATSDR. (2016). Agency for toxic substances and disease registry's priority list of hazardous substances. Retrieved from http://www.atsdr.cdc.gov/SPL/index.html

ATSDR. (2017). *Toxicological profile for polybrominated diphenyl ethers* (PBDEs). U.S. Department of Health and Human Services, Public Health Service, Agency of Toxic Substances and Disease Registry. March 2017. Atlanta, Georgia: ATSDR. Available at: https://www.atsdr.cdc.gov/toxprofiles/tp207.pdf

Binelli, A., Sarkar, S. K., Chatterjee, M., Riva, C., Parolini, M., deb Bhattacharya, B., et al. (2007). Concentration of polybrominated diphenyl ethers (PBDEs) in sediment cores of Sundarban mangrove wetland, northeastern part of Bay of Bengal (India). *Marine Pollution Bulletin, 54*(8), 1220–1229.

Corsolini, S., Baroni, D., Martellini, T., Pala, N., & Cincinelli, A. (2019). PBDEs and PCBs in terrestrial ecosystems of the Victoria Land, Antarctica. *Chemosphere, 231*, 233–239.

Eljarrat, E., De La Cal, A., Larrazabal, D., Fabrellas, B., Fernandez-Alba, A. R., Borrull, F., et al. (2005). Occurrence of polybrominated diphenylethers, polychlorinated dibenzo-p-dioxins, dibenzofurans and biphenyls in coastal sediments from Spain. *Environmental Pollution, 136*(3), 493–501.

EPA, U. (2017). *Technical fact sheet-Polybrominated diphenyl ethers (PBDEs).* United States Environmental Protection Agency.

Eriksson, U., Jakobsson, E., Fredriksson, A., & Bjorklund, C. (2001). Brominated flame retardants-a novel class of persistent environmental pollutants. *Journal of Toxicology and Environmental Health, Part A, 63*(3), 1–18.

Fang, L., Huang, J., Yu, G., & Wang, L. (2008). Photochemical degradation of six polybrominated diphenyl ether congeners under ultraviolet irradiation in hexane. *Chemosphere, 71*(2), 258–267.

Hakk, H., Larsen, G., & Klasson-Wehler, E. (2002). Tissue disposition, excretion and metabolism of 2, 2', 4, 4', 5-pentabromodiphenyl ether (BDE-99) in the male Sprague-Dawley rat. *Xenobiotica, 32*(5), 369–382.

Hoang, A. Q., Aono, D., Watanabe, I., Kuwae, M., Kunisue, T., & Takahashi, S. (2021). Contamination levels and temporal trends of legacy and current-use brominated flame retardants in a dated sediment core from Beppu Bay, southwestern Japan. *Chemosphere, 266*, 129180.

IPEN, & POPs. (2006). Country situation on persistent organic pollutants (POPs) in India, New Delhi, India.

IPEN, & POPs. (2018). Persistent organic pollutants (POPs) in India: Country situation report.

Johnson-Restrepo, B., Kannan, K., Rapaport, D. P., & Rodan, B. D. (2005). Polybrominated diphenyl ethers and polychlorinated biphenyls in human adipose tissue from New York. *Environmental Science & Technology, 39*(14), 5177–5182.

Kausar, A., Rafique, I., Anwar, Z., & Muhammad, B. (2016). Recent developments in different types of flame retardants and effect on fire retardancy of epoxy composite. *Polymer-Plastics Technology and Engineering, 55*(14), 1512–1535.

Koren, G., Carnevale, A., Ling, J., Ozsarfati, J., Kapur, B., & Bagli, D. (2019). Fetal exposure to polybrominated diphenyl ethers and the risk of hypospadias: focus on the congeners involved. *Journal of Pediatric Urology, 15*(4), 405–e1.

Kumari, K. (Ed.). (2021). *Persistent organic pollutants: Gaps in management and associated challenges* (1st ed.). CRC Press. https://doi.org/10.1201/9781003046806

Kumari, K., Sharma, J. K., Kanade, G. S., Kashyap, S. M., Juwarkar, A. A., & Wate, S. R. (2014). Investigation of polybrominated diphenyl ethers in old consumer products in India. *Environmental Monitoring and Assessment, 186*, 3001–3009.

Labunska, I., Brigden, K., Santillo, D., Kiselev, A. & Johnston, P. (2010). *Russian Refuse #2: an update on PBDEs and other contaminants detected in St-Petersburg area, Russia*. Greenpeace Research Laboratories Technical Note 04/2010.

Labunska, I., Harrad, S., Santillo, D., Johnston, P., & Yun, L. (2013). Domestic duck eggs: An important pathway of human exposure to PBDEs around e-waste and scrap metal processing areas in Eastern China. *Environmental Science & Technology, 47*(16), 9258–9266.

Li, T. Y., Zhou, J. F., Wu, C. C., Bao, L. J., Shi, L., & Zeng, E. Y. (2018). Characteristics of polybrominated diphenyl ethers released from thermal treatment and open burning of e-waste. *Environmental Science & Technology, 52*(8), 4650–4657.

Mai, C., Luo, C., Yang, S., et al. (2005). Distribution of polybrominated diphenyl ethers in sediments of the Pearl River Delta and adjacent South China Sea. *Environmental Science & Technology, 39*(10), 3521–3527.

Malik, J. K., Aggarwal, M., Kalpana, S., & Gupta, R. C. (2022). Chlorinated hydrocarbons and pyrethrins/pyrethroids. In *Reproductive and developmental toxicology* (pp. 641–664). Academic Press.

Mazdai, M., Nohynek, L. J., Miettinen, J., & Peltonen, K. (2003). Polybrominated diphenyl ethers (PBDEs) in human milk and serum in Finland. *Environmental Health Perspectives, 111*(11), 1363–1368.

Mir, R. H., Sawhney, G., Pottoo, F. H., Mohi-Ud-Din, R., Madishetti, S., Jachak, S. M., et al. (2020). Role of environmental pollutants in Alzheimer's disease: A review. *Environmental Science and Pollution Research, 27*, 44724–44742.

Moon, H. B., Kannan, K., Choi, M., & Choi, H. G. (2007). Polybrominated diphenyl ethers (PBDEs) in marine sediments from industrialized bays of Korea. *Marine Pollution Bulletin, 54*(9), 1402–1412.

National Center for Biotechnology Information. (2023). PubChem Annotation Record for, Pentabromodiphenyl ethers, Source: Hazardous Substances Data Bank (HSDB). Retrieved 12 Oct 2023 from https://pubchem.ncbi.nlm.nih.gov

NIOSH, & NOES. (2010, November 15). National Occupational Exposure Survey conducted from 1981–1983. Estimated numbers of employees potentially exposed to specific agents by 2-digit standard industrial classification (SIC). http://www.cdc.gov/noes/

Ockenden, W. A., Breivik, K., Meijer, S. N., Steinnes, E., Sweetman, A. J., & Jones, K. C. (2003). The global re-cycling of persistent organic pollutants is strongly retarded by soils. *Environmental Pollution, 121*(1), 75–80.

RPA (2000). Risk reduction strategy and analysis of advantages and drawbacks for pentabromodiphenyl ether. Stage 4 Report. Risk & Policy Analysis Limited, London

Palm, A. (2001). The environmental fate of polybrominated diphenyl ethers in the centre of Stockholm-Assessment using a multimedia fugacity model. *Sweden: Master of Science thesis. IVL B 1400*, 1–53. https://www.ivl.se/download/18.694ca0617a1de98f473b88/1628417773458/FULLTEXT01.pdf

Schecter, A., Johnson-Welch, S., Tung, K. C., Harris, T. R., Päpke, O., & Rosen, R. (2006). Polybrominated diphenyl ether (PBDE) levels in livers of US human fetuses and newborns. *Journal of Toxicology and Environmental Health, Part A, 70*(1), 1–6.

She, J., Holden, A., Sharp, M., Tanner, M., Williams-Derry, C., & Hooper, K. (2007). Polybrominated diphenyl ethers (PBDEs) and polychlorinated biphenyls (PCBs) in breast milk from the Pacific Northwest. *Chemosphere, 67*(9), S307–S317.

Söderström, G., Sellström, U., de Wit, C. A., & Tysklind, M. (2004). Photolytic debromination of decabromodiphenyl ether (BDE 209). *Environmental Science & Technology, 38*(1), 127–132.

Stubbings, W. A., & Harrad, S. (2014). Extent and mechanisms of brominated flame retardant emissions from waste soft furnishings and fabrics: A critical review. *Environment International, 71*, 164–175.

Tiwari, M., Sahu, S. K., Bhangare, R. C., Ajmal, P. Y., & Pandit, G. G. (2018). Polybrominated diphenyl ethers (PBDEs) in core sediments from creek ecosystem: Occurrence, geochronology, and source contribution. *Environmental Geochemistry and Health*. https://doi.org/10.1007/s10653-018-0125-9

UNEP (United Nations Environment Programme). (2012). *The Stockholm Convention on Persistent Organic Pollutants (POPs). Persistent Organic Pollutants: A Global Issue*. Chapter 5: Polybrominated Diphenyl Ethers (PBDEs), pp. 106–129

U.S. EPA. (2014). Report on the 2013 U.S. Environmental Protection Agency (EPA) International Decontamination Research and Development Conference. Research Triangle Park, NC, November 05–07, 2013. U.S. Environmental Protection Agency, Washington, DC, EPA/600/R-14/210.

Van der Veen, I., & de Boer, J. (2012). Phosphorus flame retardants: Properties, production, environmental occurrence, toxicity and analysis. *Chemosphere, 88*(10), 1119–1153.

Vane, C. H., Ma, Y. J., Chen, S. J., & Mai, B. X. (2010). Increasing polybrominated diphenyl ether (PBDE) contamination in sediment cores from the inner Clyde Estuary, UK. *Environmental Geochemistry and Health, 32*, 13–21.

Wang, X., Xi, B., Huo, S., Sun, W., Pan, H., Zhang, J., et al. (2013). Characterization, treatment and releases of PBDEs and PAHs in a typical municipal sewage treatment plant situated beside an urban river, East China. *Journal of Environmental Sciences, 25*(7), 1281–1290.

Wu, J., Xie, Y., Fang, Z., Cheng, W., & Tsang, P. E. (2016). Effects of Ni/Fe bimetallic nanoparticles on phytotoxicity and translocation of polybrominated diphenyl ethers in contaminated soil. *Chemosphere, 162*, 235–242.

Wurl, O., & Obbard, J. P. (2005). Organochlorine pesticides, polychlorinated biphenyls and polybrominated diphenyl ethers in Singapore's coastal marine sediments. *Chemosphere, 58*(7), 925–933.

Wurl, O., Lam, P. K. S., & Obbard, J. P. (2006). Occurrence and distribution of polybrominated diphenyl ethers (PBDEs) in the dissolved and suspended phases of the sea-surface microlayer and seawater in Hong Kong, China. *Chemosphere, 65*(9), 1660–1666.

Hexachlorobutadiene (HCBD)

Kanchan Kumari and Deepak Marathe

1 Introduction

Hexachlorobutadiene (HCBD) is classified as a halogenated aliphatic compound, commonly generated as a secondary product in the production process of chlorinated aliphatic compounds. The official categorization of the compound is referred to as "halogenated aliphatic." The chemical molecule HCBD exhibits characteristics like environmental persistence, bioaccumulation potential, and significant detrimental effects on aquatic and terrestial organisms. The substance in question possesses the capacity to induce substantial detrimental impacts on both human health and the environment in the event of long-distance transportation. Moreover, it has been classified as a potential human carcinogen with uncertain occurrence.

The use of this material as a solvent for various chlorine-containing compounds is prevalent in the chemical industry. Hexachlorobutadiene is generated as a byproduct during the chlorinolysis process of butane derivatives in the manufacturing of carbon tetrachloride and tetrachloroethene. Consequently, the aforementioned process leads to the generation of a secondary chemical compound commonly referred to as hexachlorobutadiene. Due to the substantial magnitude of production for both of these commodities, there tends to exist an ample supply of HCBD to adequately fulfill the demands of the industrial sector (Gaonkar et al., 2022).

K. Kumari (✉)
CSIR-National Environmental Engineering Research Institute (NEERI), Kolkata Zonal Center, Kolkata, West Bengal, India
e-mail: k_kumari@neeri.res.in

D. Marathe
CSIR-National Environmental Engineering Research Institute (NEERI), Nagpur, India

K. Kumari, *Pollutants of Global Concern*, Emerging Contaminants and Associated Treatment Technologies, https://doi.org/10.1007/978-3-031-50996-4_17

2 Technical Details of Hexachlorobutadiene

The various properties of HCBD such as boiling point, solubility, octanol water, and air partition coefficient have been described in detail in Table 1.

Table 1 Physicochemical properties of hexachlorobutadiene (PubChem, n.d.)

Details	Description (hexachlorobutadiene)
Structure	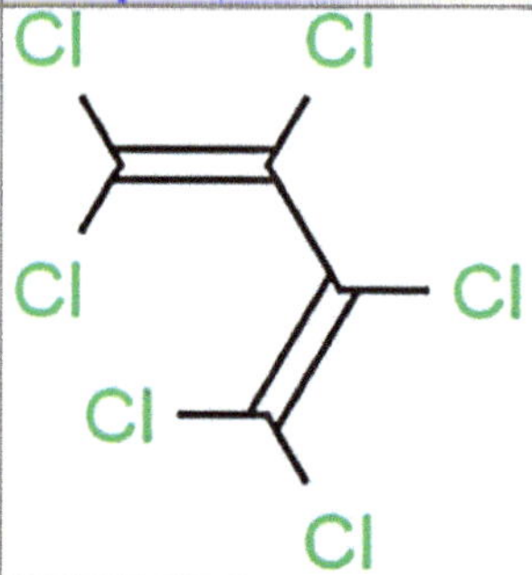
IUPAC name	1,1,2,3,4,4-hexachlorobuta-1,3-diene
Molecular formula	C_4Cl_6 or $CCl_2{=}CClCCl{=}CCl_2$
CAS no.	87-68-3
Molecular weight	260.8 g/mol
Color/form	Clear, colorless liquid
Odor	Mild, turpentine-like odor
Boiling point	419 °F at 760 mm Hg
Melting point	−6 °F
Flash point	90 °C
Solubility	Less than 0.1 mg/mL at 72 °F
Density	1.675 at 59.9 °F
Vapor pressure	0.15 mm Hg at 68 °F
Octanol/water partition coefficient	4.78 (LogP)
Octanol-air partition coefficient	–
Henry's law constant	0.01 atm-m^3/mole
Decomposition	When heated to decomposition, it emits very toxic fumes of chlorine

3 Status of Usage Within the Country and Emission of HCBD

Hexachlorobutadiene, often known as HCBD, is generated as a secondary product in the manufacturing procedures of carbon tetrachloride and tetrachloroethane. The aforementioned byproduct is sometimes known as a secondary product. The chlorinolysis process, occurring during plant metabolism, leads to the generation of the aforementioned byproduct. Chlorinolysis is a chemical procedure involving the exposure of hydrocarbons to chlorine gas inside a pyrolytic setting. Consequently, the hydrocarbons initiate a cascade of radical reactions. In India, several companies were engaged in the manufacturing of carbon tetrachloride. Notable examples are Gujarat Alkalies & Chemicals Limited located in Dahej, Adarsh Chemicals based in Chennai, and Nutan Chemicals situated in Pune (Toxic Link, 2018).

Release of HCBD

No data has been documented on the dissemination of Hexachlorobutadiene in India (Toxic Link, 2018).

Regulation

The Ministry of Environment, Forest, and Climate Change implemented a regulation on March 5, 2018 (Table 3), which prohibits the use of HCBD, as indicated in an official declaration (Marathe et al., 2021). The execution of this notice, referred to as G.S.R. 207 (E), was carried out in compliance with section 3 and section 6 of the Environment (Protection) Act, 1986 (29 of 1986), together with rule 13 of the Environment (Protection) Rules, 1986. Consequently, a legislative action was undertaken to enforce limitations on the production, commercial activities, use, importation, and exporting of HCBD. The implementation of an exemption clause serves the purpose of being applicable in situations where chemicals are utilized within the framework of conducting research. This clause permits the utilization, commercialization, or importation of chemicals specified in regulation 2, in required quantities for the objectives of research and development activities conducted in Central Universities, Council of Scientific and Industrial Research Laboratories, government institutions, other research institutions, or accredited laboratories in either the public or private sectors. In order to be eligible for this exemption, it is necessary to acquire authorization from the Ministry of Environment, Forest, and Climate Change in India, an entity that operates within the purview of the Indian government. The NIOSH and EPA guidelines for HCBD's occupational exposure in air and drinking water respectively have been outlined in Table 2. Table 3 represents the usages, sources and regulation of POP (HCBD) in India which was banned for manufacture, use , import and export since March, 2018. It was listed in POPs category in Stockholm Convention in the year 2015 under Annexure A & C (Kumari, 2021).

Table 2 Standards and guidelines for HCBD (ATSDR)

Matrix	Agency	Age group	Limits
Air (occupational exposure)	NIOSH	Adult	0.02 ppm in air for an 8-h workday over a 40-h work week
Drinking water	EPA	Adult	<0.4 mg/L for longer periods (7 years)
		Children	<0.1 mg/L for longer periods (7 years)

Table 3 Uses, sources of POPs, and regulation in India (Kumari, 2021)

Identified POP	Uses and sources	Annex	Year of listing by SC	Regulation in India
Hexachlorobutadiene (HCBD)	Industrial chemical, used as solvent, in making of rubber and lubricants; also produced unintentionally	A & C	2015	Banned for manufacture, use, import, export, etc. since March 5, 2018

Usages of HCBD (Toxic Link, 2015)

HCBD is a chemically synthesized molecule classified within the chlorinated hydrocarbons category. The formation of a byproduct can occur during the manufacture and combustion processes of certain chlorinated chemicals, such as heptachlor. Nevertheless, it is worth noting that there is currently no deliberate production or substantial use of this substance in industrial or commercial contexts. The compound HCBD has garnered attention and regulatory measures due to its recognized toxicity and long-lasting presence in the environment. The substance HCBD is classified as a persistent organic pollutant (POP) and is subject to regulation under the Stockholm Convention, a global agreement designed to manage and eradicate the production and use of persistent organic pollutants. The accumulation of HCBD in the environment has the potential to have detrimental effects on ecosystems and human health. Due to its classification as a byproduct and its detrimental characteristics, HCBD lacks any valid or endorsed use. This chemical is commonly seen as one that should be actively avoided and subject to regulation to mitigate its potential release into the environment. In the event of suspected exposure to HCBD or concerns regarding its presence, it is imperative to seek guidance from pertinent environmental agencies or specialists to guarantee appropriate management and disposal procedures. Various usages of hexachlorobutadiene are represented in Fig. 1. HCBD continue to be inadvertently discharged by industrial operations, including instances occurring during waste disposal practices. The pertinent sources encompass the manufacturing of chlorinated hydrocarbons, the synthesis of magnesium, and incineration procedures.

HCBD in the Environment

HCBD, have been detected in many biotic and abiotic settings, including surface waters, drinking water, ambient air, aquatic organisms, terrestrial fauna, and even

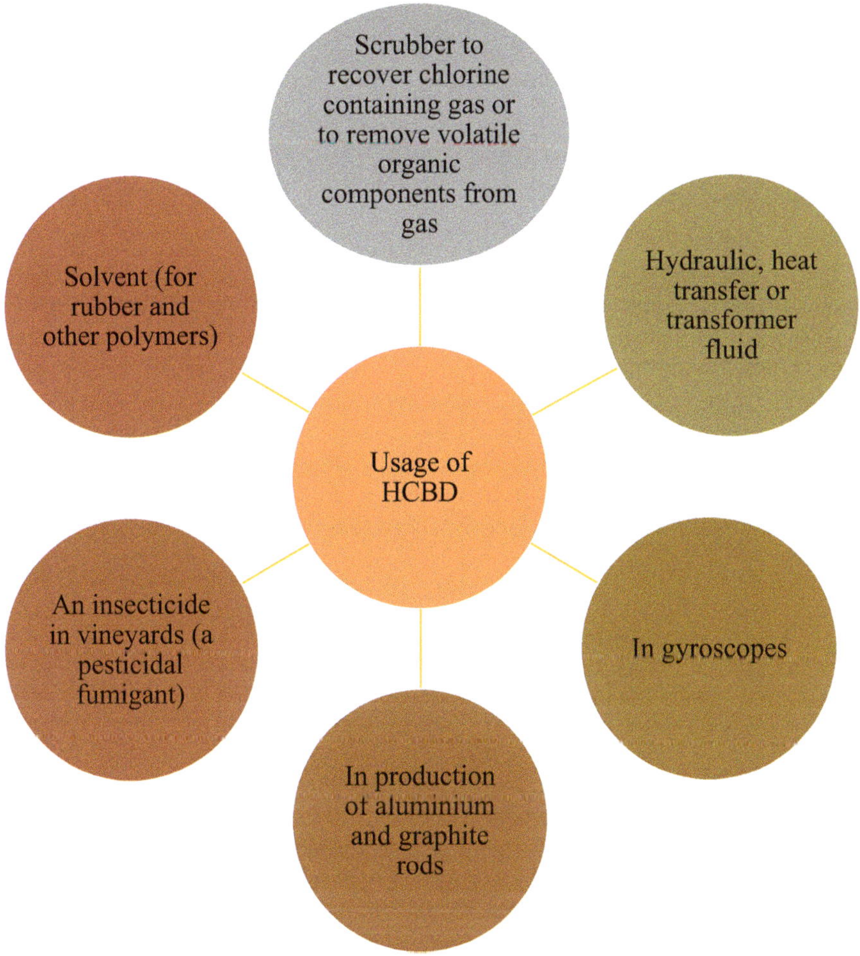

Fig. 1 Usage of heptachlorobutadiene

geographically isolated areas like the Arctic. The chemical, HCBD have demonstrated a diverse array of biological and chemical impacts, encompassing the capacity to impede the proliferation of bacteria, fungus, and yeast. HCBD have been linked to a range of potential health risks, such as an elevated risk of cancer, diabetes, and cardiovascular disease. The half-lives of HCBD in aquatic settings are expected to vary in duration, ranging from 3 days to 1 year. Projected half-life of aerobic biodegradation in aquatic conditions indicates that the anticipated durations of breakdown in soil might vary between 4 and 26 weeks. The periods of degradation can occur at any given moment. It is fair to expect that these phases of degradation will persist throughout time. Current available data on the half-life of sediment are inadequate to fully understand the role of sediments as a reservoir for HCBD. Nevertheless, sediments serve as a storage medium for HCBD. Although sediments have the potential to serve as a reservoir for HCBD, this assertion holds true. Various literature has demonstrated that HCBD, as documented in a previous study (NEW POPs 2015), had the ability to migrate from its disposal sites and disperse throughout various regions globally.

4 Health Effects of HCBD

Previous investigations have demonstrated that HCBD had the potential to induce systemic toxicity in those who encounter it by oral, inhalation, or cutaneous routes of exposure. The potential consequences encompass hepatic steatosis, necrotizing nephritis impacting epithelial tissue, arterial hypotension, myocardial dystrophy, thoracic discomfort, abnormalities in the upper respiratory tract, disturbances in sleep patterns, hand tremors, sensations of nausea, impaired olfactory perception, depression of the central nervous system, and cyanosis (Toxic Link, 2015).

The classification of HCBD's carcinogenicity by United States Environmental Protection Agency is designated as a Possible Human Carcinogen under Group C.

4.1 Status of HCBD in Different Countries

The use and production of this substance have been ceased in the UNECE member states, and there is a lack of accessible data about its current adoption in non-UNECE countries. Both Canada and Japan have implemented legislation that criminalizes the use of HCBD. The intentional manufacturing of the aforementioned drug has been ceased throughout the whole UNECE territory, encompassing both the United States (where manufacture ceased approximately in 1970) and Canada. During the latter part of the 1970s, the European-based Corporation responsible for the production of the aforementioned commodity terminated its activities. Nevertheless, it is crucial to acknowledge that empirical research conducted in China and Taiwan has substantiated the ongoing presence of industrial activity in contemporary times (Toxic Link, 2015).

As per the screening and response level (NPRPS, 2015), the level of hexachlorobutadiene is 5000 mg/kg as mentioned HW rules (2008) (Schedule II); however, the level of hexachlorobutadiene is not reported for screening and response level in different sectors such as soil (agriculture, resident, commercial, and industrial), groundwater, and surface water.

5 In-House (Countrywide) Scientific Work Done on the Chemical

No relevant information found.

5.1 *Relevance of the International Scientific Data as per Indian Conditions*

The study conducted by Tang et al. (2016) examines the occurrence of organochlorine pesticides (OCPs) and hexachlorobutadiene (HCBD) in the soil, flora, and terrestrial animals in a designated area located in Southwest China. The region exhibits a significant historical association with the manufacturing of pesticides. The researchers directed their inquiry on a specific geographic region that has historically served as a site for pesticide production. The study conducted revealed an absence of consistent distribution patterns in HCBD concentrations across different environmental compartments. The investigation unveiled diverse levels of HCBD concentrations across many environmental compartments. Obtained values of soil samples varied between 0.02 and 5.59 ng g^{-1} dw. Within the realm of plant biology, the observed quantities exhibited a wide range, spanning from 0.03 to 24.6 nanograms per gram of dry weight. The contents in animals varied between 1.65 and 3.80 ng g^{-1} lw. The aforementioned information was derived from studies conducted on soil, plant, and animal samples, respectively. The recorded measurements were rather low and did not exceed any preset criteria that would have resulted in discernible adverse effects on organisms. The concentrations of HCBD detected in the soil samples examined in this investigation were significantly lower in comparison to the recorded levels observed in soils from sites associated with the manufacturing of perchloroethylene and trichloroethylene in the United States. The United States Environmental Protection Agency (USEPA) collated and disseminated the pertinent data in 1976.

Zhang et al. (2018) highlight the limited accessibility of data pertaining to the unintentional release of HCBD from various sources, with a particular focus on its emission in the context of waste incineration. The precise detection and quantification of HCBD levels in fly ash have significant significance. Fly ash is a notable byproduct that is produced via the process of trash incineration. This paper presents a scientific approach for quantifying HCBD in fly ash, a residual substance generated by the incineration of waste items. The technology encompasses the utilization of ultrasonic extraction in conjunction with a purification methodology employing a column filled with silica gel and Florisil. This represents a vital component of the operation. Subsequently, the gathered specimen is subjected to gas chromatography-mass spectrometry analysis with the purpose of identifying any existing compounds. This approach is utilized to ascertain the presence or absence of the target material. Zhang et al. (2018) examined six fly ash samples which are collected from various waste incineration plants situated in the eastern region of China. The findings of the investigation revealed that the measured quantities of HCBD in the samples exhibited a range of 1.39–97.6 nanograms per gram (ng g^{-1}). The measured values exhibited similarities to the concentrations of pentachlorobenzene (PeCB), ranging from 1.22 to 150 ng g^{-1}, and hexachlorobenzene (HCB), ranging from 0.82 to 120 ng g^{-1}. The aforementioned findings underscore the need of considering the residual presence of HCBD, PeCB, and HCB in fly ash generated from the incineration of waste.

The study's findings offer more data indicating that the incineration of waste in China inadvertently results in the occurrence of HCBD. The offered content is insufficient in terms of information and contextual details, hence limiting its potential for academic improvement (Zhang et al., 2018).

Concerns have been raised over the possible implications of HCBD on Arctic ecosystems due to its high toxicity and ability to disperse across great distances. The presence of HCBD in the Arctic environment was detected, prompting Balmer et al. to conduct a comprehensive analysis that synthesized all existing information about its prevalence. The scope of this investigation encompassed several subjects, such as atmospheric conditions, terrestrial habitats, freshwater systems, and marine ecosystems, along with the organisms inhabiting these environments. HCBD is a chemical compound that has been detected in several Arctic environmental media. However, there is a limited availability of published literature pertaining to its presence in the region. The quantification of HCBD in air samples from the Arctic region was conducted with monitoring stations located in Finland and Canada for the purpose of sample collection. Nevertheless, there is a lack of available information about the presence of HCBD in several non-living environmental matrices, including soils, sediments, glacier ice, freshwaters, and seas. This phenomenon arises due to the inherent inability of these matrices to sustain biological existence. The biota inhabiting both terrestrial and marine environments in the Arctic region has been shown to exhibit low levels of HCBD. This observation aligns with previous laboratory studies that suggest HCBD possesses the potential for bioaccumulation but not bio magnification. Evidence indicates that terrestrial birds, mammals, and seabirds exhibit elevated concentrations of HCBD in their organisms relative to fish and other marine organisms. This phenomenon is evidenced by the observation that fish and marine organisms have comparatively reduced levels. Therefore, further investigation is required in relation to this issue. Despite the existing dearth of information on the geographical and temporal distribution of HCBD levels in the Arctic region, it is imperative to undertake future monitoring efforts to assess HCBD concentrations in this area. The monitoring process will have a significant role in assessing the effects of the recently implemented worldwide limitations in line with the Stockholm Convention on Persistent Organic Pollutants (POPs) (Balmer et al., 2019).

In their study, Juang et al. (2009) successfully detected the presence of HCBD in the atmospheric environment above the Fong Shan stream located in Taiwan. The researchers presented the results of their investigation to the assembled audience. The maximum observed concentration recorded was 716.517 g/m^3, but the mean concentration was estimated to be 225.844 g/m^3. Air samples were collected from a manufacturing site located in China, which was found to be a source of polychlorinated ethylene (PCE) emissions. Two alternative methodologies were employed to do the task of quantifying the concentration of HCBD present in the environment. One of the ways employed was the GC-MS method, while the other strategy utilized the GC-MS/MS method. Nevertheless, the acetylene approach proposed a concentration of 1170 grams per cubic meter in the vicinity directly adjacent to the facility. The CTC procedure yielded a higher concentration, measuring 5530 g/m^3, whereas the acetylene methodology suggested a concentration of 1170 g/m^3. The regions situated in the downwind direction exhibited the most elevated levels of the pollutant, as determined by the studies conducted by Wang et al. (2018) and Zhang et al. (2016), with an average concentration of 305 g/m^3 in those areas. Each of these projections exceeded the predetermined maximum value.

According to the study conducted by Juang et al. (2009), in Taiwan, the regulations governing occupational safety and hygiene provide a permissible threshold for workplace air quality, specifically setting the highest allowable concentration as 210 micrograms per cubic meter. The personnel should not be exposed to an amount exceeding this limit at any given moment. This concentration is the highest level that may be considered as safe for consumption (Kong et al., 2020).

Previous study has showed that the concentrations of HCBD in various aquatic environments worldwide can vary significantly, ranging from less than 0.05 to 404.85 g/L (grams per liter). These ecosystems encompass a range of water sources, such as drinking water sources, groundwater, rivers, reservoirs, wastewater, and the effluents of sewage treatment facilities. Additionally, they encompass diverse water forms and settings that are centered on water. According to the findings of a study conducted by Chen et al. (2015), the concentrations of HCBD in drinking water sources across five significant river basins in China varied between 0.10 and 1.23 g L1. The average quantity reported was 0.61 g/L. This paper presents the latest research findings pertaining to the topic matter. The significance of this discovery lies in its provision of clear proof that the amounts of this chemical compound surpass previous estimations. Based on the findings, it was observed that the concentrations of HCBD detected in the drinking water above the maximum threshold set by the World Health Organization (WHO) at 0.6 micrograms per litre. This phenomenon occurred because of the identification of HCBD levels in the drinking water, which were shown to be present in the aforementioned water source. The origin of this concept might be attributed to the World Health Organization (WHO). Furthermore, Laska et al. (1976) and Nikolaou et al. (2002) discovered that HCBD was detected in the river waters of Greece at a concentration of 3.9 g L1 and in the river waters of the Mississippi River at a concentration ranging from 0.7 to 1.5 g L1, respectively. Both of these investigations were done inside the geographical boundaries of Greece. Both studies focused their emphasis on the Mississippi River as the primary region of their inquiry. It is of utmost significance to highlight that the aforementioned levels surpass the specified environmental threshold for inland surface water, as determined by the European Commission and currently fixed at 0.1 micrograms per liter. The establishment of this restriction was undertaken by the European Commission. The implementation of this limitation is intended to mitigate the possible introduction of hazardous substances into the environment. Furthermore, the introduction of industrial effluent into natural aquatic systems has been linked to a substantial increase in the concentration of HCBD detected in the water (Wang et al., 2012). Researchers based in China conducted the aforementioned discovery. This chemical has a high level of toxicity toward aquatic creatures. Researchers from China achieved the discovery of this consequence. The research conducted by Wang et al. (2012) revealed significant concentrations of HCBD in various water sources within the Dongjiang River basin in China. This region is renowned for its numerous industrial establishments. Specifically, the study measured HCBD concentrations of 18.04 g/L, 349.02 g/L, and 404.85 g/L in river water, groundwater, and reservoir water, respectively. It was found that these concentrations were widespread over the whole region. The measurements were collected from several locations throughout the basin for the purpose of evaluation.

Besides sludge and leachates, it has been demonstrated that HCBD is also detected in sediments and soils, alongside leachates. The study conducted by Zhang et al. (2014) reported a range of HCBD concentrations in sludge, with values spanning from 0.03 to 74.3 g/kg of dry weight (DW). The aforementioned data was obtained by the examination and evaluation of the sludge. Based on the research conducted by Marek et al. (2011), the examination revealed a variety of chemical concentrations detected in the leachates, spanning from 0.008 to 0.064 g/L. In the course of their respective investigations, the scientific study groups led by Lee et al. (2000) and Pinto et al. (2016) discovered a variety of quantities of HCBD in sediments. These concentrations varied from non-detectable (n.d.) to 47.4 grams per kilogram (g/kg) across the span of their individual studies. The aforementioned information was discovered in the course of the researchers' investigation. Multiple studies have reported different concentrations of HCBD in soils along the Mississippi River in Louisiana and various locations in China. The reported concentrations range from 21.6 to 793.4 g/L (Laska et al., 1976) in Louisiana and from 0.02 to 8.47 g/kg (dry weight) in China (Zhang et al., 2014; Tang et al., 2014, 2016; Sun et al., 2018).

Tang et al. (2014) reported the presence of HCBD in diverse aquatic species, terrestrial animals, and plants. This phenomenon can be attributed to the propensity of HCBD to undergo bioaccumulation and bio magnification. These discoveries have emerged because of the global ubiquity of HCBD. Based on the findings of a research conducted in 1976 by Laska et al., the concentrations of HCBD detected in mosquito fish specimens collected from the Mississippi River varied between 112.8 and 827.3 g/kg, contingent upon the particular species examined. Tang et al. (2014) identified the presence of HCBD in herbivorous insects, earthworms, and Chinese toads. The researchers conducted the experiment in a region of East China renowned for its agricultural activities. The levels of HCBD concentration exhibited notable variability, spanning from 1.3 to 8.2 micrograms per kilogram of liquid weight. In a study conducted in 2017, Tang et al. investigated the topic at hand. According to a study conducted by Tang et al. HCBD was detected in plants originating from a specific region in Southwest China, which had a history of pesticide production. The concentrations of HCBD ranged from 0.03 to 24.6 g/kg (dry weight) in these plants.

6 Remedial Action Taken for Management of POPs

In India, there exist manufacturing methods that result in the emission of HCBD, a chemical known for its carcinogenic qualities. It is imperative to undertake measures aimed at minimizing the release of HCBD by implementing technological advancements.

Insufficient data is available about the cleanup of HCBD in the context of India. The following is an overview of the remediation techniques (Table 4) employed by different nations for the treatment of HCBD.

Replacement Within the territory of the United Nations Economic Commission for Europe (UNECE), encompassing countries such as the United States and Canada, there exists a notable lack of understanding regarding the deliberate manufacturing and utilization of HCBD. This knowledge gap persists both in contemporary times and during the last three decades. However currently the majority of

Table 4 Summary of the chemical methods to remove HCBD from aquatic environment (Kong et al., 2020)

Mechanism	Methods	Conditions	Removal	References
Oxidation	O_3	O_3 = 317 mg/L	90.7%	Derco et al. (2013)
		HCBD = 82.6 ng/L, 60 min		
		O_3 = 22.3 mg/L, 20 min	87%	Derco et al. (2015)
		25 °C, pH 7.95–8.05	16%	Lee et al. (2016)
		O_3 = 0.5 mg/L		
		HCBD = 9 ∝ g/L		
	O_3/UV	O_3 = 22.3 mg/L, 20 min	≈ 90%	Derco et al. (2015)
	O_3/H_2O_2	25 °C, pH 7.95–8.05	>30%	Lee et al. (2016)
		O_3 = 0.5 mg/L		
		H_2O_2 = 0.25 mg/L		
		HCBD = 9 ∝ g/L		
Photolysis	UV	25 °C, pH 7.95–8.05	≈ 100%	Lee et al. (2016)
		UV dose: 6000 J m^{-2}		
	UV/ H_2O_2	25 °C, pH 7.95–8.05	≈ 85%	Lee et al. (2016)
		UV dose: 6000 J m^{-2}		
		H_2O_2 = 5.0 mg/L		
		HCBD = 9 ∝ g/L		
Adsorption, oxidation, and reduction	O_3/nZVI	O_3 = 22.3 mg/L, 3 min	100%	Derco et al. (2015)
Adsorption and reductive dechlorination	nPd/ μZVI	25 °C, 180 min, nPd/ μZVI = 600 mg/L, HCBD = 10 mg/L	100%	Rodrigues et al. (2017)
	nZVI	4 h	≈ 73%	Derco et al. (2015)

HCBD unintentionally manufactured as a waste byproduct is destroyed via incineration by the chemical manufacturer in US. There is a need to develop a plan to eliminate HCBD to the extent practicable through alternative chlorinated solvent manufacturing. This discovery provides support for the concept that a process of substitution has occurred, resulting in the emergence of feasible alternatives.

7 Recommendations

The chemical process referred to as chlorinolysis is accountable for the formation of carbon tetrachloride and tetrachloroethane. Hexachlorobutadiene, generally referred to as HCBD, is a notable derivative resulting from this chemical process. The manufacturing of carbon tetrachloride in India is reported to be carried out by many companies, including Gujarat Alkalies & Chemicals Limited (located in Dahej), Adarsh Chemicals (based in Chennai), and Nutan Chemicals (situated in Pune). There is an urgent need to initiate research for the Environment and Green Chemistry alternatives assessment of HBCD in order to support the Industries involved. General consumer and commercial applications impacting public health may be limited by formulating regulations. Currently, there exists a dearth of substantial data pertaining to the prevalence of the aforementioned phenomenon, as well as its associated polluted locations, stockpiles, and related information. Furthermore, research study on the historical background and toxicological data pertaining to the Indian ecosystem is vital.

References

Balmer, J. E., et al. (2019). Hexachlorobutadiene (HCBD) contamination in the Arctic environment: A review. *Emerging Contaminants, 5*, 116–122.

Chen, D., Hale, R. C., La Guardia, M. J., Luellen, D., Kim, S., & Geisz, H. N. (2015). Hexabromocyclododecane flame retardant in Antarctica: Research stations as sources. *Environmental Pollution, 206*, 611–618. https://doi.org/10.1016/j.envpol.2015.08.024

Derco, J., Valičková, M., Šilhárová, K., et al. (2013). Removal of selected chlorinated micropollutants by ozonation. *Chemical Papers, 67*(12), 1585–1593.

Derco, J., Dudáš, J., Valičková, M., et al. (2015). Removal of micropollutants by ozone based processes. *Chemical Engineering and Processing, 94*(SI), 78–84.

Gaonkar, O., Roy-Basu, A., Bharat, G., & Chakraborty, P. (2022). *Current scenario of new persistent organic pollutants in India: Review on quantification techniques: Persistent organic pollutants: Gaps in management and associated challenges* (1st ed., pp. 92–123). CRC Press.

Juang, D. F., Yuan, C. S., Hsueh, S. C., & Chiou, L. J. (2009). Distribution of volatile organic compounds around a polluted river. *International Journal of Environmental Science and Technology, 6*(1), 91–104.

Kong, Q., Wang, Y., & Yang, X. (2020). A review on Hexachloro-1,3-butadiene (HCBD): Sources, occurrence, toxicity and transformation. *Bulletin of Environmental Contamination and Toxicology, 104*(1), 1–7. https://doi.org/10.1007/s00128-019-02744-5. Epub 2019 Nov 19.

Kumari, K. (Ed.). (2021). *Persistent organic pollutants: Gaps in management and associated challenges* (1st ed.). CRC Press. https://doi.org/10.1201/9781003046806

Laska, A. L., Bartell, C. K., & Laseter, J. L. (1976). Distribution of hexachlorobenzene and hexachlorobutadiene in water, soil, and selected aquatic organisms along the lower Mississippi River, Louisiana. *Bulletin of Environmental Contamination and Toxicology, 15*, 535–542. https://doi.org/10.1007/BF01685701

Lee, C.-L., Song, H.-J., & Fang, M.-D. (2000). Concentrations of chlorobenzenes, hexachlorobutadiene and heavy metals in surficial sediments of Kaohsiung coast, Taiwan. *Chemosphere, 41*, 889–899.

Lee, M., Merle, T., Rentsch, D., et al. (2016). Abatement of polychoro-1,3-butadienes in aqueous solution by ozone, UV photolysis, and advanced oxidation processes (O_3/H_2O_2 and UV/H_2O_2). *Environmental Science & Technology, 51*(1), 497–505.

Marathe, D., Balbudhe, S., & Kumari, K. (2021). Persistent organic pollutants: A global issue, a global response. In *Persistent Organic Pollutants* (pp. 1–32). CRC Press.

Marek, M., Plaza, G. A., Nalecz-Jawecki, G., Ulfig, K., & Markowska-Szczupak, A. (2011). Estimation of the environmental risk posed by landfills using chemical, microbiological and ecotoxicological test-ing of leachates. *Chemosphere, 82*(7), 1017–1023. https://doi.org/10.1016/j.chemosphere.2010.10.06. *(15) (PDF) Spatial distributions of hexachlorobutadiene in agricultural soils from the Yangtze River Delta region of China*. Available from: https://www.researchgate.net/publication/321152832_Spatial_distributions_of_hexachlorobutadiene_in_agricultural_soils_from_the_Yangtze_River_Delta_region_of_China. Accessed 16 Oct 2023.

Nikolaou, A. D., Golfinopoulos, S. K., Kostopoulou, M. N., Kolokythas, G. A., & Lekkas, T. D. (2002). Determination of volatile organic compounds in surface waters and treated wastewater in Greece. *Water Research, 36*, 2883–2890. https://doi.org/10.1016/S0043-1354(01)00497-3

NPRPS. (2015, December). *National Program for Rehabilitation of Polluted Sites in India. Guidance document for assessment and remediation of contaminated sites in India* (1st ed., pp. 1–564). https://cpcb.nic.in/uploads/hwmd/MoEFCC_guidelines_contaminatedsites.pdf

Pinto, M. I., Vale, C., Sontag, G., & Noronha, J. P. (2016). Pathways of priority pesticides in sediments of coastal lagoons: A case study of Obidos Lagoon, Portugal. *Marine Pollution Bulletin, 106*(1–2), 335–340.

PubChem. https://pubchem.ncbi.nlm.nih.gov/compound/Hexachloro-1_3-butadiene#section=Chemical-and-Physical-Properties

Rodrigues, R., Betelu, S., Colombano, S., et al. (2017). Reductive dechlorination of hexachlorobutadiene by Pd/Fe microParticles suspension in dissolved lactic acID polymers: Degradation mechanism and kinetics. *Industrial and Engineering Chemistry Research, 56*(42), 12092–12100.

Sun, J., Pan, L., Zhan, Y., & Zhu, L. (2018). Spatial distributions of hexachlorobutadiene in agricultural soils from the Yangtze River Delta region of China. *Environmental Science and Pollution Research International, 25*(4), 3378–3385. https://doi.org/10.1007/s11356-017-0707-6. Epub 2017 Nov 18.

Tang, B., Luo, X. J., Zeng, Y. H., & Mai, B. X. (2014). Polychlorinated biphenyls and their methylsulfonyl metabolites in fish from an electronic waste recycling site in South China: Tissue distribution and human dietary exposure. *Environmental Science (China), 12*, 4655–4662.

Tang, Z., Huang, Q., Nie, Z., et al. (2016). Levels and distribution of organochlorine pesticides and hexachlorobutadiene in soils and terrestrial organisms from a former pesticide-producing area in Southwest China. *Stochastic Environmental Research and Risk Assessment, 30*, 1249–1262. https://doi.org/10.1007/s00477-015-1144-z

Toxic Link. (2015, October). New POPs: Pentachlorophenol, chlorinated naphthalenes, hexachlorobutadiene, toxics link factsheet number 50.

Toxic Link. (2018). https://ipen.org/sites/default/files/documents/POPs%20Country%20Situation%20report.pdf

Wang, R. S., Xu, Q. J., Zhang, X., Wei, Q. S., & Yan, C. Z. (2012). Health risk assessment of heavy metals in typical township water sources in DongjianG River basin. *Huan Jing Ke Xue, 33*(9), 3083–3088. Chinese.

Wang, L., Bie, P., & Zhang, J. (2018). Estimates of unintentional production and emission of hexachlorobutadiene from 1992 to 2016 in China. *Environmental Pollution, 238*, 204–212.

Zhang, H., Wang, Y., Sun, C., Yu, M., Gao, Y., Wang, T., Liu, J., & Jiang, G. (2014). Levels and distributions of hexachlorobutadiene and three chlorobenzenes in biosolids from wastewater treatment plants and in soils within and surrounding a chemical plant in China. *Environmental Science & Technology, 48*(3), 1525–1531. https://doi.org/10.1021/es405171t. Epub 2014 Jan 17.

Zhang, Z., Zhang, Y., Wang, X., Lü, S., Huang, Z., Huang, X., Yang, W., Wang, Y., & Zhang, Q. (2016). Spatiotemporal patterns and source implications of aromatic hydrocarbons at six rural sites across China's developed coastal regions. *Journal of Geophysical Research, [Atmospheres], 121*, 6669–6687. https://doi.org/10.1002/2016jd025115

Zhang, H., Jiang, L., Zhou, X., Zeng, T., He, Z., Huang, X., Chen, J., & Song, S. (2018). Determination of hexachlorobutadiene, pentachlorobenzene, and hexachlorobenzene in waste incineration fly ash using ultrasonic extraction followed by column cleanup and GC-MS analysis. *Analytical and Bioanalytical Chemistry, 410*(7), 1893–1902. https://doi.org/10.1007/s00216-018-0849-5. Epub 2018 Feb 2.

Penta-Chlorobenzene (PeCB)

Kanchan Kumari, Deepak Marathe, and Md. Shahbaz

1 Introduction

Penta-Chlorobenzene (PeCBs) falls under the broader category of Chlorobenzene compounds. PeCB is categorized within the Chlorobenzene family, notable for the presence of five chlorine atoms on the benzene ring in place of hydrogen atoms. In the past, PeCB was once part of a mixture of chlorobenzenes, utilized to lower the viscosity of PCB products for heat transfer purposes (Environment Canada, 1993; King et al., 2003). PeCB has been identified in numerous global locations and media (Hoff et al., 1992; Chan et al., 1996; Bright & Healey, 2003; Lacorte et al., 2006; Oren et al., 2006; Wenzel et al., 2006). Several countries have been monitoring its release over the years as part of their preparations for national PeCB regulation. PeCB may also be released from certain fuel combustion processes that contain chlorine, either in organochlorine or chloride form, as suggested by Carpenter in 2006.

While PeCB is relatively less toxic for human consumption, its influence on aquatic ecosystems raises concerns for aquatic life. Current PeCB emissions into the environment are approximately estimated at 121,000 kilograms per year, as reported by Bailey et al. in 2009.

Penta-Chlorobenzene, often referred to as PeCB, has found extensive use in various industries and products such as PCBs, dyestuff carriers, fungicides, and

K. Kumari (✉) · M. Shahbaz
CSIR-National Environmental Engineering Research Institute (NEERI), Kolkata Zonal Center, Kolkata, West Bengal, India
e-mail: k_kumari@neeri.res.in

D. Marathe
CSIR-National Environmental Engineering Research Institute (NEERI), Nagpur, India

K. Kumari, *Pollutants of Global Concern*, Emerging Contaminants and Associated Treatment Technologies, https://doi.org/10.1007/978-3-031-50996-4_18

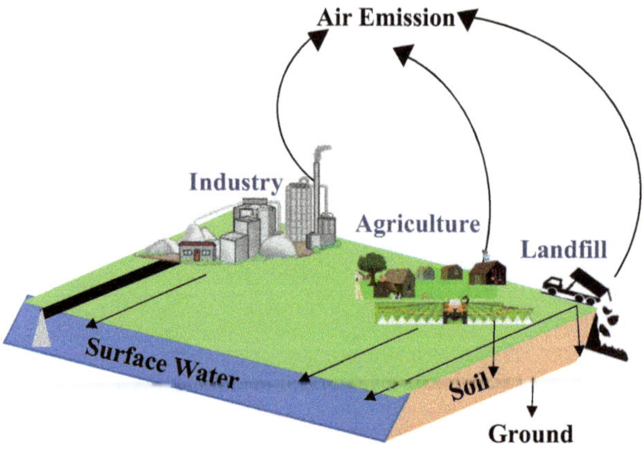

Fig. 1 Different sources of exposure in environment

flame-retardants. It is expected that PeCB will continue to serve as a chemical intermediary, particularly in the synthesis of quintozene.

Accidental formation of PeCB can arise from combustion, thermal processes, and industrial activities (Fig. 1). It might also be present as impurities in substances like solvents or pesticides. While belonging to the Chlorobenzene family characterized by a benzene ring with chlorine atoms replacing hydrogen atoms, PeCB stands out due to its distinct benzene ring structure.

The increasing presence of Persistent Organic Pollutants (POPs) in various environmental matrix has presented a global concern. This concern prompted the creation of the Stockholm Convention (SC) in 2001, as part of the United Nations Environment Programme (Kumari et al., 2021).

PeCB's substantial environmental presence, prolonged persistence, and potential for long range transport raise concerns. In addition, it poses only moderate risks to human health but a significant threat to aquatic organisms. It is mostly used for producing dyestuff carriers, fungicides, and flame-retardants and thus, PeCB's versatility extends to various applications. It is expected to retain its role as a chemical intermediary, particularly in quintozene production. Accidental formation during combustion, thermal processes, and industrial activities is common, and it can emerge as impurities in solvents or insecticides.

The molecular structure of Penta-Chlorobenzene (C_6Cl_5H) distinguishes it, marked by its formula and name. Unlike other hydrocarbons, chlorinated aromatic hydrocarbons possess a benzene ring comprising five chlorine atoms (McNeill et al., 1995). The compound's classification hinges on this unique chemical composition. Notably, it appears as a solid, displaying a crystalline structure with a white hue, setting it apart. Its distinct and intense aroma carries a sweet quality. With notable resistance to chemical reactions, its persistence is attributed to limited reactivity and resistance to degradation in the natural environment.

It was previously utilized as a pesticide, particularly for seed treatment against fungal infections in crops, but later on the use of Penta-Chlorobenzene in pesticides has diminished. Its potential for bio-accumulation and long persistence in the environment prompted strict regulations, even leading to its discontinuation in certain regions due to its adverse effects on various animal species and potential risks to human health.

Many research reported the threats of this chemical which poses major impacts to both human well-being and the environment. Its tendency to accumulate in the fat tissues of organisms can result in skin and eye irritation, respiratory problems, and cancer risks. Regulatory measures have been put in place to limit production, usage, and its spread in ecosystems. Its hazardous nature and related environmental concerns have contributed to a noticeable decline in its use and production.

This study emphasizes the significance of assessing the potential long-term consequences of chemical compounds on human health and the environment.

2 Technical Details of Penta-Chlorobenzene

Penta-Chlorobenzene is a chemical compound which is also known as 1,2,3,4,5-Penta-Chlorobenzene. It has the molecular formula C_6HCl_5 and a CAS number of 608-93-5 and a molecular weight of 250.3 g/mol.

It appears as a white crystalline solid and emits a pleasant aroma. Its boiling point is 531 °F at 760 mm Hg, while the melting point is 187 °F. The compound's solubility is less than 1 mg/mL at 72 °F, and its density is 1.8342. At 25 °C, it has a vapor pressure of 0.002 mm Hg. The octanol/water partition coefficient is 5.17 (LogP), and the octanol-air partition coefficient is 6.49.

The Henry's Law Constant is 7.03e-04 atm-m^3/mole. Penta-Chlorobenzene remains stable under recommended storage conditions but decomposes when heated, emitting fumes of hydrogen chloride (Table 1). It is not corrosive in nature.

The molecule's stability and low reactivity are as result of the resonance stabilization of the benzene ring and the electronegativity of chlorine atoms. These properties provide protection against environmental degradation.

Recognizing these risks, numerous countries have implemented bans on PeCBs, and classified in the category of Persistent Organic Pollutants (POPs) under the Stockholm Convention. These substances are known for their long-lasting environmental persistence and associated health threats. Proper handling of PeCBs is essential to prevent environmental contamination (Fig. 1), in accordance with local hazardous waste disposal regulations (Misra & Pandey, 2005).

Contemporary practices have safely and sustainably replaced PeCBs in various applications. In essence, Penta-Chlorobenzene displays a distinctive molecular structure, physical characteristics, and chemical behavior. The historical context of its use underscores the necessity of comprehending the chemical's effects on health and the environment. Current actions align with ethical and regulatory standards, safeguarding both human well-being and the ecosystem.

Under standard conditions, Penta-Chlorobenzene takes the form of a white crystalline solid, melting at 273–275 °C (527–527 °F), and exhibiting a notably high boiling point of 437 °C (819 °F).

Table 1 Technical details and physicochemical properties of Penta-Chlorobenzene (NCBI, 2023)

Details	Description (Penta-Chlorobenzene)
Structure	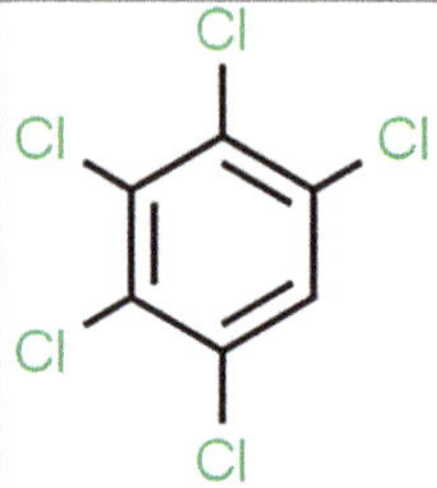
IUPAC name	1,2,3,4,5-Penta-Chlorobenzene
Molecular formula	C_6HCl_5
CAS no.	608-93-5
Molecular weight	250.3 g/mol
Color/form	White crystalline solid
Odor	Pleasant aroma
Boiling point	531 °F at 760 mm Hg
Melting point	187 °F
Solubility	Less than 1 mg/mL at 72 °F
Density	1.8342
Vapor pressure	0.002 mm Hg at 25 °C
Octanol/water partition coefficient	5.17 (LogP)
Octanol-air partition coefficient	6.49
Henry's law constant	7.03e-04 atm-m^3/mole
Stability/shelf life	Stable under recommended storage conditions
Decomposition	When heated to decompose, it emits fumes of hydrogen chloride

Boasting a substantial molecular weight, Penta-Chlorobenzene possesses a density of 2.04 g/cm^3. This hydrophobic compound demonstrates limited solubility in water while readily dissolving in nonpolar solvents like benzene and toluene. The hexachlorobenzene (HCB) also gets degraded to Penta-Chlorobenzene (PeCB) by a reductive dechlorination process in soil under the influence of microbial activity (Fig. 2).

Technical insights into Penta-Chlorobenzene encompass its distinct chemical structure, physical characteristics, and notable reactivity. This pesticide serves as a lesson in comprehending the impact of chemical compounds on human health and the environment. Modern values and ethics find expression in responsible business practices that prioritize both human well-being and the planet's sustainability.

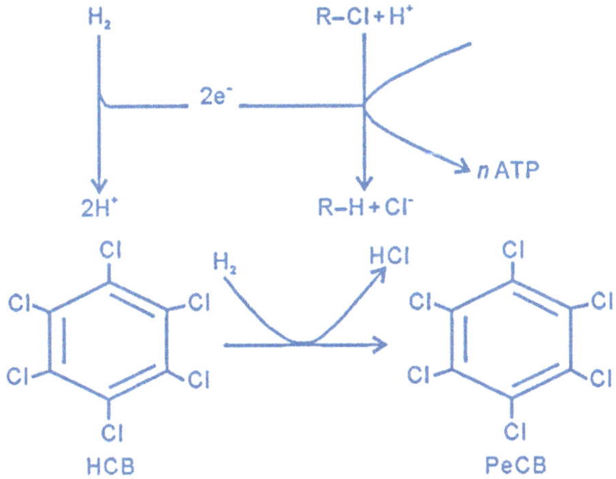

Fig. 2 Process of dechlorination of hexachlorobenzene to Penta-Chlorobenzene. (Brahushi et al., 2004)

3 Status of Usage Within the Country

There is currently a lack of comprehensive data regarding the production and accumulation of Penta-Chlorobenzene (IPEN, POPs, 2018). The ban on Penta-Chlorobenzene, a chemical compound, was enforced through a notification issued by the Ministry of Environment, Forest and Climate Change, Government of India on March 5, 2018. This communication, identified as G.S.R. 207 (E), aligns with the stipulations of section 3 and section 6 of the Environment (Protection) Act, 1986 (29 of 1986), along with rule 13 of the Environment (Protection) Rules, 1986. Consequently, this directive has resulted in the prohibition of various activities related to Penta-Chlorobenzene, encompassing its manufacture, trade, usage, import, and export.

3.1 In-House (Countrywide) Scientific Work Done on the Chemical

No relevant information found in Indian context.

3.2 *Relevance of the International Scientific Data as per Indian Conditions*

Zhang et al. (2018) in their investigation developed a direct and efficient method to quantify three Persistent Organic Pollutants (POPs)—hexachlorobutadiene (HCBD), Penta-Chloro benzene (PeCB), and hexachlorobenzene (HCB)—within fly ash (Fig. 3). This procedure involves ultrasonic extraction, silica gel—Florisil column purification, followed by measurement via gas chromatography-mass spectrometry (GC-MS). The recorded PeCB levels ranged from 1.22 to 150 mg g^{-1}, indicating heightened PeCB concentrations in fly ash samples sourced from industrial waste incineration. This discovery underscores the necessity for continuous vigilance and monitoring regarding this concern. Further, US EPA Regional Screening Levels for Chemical Contaminant—PeCB, has been outlined in Fig. 4 wherein the soil screening level (SSL) in resident and industrial soil along with risk based SSL and toxicity values have been provided. The unified use of such regional screening levels is in screening of chemicals at contaminated sites for national consistency.

3.3 *Exemption (Use of Chemicals for Scientific Purpose)*

According to established regulations, specific chemicals enumerated in rule 2 are permissible for use, trade, or import in appropriate quantities, exclusively intended for research and development endeavors within Central Universities, Council of Scientific and Industrial Research Laboratories, officially recognized government institutions, or designated research facilities in the public or private sector. However,

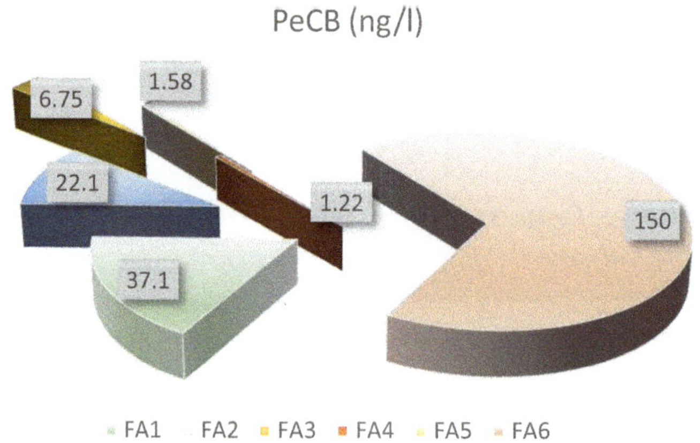

Fig. 3 Measured concentrations of Penta-Chlorobenzene (PeCB) in different waste incineration fly ash (FA) samples. (Zhang et al., 2018)

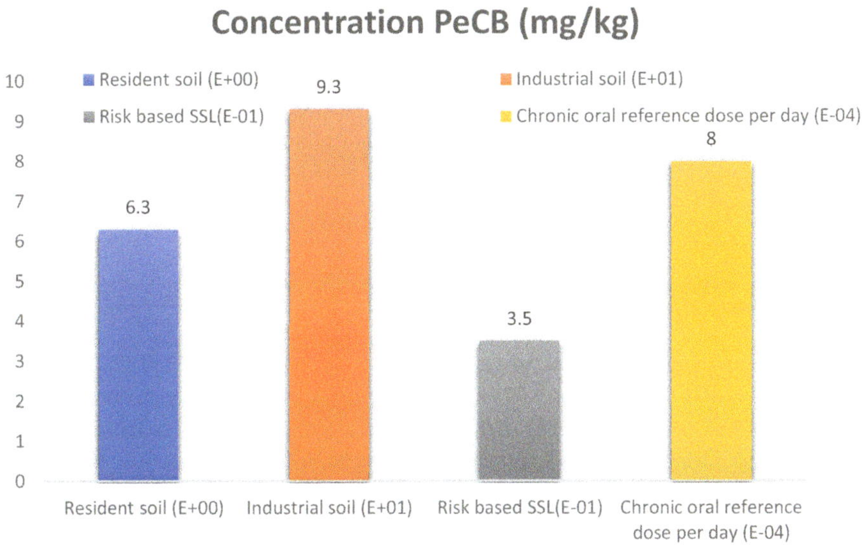

Fig. 4 US EPA regional screening levels for chemical contaminants. (US EPA, 2004)

it is crucial to obtain prior authorization from the Ministry of Environment, Forest and Climate Change, a governmental body in India, before embarking on such initiatives (IPEN POPs, 2018).

Penta-Chlorobenzene, a recognized Persistent Organic Pollutant (POP) listed in Annex A and C, was primarily employed in pesticides and industrial chemicals. It was added to the list in 2009. From March 5, 2018 onward, the manufacturing, usage, import, export, and related activities of this substance have been prohibited in India in accordance with regulations.

3.4 *Producers, Trade Names, Uses, and Effects of PeCB*

Initially, Penta-Chlorobenzene (PeCB) was intentionally produced for the synthesis of Pentachloronitrobenzene (quintozene), as a pesticide. However, its current sources are unintentional, originating from various origins like PCBs, chlorinated solvents, pesticides, chemical manufacturing, aluminum casting, waste combustion (including open burning), ore treatment for metal production (magnesium, copper, niobium, tantalum), titanium dioxide production, wood treatment plants, and hazardous waste incineration.

PeCB's impact on humans is moderately toxic, as indicated by animal studies showing effects such as reduced thyroxin levels, abnormal sperm, and histopathological changes in kidneys.

Furthermore, Penta-Chlorobenzene displays high toxicity to aquatic organisms, with the potential to cause lasting adverse effects in aquatic environments. The chemical compound, known as Penta-Chlorobenzene (PeCB), has been extensively used across various sectors due to its distinctive chemical properties. However, its utilization has significantly decreased over time due to increasing concerns about its potential environmental and health consequences. A deeper exploration of its historical applications and unintended generation provides a more comprehensive understanding of this phenomenon (Huesemann & Huesemann, 2011).

3.5 The Historical Applications and Utilizations

PeCB, effectively lowering the viscosity of polychlorinated biphenyl (PCB) products used in heat transfer applications, leading to improved thermal conductivity and heat transfer (EFSA, 2018). PeCB, also referred to as p-Chlorobenzene, has found application as a dyestuff carrier in the textile industry, facilitating even color dispersion on textiles and substrates. Additionally known as Penta-Chlorobenzene, PeCB was employed as an agricultural fungicide to protect crops from fungal diseases, but its persistent nature and potential ecological impact led to reduced usage for this purpose. PeCB has been employed as a chemical intermediate in the manufacturing of quintozene, a soil fungicide used for controlling soil-borne diseases in agricultural crops. Even so, the unintentional production and environmental emission of PeCB can occur during combustion processes, especially in thermal and industrial operations, leading to the formation of polychlorinated biphenyls (PCBs) as a byproduct (Vorkamp, 2016). It can also be found as an impurity in certain products such as solvents and insecticides, resulting from manufacturing processes or contamination.

Activities like garbage incineration and domestic waste combustion hold the potential to release polychlorinated biphenyls (PCBs), including PeCB, into the atmosphere and surrounding ecosystems. PeCB enters the environment significantly through industrial waste streams, originating from industries such as pulp and paper mills, iron and steel mills, and petroleum refineries, posing a risk of environmental contamination.

Despite undergoing treatment, wastewater facilities might retain PeCB in activated sludge, which can then potentially spread into the environment through disposal or utilization as fertilizer. While PeCB's historical applications reflect its chemical properties and versatility, concerns about its adverse environmental and health effects have led to regulations and guidelines governing its production and usage. Recognizing unintended generation and release routes highlights the need for effective waste management practices and the adoption of ecologically sustainable alternatives across various sectors.

4 Adverse Effects

As per the European Union's classification criteria, Penta-Chlorobenzene (PeCB) is categorized as "Harmful if ingested" and "Highly toxic to aquatic organisms, potentially leading to persistent negative impacts on the aquatic ecosystem." Both classifications emphasize PeCB's potential to cause enduring harm to aquatic ecosystems (Wright et al., 2013).

In one study, nursing puppies exhibited tremors after their mothers were administered a dose of 12.5 mg/kg body weight, with symptoms emerging between 4 and 14 days after litter delivery. At a dose of 6.3 milligrams per kilogram of the mother's body weight, no observable effects were seen in the offspring. In another study, pregnant rats were exposed to PeCB doses of 50, 100, and 200 mg/kg body weight per day during days 6–15 of pregnancy. Although the number of viable fetuses remained constant, observed that the highest dosage group exhibited a decrease in average fetal weight.

Current research suggests no notable discrepancy in sensitivity between marine and freshwater species concerning PeCB. Based on outcomes from acute and subchronic animal tests, it is reasonable to conclude that PeCB presents a moderate degree of toxicity to humans. Empirical investigations involving aquatic organisms confirmed PeCB's toxicity to a diverse range of aquatic species, leading to this revelation.

4.1 Probable Routes of Human Exposure

Occupational exposure to PeCB can occur through both inhalation and skin contact, posing significant risks (WHO, 1993). This chemical is classified as a carcinogen, with potential for spreading via occupational exposure. Consumer products can also pose exposure risks to individuals through inhalation, ingestion of food and beverages, as well as skin contact. These findings are derived from the latest monitoring information available (Wambaugh et al., 2019).

5 Remedial Action Taken for Management of POPs

No pertinent material specific to the Indian context was discovered. Polychlorinated biphenyls, commonly referred to as PCBs, have attracted significant research attention, all of which has generated data indicating a notable tendency toward biodegradation. However, due to limited knowledge about the aerobic bacteria's breakdown of the chemical, it's plausible to assume that anaerobic conditions are essential for the decomposition process (Environmental Protection Agency, 2010). Brahushi et al. (2004) and Pavlostathis and Prytula (2000) studies have conducted extensive studies on the topic of PeCB's biological dechlorination. Nevertheless, prior research has empirically shown that PeCB can persist in soil and water environments for prolonged periods.

Various dehalogenation methods exist, each with potential to enhance the breakdown rate of Penta-Chlorobenzene. Atkinson et al. (1992) described a method utilizing hydroxyl radicals in the Earth's atmosphere. Gach et al. examined another approach, cobalt salen-catalyzed reduction, in 2008. Additionally, Takaoka et al. (2007) explored nitric acid-modified activated carbon in a packed-bed reactor at 350 °C.

Considering its positive environmental impact and cost-effectiveness, microbial decomposition should be prioritized where feasible. Microorganisms must not only metabolize Persistent Organic Pollutants (POPs), but also resist their harmful effects. White rot fungus species, such as *Phanerochaete chrysosporium*, *Trametes versicolor*, and *Pleurotus ostreatus*, are recognized for this resistance.

Madaj et al. (2018) demonstrated that certain enzymes are effective in degrading nitroaromatic pollutants like 2,4,6-trinitrotoluene (TNT) and 3,5-dinitrophenol (DNP). Additionally, these organisms can be rapidly cultured, making them appealing for chloro organic Persistent Organic Pollutant degradation. The provided user content doesn't necessitate significant academic revision.

5.1 Replacement

Production of PeCB has been halted in the main manufacturing countries for multiple decades, driven by the rise of more effective and economically feasible substitutes. In order to significantly decrease the unintentional production of PeCB, it becomes crucial to adopt the application of Best Available practices (BAP) and Best Environmental Practices (BEP).

6 Recommendations

In spite of the explicit ban on intentional production of PeCBs according to the Stockholm Convention, it is commonly believed that most of these substances actually stem from unintended sources of generation. Currently, there exists a significant gap in available information concerning the production and storage of Penta-Chlorobenzene. It becomes crucial to develop an all-encompassing surveillance mechanism for the purpose of tracing inadvertent release of PeCBs arising from the practice of waste incineration.

References

Atkinson, R., Aschmann, S. M., Arey, J., & Shorees, B. (1992). Formation of OH radicals in the gas phase reactions of O3 with a series of terpenes. *Journal of Geophysical Research: Atmospheres, 97*(D5), 6065–6073.

Bailey, R. E., van Wijk, D., & Thomas, P. C. (2009). Sources and prevalence of pentachlorobenzene in the environment. *Chemosphere, 75*(5), 555–564.

Brahushi, F., Dörfler, U., Schroll, R., & Munch, J. C. (2004). Stimulation of reductive dechlorination of hexachlorobenzene in soil by inducing the native microbial activity. *Chemosphere, 55*, 1477–1484. https://doi.org/10.1016/j.chemosphere.2004.01.022

Bright, D. A., & Healey, N. (2003). Contaminant risks from biosolids land application: Contemporary organic contaminant levels in digested sewage sludge from five treatment plants in Greater Vancouver, British Columbia. *Environmental Pollution, 126*(1), 39–49.

Carpenter, D. O. (2006). Polychlorinated biphenyls (PCBs): Routes of exposure and effects on human health. *Reviews on Environmental Health, 21*(1), 1–24.

Chan, H. M., El Khoury, M., Sedgemore, M., Sedgemore, S., & Kuhnlein, H. V. (1996). Organochlorine pesticides and polychlorinated biphenyl congeners in Ooligan grease: A traditional food fat of British Columbia First Nations. *Journal of Food Composition and Analysis, 9*(1), 32–42.

EFSA Panel on Contaminants in the Food Chain (CONTAM), Knutsen, H. K., Alexander, J., Barregård, L., Bignami, M., Brüschweiler, B., et al. (2018). Risk for animal and human health related to the presence of dioxins and dioxin-like PCBs in feed and food. *EFSA Journal, 16*(11), e05333.

Environment Canada. (1993). *Priority substances list assessment report: Pentachlorobenzene. 26, Cat. No. EN40-215/26E, Canadian Environmental Protection Act, Ottawa, Ontario.*

Gach, P. C., Karty, J. A., & Peters, D. G. (2008). Catalytic reduction of hexachlorobenzene and pentachlorobenzene by cobalt (I) salen electrogenerated at vitreous carbon cathodes in dimethylformamide. *Journal of Electroanalytical Chemistry, 612*(1), 22–28.

Hoff, R. M., Muir, D. C. G., & Grift, N. P. (1992). Annual cycle of polychlorinated biphenyls and organohalogen pesticides in air in southern Ontario. 1. Air concentration data. *Environmental Science & Technology, 26*(2), 266–275.

Huesemann, M., & Huesemann, J. (2011). *Techno-fix: Why technology won't save us or the environment.* New Society Publishers.

IPEN. (2018, June). *Persistent organic pollutants (POPs) in India: Country situation report by toxics link.*

King, H. R., Pankhurst, N. W., Watts, M., & Pankhurst, P. M. (2003). Effect of elevated summer temperatures on gonadal steroid production, vitellogenesis and egg quality in female Atlantic salmon. *Journal of Fish Biology, 63*(1), 153–167.

Kumari, K., Swamy, S., & Singh, A. (2021). Global monitoring plan on persistent organic pollutants (POPs). In *Persistent organic pollutants: Gaps in management and associated challenges* (p. 227). CRC Press.

Lacorte, S., Raldúa, D., Martínez, E., Navarro, A., Diez, S., Bayona, J. M., & Barceló, D. (2006). Pilot survey of a broad range of priority pollutants in sediment and fish from the Ebro river basin (NE Spain). *Environmental Pollution, 140*(3), 471–482.

Madaj, R., Sobiecka, E., & Kalinowska, H. (2018). Lindane, kepone and pentachlorobenzene: Chloropesticides banned by Stockholm convention. *International journal of Environmental Science and Technology, 15*, 471–480.

McNeill, I. C., Memetea, L., & Cole, W. J. (1995). A study of the products of PVC thermal degradation. *Polymer Degradation and Stability, 49*(1), 181–191.

Misra, V., & Pandey, S. D. (2005). Hazardous waste, impact on health and environment for development of better waste management strategies in future in India. *Environment International, 31*(3), 417–431.

National Center for Biotechnology Information (2023). *PubChem compound summary for CID 35823, 2,3′,4,4′,5-Pentachlorobiphenyl*. Retrieved October 16, 2023, from https://pubchem.ncbi.nlm.nih.gov/compound/1_2_4-trichloro-5-_3_4-dichlorophenyl_benzene

Oren, A., Aizenshtat, Z., & Chefetz, B. (2006). Persistent organic pollutants and sedimentary organic matter properties: A case study in the Kishon River, Israel. *Environmental Pollution, 141*(2), 265–274.

Pavlostathis, S. G., & Prytula, M. T. (2000). Kinetics of the sequential microbial reductive dechlorination of hexachlorobenzene. *Environmental Science & Technology, 34*(18), 4001–4009.

Takaoka, M., Yokokawa, H., & Takeda, N. (2007). The effect of treatment of activated carbon by H2O2 or HNO3 on the decomposition of pentachlorobenzene. *Applied Catalysis B: Environmental, 74*(3–4), 179–186, ISSN 0926-3373. https://doi.org/10.1016/j.apcatb.2007.02.009

U.S. EPA. (2010). *U.S. Environmental Protection Agency (EPA) decontamination research and development conference*. U.S. Environmental Protection Agency, Washington, DC, EPA/600/R-11/052, 2011.

United States. Environmental Protection Agency. Office of Pesticide Programs. (2004). *Overview of the ecological risk assessment process in the Office of Pesticide Programs, US Environmental Protection Agency: Endangered and threatened species effects determinations*. DIANE Publishing.

Vorkamp, K. (2016). An overlooked environmental issue? A review of the inadvertent formation of PCB-11 and other PCB congeners and their occurrence in consumer products and in the environment. *Science of the Total Environment, 541*, 1463–1476.

Wambaugh, J. F., Bare, J. C., Carignan, C. C., Dionisio, K. L., Dodson, R. E., Jolliet, O., et al. (2019). New approach methodologies for exposure science. *Current Opinion in Toxicology, 15*, 76–92.

Wenzel, K. D., Hubert, A., Weissflog, L., Kühne, R., Popp, P., Kindler, A., & Schüürmann, G. (2006). Influence of different emission sources on atmospheric organochlorine patterns in Germany. *Atmospheric Environment, 40*(5), 943–957.

World Health Organization. (1993). *Polychlorinated biphenyls and terphenyls*. World Health Organization.

Wright, S. L., Thompson, R. C., & Galloway, T. S. (2013). The physical impacts of microplastics on marine organisms: A review. *Environmental Pollution, 178*, 483–492.

Zhang, H., Jiang, L., Zhou, X., Zeng, T., He, Z., Huang, X., Chen, J., & Song, S. (2018). Determination of hexachlorobutadiene, pentachlorobenzene, and hexachlorobenzene in waste incineration fly ash using ultrasonic extraction followed by column cleanup and GC-MS analysis. *Analytical and Bioanalytical Chemistry, 410*, 1893. https://doi.org/10.1007/s00216-018-0849-5

Pentachlorophenol (PCP): Its Salts and Esters

Kanchan Kumari and Rachna Jain

1 Introduction

Pentachlorophenol (PCP) is a persistent organic compound, which has been initially introduced in 1930, and is now used as a well-known wood preservative. It is an aromatic hydrocarbon belonging to chlorophenol (CP) family.

Along with this it is also been used as crop protectant, disinfecting agent as well as an antifouling agent (Kumari et al., 2022). It is also used in certain non-edible agricultural practices, building related work, in water cooling towers and paper and pulp industries. Recently PCP has been banned for the packaging of food material due to its highly toxic and less biodegradable nature (Fiege et al., 2000). In its purest form it is colourless crystal with relative volatile nature, less water solubility but dissoluble in "organic solvents, such as alcohol, ether and benzene". Eighty-six per cent pure PCP is commercially available for use (NPIRS, 2021).

Trade and Other Names PCP has been sold with different trade names, namely, "Dowicide, PCP, Penchlorol, Penta, Penta Plus, Pentachloral, Pentacon, Penwar, Priltox, Santobrite, Santophen, Sinituho and Weedone".

K. Kumari (✉) · R. Jain
CSIR-National Environmental Engineering Research Institute (NEERI), Kolkata Zonal Centre, Kolkata, West Bengal, India
e-mail: k_kumari@neeri.res.in

K. Kumari, *Pollutants of Global Concern*, Emerging Contaminants and Associated Treatment Technologies, https://doi.org/10.1007/978-3-031-50996-4_19

2 Technical and Physicochemical Properties of PCP and Its Esters

PCP decomposes >200 °C and produces toxic by-products, i.e. fumes and gases that include hydrogen chloride, dioxins, chlorinated phenols which react pugnaciously with strong oxidants and water, resulting in fire and explosion threat. Technical details of Pentachlorophenol, its salts and ethers (PCP) are given in Table 1, whereas Table 2 outlines the structural formula.

In its purest form, PCP is moderately volatile; crystal like white/tan white structure. Technical grade PCP is 86% pure with 10% other chlorophenols and 4% in the inert form. PCP and its esters (Na-PCP and PCP-L) are commercially available in form of solid slabs, shavings, grinded powder or as a dilatable liquid (UNEP/POPS/POPRC.9/13/Add.3).

3 Adverse Effects

3.1 Toxicological Characteristics

Classification

- WHO—Class Ib (highly hazardous: WHO, 1994).
- EU—Carcinogen (Cat.3): highly toxic; irritant.
- IARC—No evaluation of carcinogenicity because data are inadequate.
- In general, PCP is highly toxic to mammals and birds. It is the most toxic compound of the chlorophenols family tested yet. Some of the acute effects of exposure to commercial PCP are attributable to micro-contaminants present in the technical preparation (IPCS, 1987).
- Uptake—PCP is readily absorbed when applied to the skin.

Table 1 Technical details and physicochemical properties of Pentachlorophenol, its salts and esters (PCP)

POPs name	Density	Solubility	Vapour pressure	Octanol/ water partition coefficient	Octanol-air partition coefficient	Henry's Law Constant
Pentachlorophenol, its salts and esters (PCP)	1.98 at 59 °F	Less than 1 mg/mL at 68 °F	0.00011 mm Hg at 68 °F, 40 mm Hg at 412.2 °F	5.12 (LogP)	–	2.45e-08 atm-m^3/mole

Table 2 Structural formulas of PCP, its salt Na-PCP and ester PCP-L

	Pentachlorophenol	Sodium pentachlorophenate	Pentachlorophenyl laurate
Chemical name and abbreviation	2,3,4,5,6-pentachlorophenol (PCP)	Na-PCP	PCP-L
CAS number	87-86-5	131-52-2 and 27735-64-4 (as monohydrate)	3772-94-9
Molecular formula	C_6HCl_5O	C_6Cl_5ONa	$C_{18}H_{23}Cl_5O_2$
Molecular mass	266.34 g/mol	288.32 g/mol	448.64 g/mol
Structural formulas of PCP, its salt and ester, as well as the main transformation product	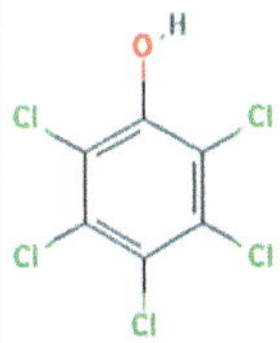	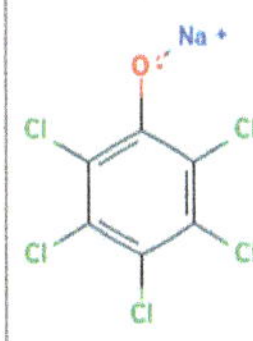	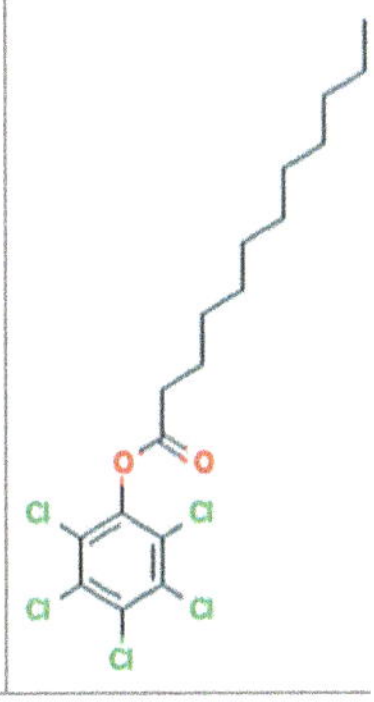

3.2 *Acute Toxicity*

Oral LD_{50}: 27–205 mg/kg (tests with a variety of species) (IPCS, 1987).
Inhalation Rabbit (35.2%) LD_{50}: 201 mg/kg.
Dermal LD_{50}: 80–350 mg/kg (tests with a variety of species) (IPCS, 1987), causes irritation on skin, eyes and mucous membranes (US PHS, 1989; IPCS, 1987).

3.3 *Short-Term Toxicity*

Reproduction

PCP and its dioxin and hexachlorobenzene exhibit foetotoxicity and teratogenicity when administered during early gestation phase into female human body (Eisler, 1989). NOEL for teratogenicity, foetotoxicity and embryotoxicity in rats was observed to be 10 mg/kg. The NOEL for subchronic inhalation as determined from animal-based study was observed to be 1.25 mg/kg/day (25 ppm, LDT).

Chronic Toxicity

Carcinogenic assessment carried by an US EPA group stated PCP as non-oncogenic (Williams, 1982). Similar observations were reported by Innes et al. (1969), with no significant increase in tumour incidences in mice.

Mutagenicity
Insufficient human experimentations and data reports PCP as non-mutagenic (Williams, 1982). Whereas, other PCPs have reported immune toxic and neurotoxic effects on mice, rats, chicken and cattle (IPCS, 1987).

Epidemiological Data
Limited epidemiological studies suggest adverse effect of PCP on liver, kidney, skin, blood, lungs and central nervous system. Analysis may vary as per sample size, regular follow up and exposure duration.

4 Production and Usage

Worldwide production of pentachlorophenol was estimated to be 90,000 tonnes in 1981. Figure 1 represents the production of PCP in the USA, Canada and China in different years. In most of the European countries, namely, Poland, Germany, the Netherlands, Switzerland, the UK, Spain and France production ceased after 1992. Spain stopped production in 2003. Canada imported 472 tons from the USA and Mexico in 2007 (CAREX Canada, 2009).

Until 1986, 97% of the total production was used as wood preservative, 1% as biocide and remaining for other smaller applications (IARC, 1991). Currently, it has been listed as a restricted use pesticide and hence its use as wood preserving agent, herbicide, biocide in indoor places is prohibited (ATSDR, 2001). The use of pentachlorophenol (and its sodium salt) is currently restricted to industrial areas in Canada and USA (such as utility poles, cross arms, railroad cross ties, wooden pilings, fence posts and lumber/timbers for construction); products containing pentachlorophenol remain registered for heavy-duty wood preservation, primarily to treat utility poles and cross arms (CAREX Canada, 2009; EPA, 2010). Worldwide usage of PCP and its esters has been described in Table 3.

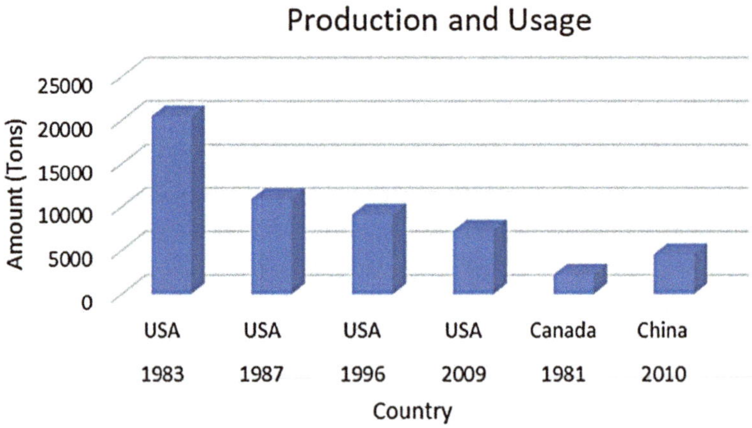

Fig. 1 ATSDR (2001), United Nations (2010) and CAREX Canada (2009)

Table 3 Worldwide usage of PCP and its esters

Country	Usage	References
USA	Used in paints, canvas, ropes, insulation, adhesives, leather and brick walls as wood preservative, algicide, defoliant, herbicide, germicide, molluscicide, fungicide, etc.	NTP (1999) and EPA (2010)
Tuscany, Italy and Sweden	Tanneries	Seniori-Costantini et al. (1989), Mikoczy et al. (1994) and Mikoczy and Hagmar (2005)
China	Molluscicide	Zheng et al. (2012)
	Pentachlorophenol laurate used in textiles and other fabrics	United Nations (2010)
Canada and USA	As railroad cross ties, wooden pilings, utility poles, fence posts, cross arms in industries	CAREX Canada (2009) and EPA (2010)
European Union (mainly France, Portugal and Spain)	As a waterproof material in millwork to prevent microbial growth. As Anti-sap stain in green lumber	IARC (1999), OSPAR (2004) and INERIS (2011)
North America	As wood preservative	United Nations (2010)

5 Sources of Environmental Releases for PCP

High adsorption coefficient lets PCP residues into soil and sediment. PCP is reasonably importunate with approximate half-life (t1/2) between 45 and 70 days in soil matrix of open fields. Microbial processes reduced the half-life to 15–48 days. In the presence of high organic matter and anaerobic soil, conditions of PCP reduced very swiftly when incubated at high temperature. Its defoliant and desiccant makes them highly toxic towards plants. PCP gets accumulated into the food chain of aquatic organisms by uptake of contaminated water and food (WHO, 1987).

Major release or disposal points of PCP include the following:

- PCP treatment facilities.
- Evaporation during in service life of PCP-treated products and stockpiles.
- Waste handling processes where PCP-treated products are sent to landfill or incinerated.
- Contaminated sites resulting from historical use/production of PCP.
- Improper practices (e.g. spills from industrial holding ponds from wood treatment facilities).
- Re-volatilisation from adsorbed residues.
- Domestic or forest fires.

6 Environmental Detection and Exposure

PCP and its metabolites have been detected in different environmental matrix, namely, air, soil, water and its sediments. Human exposure occurs through foul air, contaminated water, food or through dermal contact to PCP-coated products. Although due to restricted use, chances of exposure lessen but occupational exposure at manufacturing units by inhalation or skin absorption is quite possible. Acute inhalation of PCP exposure in humans may have an effect on the cardiovascular system, blood, liver and eyes. PCP is thought to be carcinogenic in nature.

The main contaminants include other polychlorinated phenols, polychlorinated dibenzo-p-dioxins and polychlorinated dibenzo furans (IARC, 2019).

6.1 Exposure

Occupational Exposure

It majorly occurs to the workers that are employed in PCP manufacturing units as well as during making different formulations of PCP. Agricultural exposure occurs while spraying these formulations onto fields as pesticides, exposure to workers occurs doing wood preservation and other painting works where PCP has been used. Workers in waste management and incineration units also get exposed to this hazardous chemical. Due to its volatile and adsorbing nature, it readily absorbs into human body and hence significant amounts of these hazardous chemicals have been found in blood, plasma, serum, urine samples of workers in different occupations as listed in Table 4.

Community Exposure Exposure of PCP to public may occur even long time after cessation of its production and use through wood and leather products coated with PCP. Due to its persistent nature it, accumulates into food chain especially grains, fruits and vegetables and hence enters human body. Other sources of exposure may be land, air and soil sediments. Significant concentrations of PCP have been detected in blood plasma, umbilical cord, spinal cord, serum, etc., in different age group communities.

In Belgium, spinal fluid samples (16 individuals, aged 21–74) showed a mean exposure of 0.75 ± 0.49 μg/L with a range of 0.24–2.03 μg/L, and serum samples from the same population exhibited a mean exposure of 22 ± 16 μg/L within a range of 4 to 60 μg/L. These individuals had no correlation with serum PCP levels, but three subjects had previously used PCP-containing wood preservatives (Jorens et al., 1991). In Canada, umbilical cord plasma samples (30 individuals, aged 21–74) revealed a median exposure of 1670 pg/g wet weight, ranging from 628 to 7680 pg/g wet weight, with 100% detection. Urine samples (31 individuals, aged 36–76) from sport fish consumers around the Great Lakes showed a mean exposure

Table 4 PCP concentrations in biological samples from occupationally exposed workers

Sample matrix	Country, year	Occupation	Work tasks, type of worker or specific exposure	No. of workers	Exposure		References
					Level	Range	
Blood	USA, NR	Wood treatment	Treatment of lumber, furniture and other wood products	18	Mean, 2190–5140 μg/L	190–14,000 μg/L	Begley et al. (1977)
Blood	Portugal, NR	Wood transformation unit	Mean, 6 years of exposure	11	Mean, 2273 μg/L	133–6884 μg/L	Ferreira et al. (1997)
Plasma	Italy, NR	Wooden strip board factory	Applied PCP by brush to wooden boards	14	Mean, 288.7 μg/L; median, 67 μg/L	2–1442 μg/L	Colosio et al. (1993)
Plasma	UK, 1982–1983	Wood treatment	Remedial sprayers working on house timber treatment	108	Mean, 6.0 mmol/L[16 × 105 μg/L]	0.2–29 mmol/L[0.5 × 105–77.2 × 105 μg/L]	Jones et al. (1986)
Serum	Canada, NR	Sawmill workers	Tasks involving dermal contact with treated lumber	5	Mean, 714 ± 383 μg/L	NR	Embree et al. (1984)
Serum	USA, 1967–1973	Agricultural workers and wood processing workers	Farmers or pest control operators	280	Mean, 250 μg/L	< 10–8400 μg/L	Klemmer et al. (1980)
Serum	Finland, NR	Sawmill workers	Moving lumber that was dipped in chlorophenol solution containing 23% TCP, 74% tetrachlorophenol, 3% PCP	7	Mean, (p.m.) 0.85 μmol/L [226 μg/L]	NR	Pekari et al. (1991)
Urine	Germany, NR	Municipal waste incinerator	Municipal waste workers	53	Mean, 2.60 μg/g creatinine	0.43–8.88 μg/g creatinine	Angerer et al. (1992)
Urine	Germany, NR	Municipal waste incinerator	Unexposed	248	Mean, 3.21 μg/g creatinine	< 0.8–67.79 μg/g creatinine	

(continued)

Table 4 (continued)

Sample matrix	Country, year	Occupation	Work tasks, type of worker or specific exposure	No. of workers	Exposure		References
					Level	Range	
Urine	Germany, 1990–1993	PCP-exposed construction painters, bricklayers	40% of the painters are exposed to PCP free wood preservatives once in a week	189	Median, 2.4 μg/g creatinine	< 0.2–52 μg/g creatinine	Bader et al. (2007)
Urine	Germany, 1990–1993	PCP-exposed construction painters, bricklayers	< 10% Bricklayers are reportedly in contact of wood preservatives per month	148	Median, 1.8 μg/g creatinine	< 0.2–25 μg/g creatinine	
Urine	Spain, 1999–2011	Hazardous waste incinerator	Plant workers, working in different mechanical sections, laboratory workers, administrative/management workers	27	Annual means, 0.1–1.9 μg/g creatinine	NR	Agramunt et al. (2003) and Mari et al. (2009, 2013)
Urine	USA, NR	Wood treating plant	PCP application to the lumber—vat dipping, lumber—tank	22	Mean, 4200 μg/L	NR	Casarett et al. (1969)
Urine	Italy, NR	Wooden strip board factory	Applied PCP by brush to wooden boards, handled treated wood and other indirect exposure	32	Mean, 281.3 μg/L; median, 194 μg/L	2–324 μg/L	Colosio et al. (1993)
Urine	Italy, NR	Wooden strip board factory	Unexposed	37	Mean, 4.7 μg/L; median, 3.7 μg/L	0–27 μg/L	
Urine	Spain, 1999–2000	Hazardous waste incinerator	Plant workers, laboratory workers	22	Annual means 0.64–3.8 μg/g creatinine	NR	Domingo et al. (2001) and Schuhmacher et al. (2002)
Blood	Germany, NR	PCP production facility	PCP production area, transported, weighed raw materials for PCP production	42	5415.71 ± 3667.14 μg/L	NR	Bauchinger et al. (1982)

GSD geometric standard deviation, *LOD* limit of detection, *Na-PCP* sodium pentachlorophenate, *NA* not applicable, *NR* not reported, *PCP*, pentachlorophenol

of 1.3 μg/L with a range of <2–3.2 μg/L (Anderson et al., 1998). In Germany, urine samples (1895 individuals, aged 25–69) had a geometric mean exposure of 2.7 μg/L (1990/92) and 1 μg/L (1998), with exposure ranging from <1 to 13 μg/L (95th percentile) in 1990/92 and 50% detection in 1998. There was no difference in exposure based on the presence of wood preservatives at home (Schulz et al., 2007). Plasma samples (623 individuals, aged 0–62) exhibited a mean exposure of 2.4 μg/L and a median of 1.7 μg/L, ranging from < LOD to 59.3 μg/L (Heudorf et al., 2000). Urine samples (1295 adults, aged 18–79) showed a geometric mean exposure of 2.67 μg/L, with a 95th percentile of 12.8 μg/L (Seifert et al., 2000). In Norway, plasma samples (281 individuals, aged 48–62) displayed a mean exposure of 958 ng/L and a median of 711 ng/L, ranging from < LOD to 7686 ng/L, with 94% detection (Rylander et al., 2012). Nigeria reported blood samples (29 adults) with exposures ranging from < trace to 21.3 ppb (trace–0.23 μg/L), detected in 100% of cases (Atuma & Okor, 1985). In Spain, serum samples from children (66 individuals, aged 4) showed a mean exposure of 6.4 μg/L within a range of 1.5–35 μg/L, while serum samples from children in rural environments (131 individuals, aged 4) exhibited a mean exposure of 0.61 μg/L with a range of <0–4.7 μg/L. Adults in Spain showed serum exposure with a mean of 21.9 μg/L and a range of 2.5–116.5 μg/L, along with urine exposure with a mean of 25 μg/L and a range of 4–136 μg/L, with a correlation with tetrachlorophenol ($r = 0.883$) (Gómez-Catalán et al., 1987). In Sweden (2000–2001), breast milk samples (15 individuals, mean age 32) exhibited exposure of 20 pg/g fresh weight, with a range of 10–570 pg/g fresh weight. Maternal plasma samples showed exposure of 2830 pg/g fresh weight with a range of 1360–13,200 pg/g fresh weight, and cord blood plasma samples had exposure of 1960 pg/g fresh weight, ranging from 820 to 7580 pg/g fresh weight (Guvenius et al., 2003). In the USA, urine samples from adults (106 individuals) revealed a mean exposure of 0.7 μg/g creatinine and a median of 0.4 μg/g creatinine, with a range of <0.2–6.4 μg/g creatinine and 96% detection (Morgan, 2015). Pregnant women in the USA showed urine exposures (747 individuals) ranging from <0.9 to 29.5 μg/L (2–4% detects) (Castorina et al., 2010). Another group of 254 individuals aged 1.5–5 exhibited mean urine exposures ranging from 0.61 to 1.27 μg/L and GM exposure of 0.63–2.20 μg/L, with detection in 94% of cases (Wilson et al., 2007). Pregnant women in the USA (361 individuals) had a median urine exposure of 7.3 μg/g creatinine within a 10–90th percentile range of 1.1–67 μg/g creatinine (Berkowitz et al., 2003). Additionally, urine samples from adults in the USA (886 individuals, aged 20–59) exhibited a median exposure of 1.2 μg/g creatinine, ranging from <1 to 29 μg/g creatinine, with detection in 59% of cases (Hill et al., 1995). Urine samples (6990 individuals, aged 12–74) had a median detected exposure of 6 μg/L, with a range of <2–2670 μg/L and 72% detection (Kutz et al., 1992). Serum and urine samples from individuals in PCP-treated homes had higher exposure levels compared to those in conventional homes (means 420 vs. 40 ppb [μg/L], respectively) (Cline et al., 1989). Lastly, urine samples from children (197 individuals, aged 2–6) exhibited a median exposure of 14 μg/g creatinine, ranging from >1 (minimum NR) to 330 μg/g creatine, with 100% detection (Hill et al., 1989). In Portugal, urine samples from adults (10 individuals) displayed a mean exposure of 6 μg/L with a range of 1–31 μg/L (Ferreira et al., 1997), and serum samples from adults (10 individuals) showed a mean exposure of 15 μg/L within a range of 3–17 μg/L (Ferreira et al., 1997). Nigeria reported urine samples from adults (35 individuals) with exposures ranging from <0.025 to 0.23 ppm [25–230 μg/L], detected in 100% of cases (Atuma & Okor, 1985).

6.2 *Human Health Hazard Information*

Pentachlorophenol (PCP) significantly affects human body after ingestion. Animal experimentation research establishes its acute effect on "cardiovascular system, blood, liver, immune system, and central nervous system (CNS)" majorly through inhalation and oral exposure (McQueen, 2017).

Its acute and chronic side effects are listed as below.

6.2.1 Acute Effects

- Effects cardiovascular system, blood, liver (jaundice) and eyes (visual damage and irritation).
- Neurological effects include lethargy, tachypnoea, tachycardia, delirium, and convulsions.

6.2.2 Chronic Effects (Non-cancer)

- Inflammation of the upper respiratory tract and bronchitis, blood effects such as aplastic anaemia, effects on the kidney and liver, immunological effects and irritation of the eyes, nose and skin.
- Effects on the liver, kidney, blood, endocrine, immune system and CNS.

There is no specified Reference Concentration (RfC) for uptake and exposure regulation by EPA. Pathological studies based on liver and kidney pathology in rats have defined Reference Dose (RfD) as 0.03 milligrams per kilogram body weight per day (mg/kg/d). Beyond this concentration, PCP causes adverse health impacts.

6.2.3 Reproductive/Developmental Effects

PCP exposure leads to different reproductive and physiological problems like spontaneous abortions, unexplained infertility or menstrual disorders in women. However direct correlation could not be established due to presence of other aromatic compounds as confounding factors, i.e. lindane. Oral administration studies on rats suggest less survival rate of offsprings due to PCP exposure. PCP also exhibits maternal toxicity (depressed maternal body weight) but does not cause birth defects (Guo et al., 2016).

6.2.4 Cancer Risk

EPA has classified pentachlorophenol as a Group B2, probable human carcinogen. PCP inhalation has a reported association with certain types of cancer, namely, Hodgkin's disease, soft tissue sarcoma and acute leukaemia. However, direct correlation cannot be established due to associated exposure to other environmental toxins, namely, certain heavy metals and other persistent chemicals, which may also contributed to them. Oral administration of PCP also causes health problems. Calculated oral safe intake limit as per EPA guidelines has been observed to be 3×10^{-6} (μg/L).

Based on animal studies, a probable correlation between PCP inhalation and occurrence of tumours, i.e. cancer liver tumours (hepatocellular adenomas and carcinomas) and two uncommon tumours (adrenal medulla pheochromocytomas, hemangiosarcomas and hemangiomas) has been reported.

7 Regulations and Guidelines

7.1 International Context

PCP has been completely banned in 90 nations in the "Conference of Parties (COP) 7 of the Stockholm Convention held in Geneva in May 2015". IPEN states that 94 countries voted in favour of global prohibition, 2 opposed and 8 nations abstained themselves. However, this chemical was restricted in various countries much before it was banned in the Stockholm Convention as outlined in Table 5.

7.2 Occupational Limits for Pentachlorophenol in Air

The values of occupational limits are expressed as threshold limit value (TLV) after 8 h of time-weighted average (TWA) and a short exposure (STEL) of 15 min. The TLV recommended by American Conference of Governmental Industrial Hygienists (ACGIH) is 50 ppm (TWA) and a 15-min STEL of 100 ppm. The occupational limit prescribed by National Institute for Occupational Safety and Health (NIOSH) and limits set in different countries have been represented in Fig. 2.

Table 5 Restrictions posed in various countries

Sl. no	Country	Nature	Year
1.	Sri Lanka	Complete ban	1994
2.	Thailand	Agriculture	1995
		Public health	2000
		Industrial use	2001
3.	China	Complete ban	2015

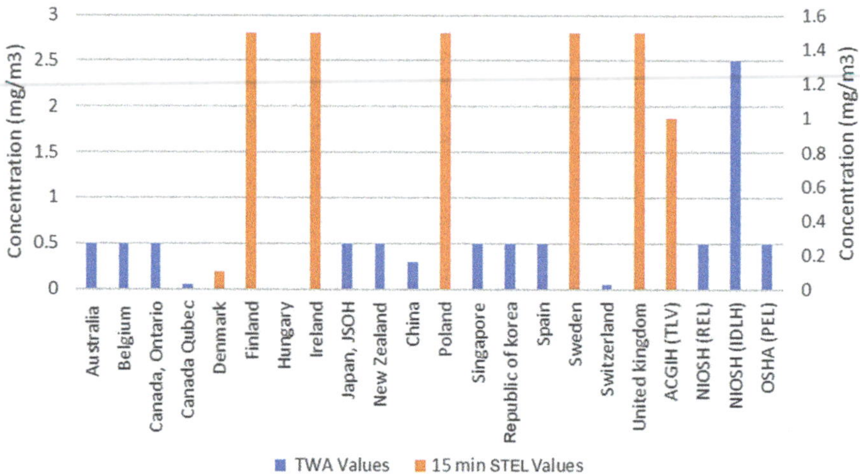

Fig. 2 Regulations and guidelines for occupational limits for pentachlorophenol

7.3 *Production and Stockpiles*

An Indian Chemical Council (ICC) report states that in Indian context, sodium ester of pentachlorophenol is majorly used as a wood preservative but nowadays its use as an antifouling agent in "distemper paints" has also been increased. Maharashtra and West Bengal are among the major contributors in the production of Na-PCP, i.e. 1800 tonnes per year.

Figure 3 represents the export data of India (Source: Toxic Link, 2018) and Fig. 4 represents top manufacturers of PCP. Maximum amount of PCP that has been exported from India is 20.08 thousand kgs to Nepal, until this date there is no data of import for this POP in India.

7.4 *Indian Context*

India opposed the ban on PCP in the Stockholm Convention meeting majorly due to two reasons—scientific infirmities in the report of Stockholm scientific committee and lack of suitable alternatives. Na-PCP is required to preserve wood and for forest

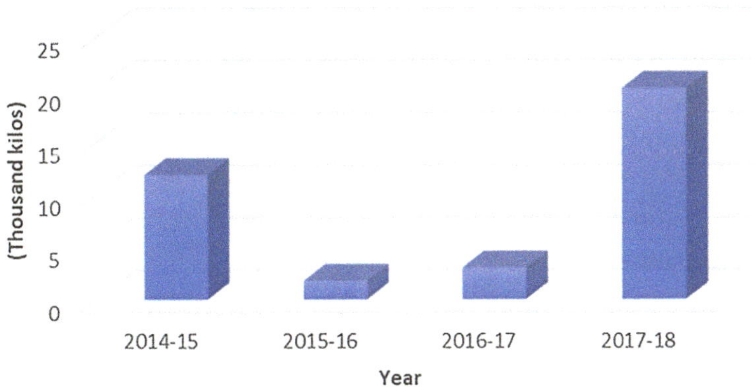

Fig. 3 Annual export of PCP and its esters from India

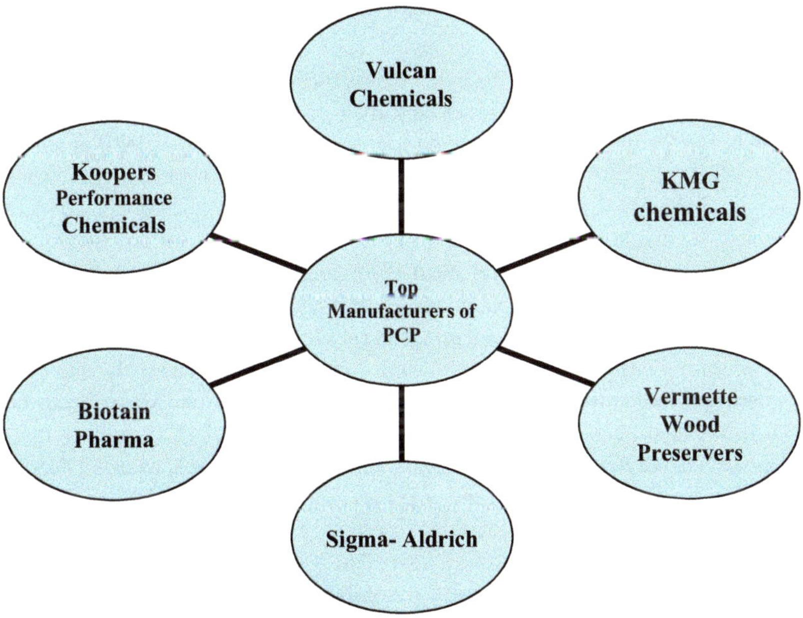

Fig. 4 Top manufacturer of Pentachlorophenol

conservation in India. They estimate that it will take at least 8–10 years in India to discover, produce and disseminate cost-effective alternatives to Na-PCP. In this context, ICC emphasised the socioeconomic significance of the wood industry in a country whose demand for timber is expected to rise from 58 million m^3 in 2005 to 153 million m^3 in 2020.

However, India imposed a ban on the use of PCP in agricultural and leather industry but still has not applied complete ban on its manufacture as well as usage. Central Insecticides Board and Registration (CIBR) had imposed ban on sodium ester of PCP (Na-PCP) for use in agricultural activities in 1998. Currently it is only used as preservative agent for wood and distemper paints during storage.

7.5 Research Work on PCP

Not much research has been done on the assessment of concentration and their adverse impacts of PCP exposure on humans and other aquatic animals. Reported literature in Indian and global context has been compiled subsequently.

7.6 In-House Work (Done in India)

Pandey (2015) studied the effect of Pentachlorophenol (PCP) on the histo-architectural alterations in the kidney and liver of Heteropneustes fossilis, a fresh water fish. For 21 days, fish were given low and high dosages of PCP at sublethal concentrations (6 g and 32 g/l/day). Histological examination revealed progressive degenerative abnormalities in the kidney, including glomerulosclerosis, an increase in Bowman's capsule space and tubular cell destruction and necrosis. Liver showed degeneration and disintegration in most cytoplasmic content, lymphatic aggregation, necrosis and rupture of hepatocytes. PCP, a widely used synthetic organochlorine compound, induces oxidative damage in rat kidneys, compromising antioxidant status, inhibiting enzymes, elevating plasma creatinine and urea levels, and causing DNA damage, indicating nephrotoxicity and renal dysfunction (Maheshwari et al., 2023). It has also been reported that oral administration of PCP induces reactive oxygen species (ROS) generation, impairs antioxidant defenses, causes DNA damage, inhibits brush border membrane enzymes, and disrupts metabolic pathways in rat intestines, highlighting its potential toxic effects on intestinal tissues (Maheshwari et al., 2022). Few degradation studies have also been reported from India. Santosh et al., (2017) in their study employed a combination of PCP-resistant Vibrio sp. (SK1) microorganisms and biochar derived from *Pistia stratiotes* to effectively remove pentachlorophenol (PCP) from Indian agricultural soil, achieving up to a 75% reduction in PCP levels within one week, with 5% biochar concentration significantly enhancing PCP stabilization, and kinetic modeling using various mathematical models provided estimations of kinetic parameters. Isolates from tannery effluent sludge in Kanpur, U.P., India, predominantly Pseudomonas and Arthrobacter species, exhibited varied pentachlorophenol (PCP) degrading capabilities. Arthrobacter showed the highest potential, degrading 55% of PCP within 30 days in lab-scale bioreactor studies (Tewari et al., 2011). Further it has been reported that a modified chitosan-based bioadsorbent, CHTAC, synthesized through crosslinking with 2-hydroxy-1-naphthaldehyde and grafting with CuCl2, exhibits significantly enhanced pentachlorophenol (PCP) adsorption capacity (94% improvement for (1) and 76% for (2)) compared to unmodified chitosan (CHT), with demonstrated efficacy in multiple adsorption-desorption cycles (Shankar et al., 2020).

7.7 Global Purview

Cooper and Jones (2008) had swotted epidemiologic studies to establish an association of PCP exposure with that of possible cancer risk and in turn extricate it from other chlorophenols and PCDD/Fs. He also stated that the association does not belong to the impurities present in PCP and its salts and esters, and therefore PCP has been categorised in Group 2B as a "possible human carcinogen" and also one of its esters 2,3,7,8-tetrachlorodibenzo-p-dioxin (2,3,7,8-TCDD) has been categorised in Group 1 as a "known human carcinogen" by IARC in 1997. Co-existence PCP and PCDD/Fs leads to collaborative risk of cancer, which should be considered as a major health concern (Cooper & Jones, 2008).

Zheng et al. (2012) has reviewed the occurrence and concentration of PCP in different human urine samples across different countries in various decades. PCP concentration decreased during 1967–2001 in different samples collected from USA and Germany, which can be very well explained in lieu of the ban imposed on PCP production in this region. In comparison to other western countries, China displayed lowest PCP concentration during the 1990s but unexpected high concentration in "Penshan County" was comparable to the levels in the USA and Germany when PCP was extensively used (Fang et al., 2010). Even the least detected Na-PCP concentration, i.e. 13.20 μg/L was higher than the 1992 established German Human Biological Monitoring Value (HBM) of 12 μg/L. HBM I (40 μg/L) and HMB II (70 μg/L) are the reference standards of environmental toxins apt to be present in human blood and urine samples collected from exposed population (Zheng et al., 2012).

In his urinary bio-monitoring analysis report Morgan et al. (2015) stated that children exposed to different persistent pesticides and other POPs (CTEPP OH) are regularly exposed to PCP once in their lifetime. Urinary PCP concentration as observed in different children's samples was found highly variable and several spot urine samples (for 48 h approx.) were collected in order to understand the variation. In addition, they identified several other specific factors like age of residence, sampling season, time spent outside and pet ownership also contributes to their enhanced exposure and variable PCP concentration in urine samples collected from children (Morgan, 2015). Table 6 enlists different studies conducted to estimate urinary PCP concentration in children ageing in the range of 2–14 years.

Table 6 Reported urinary PCP levels (ng/mL) in young children from published studies worldwide

Country	Location	Year	N	Age (years)	Median	95th	Maximum	References
Germany	National (GerES II)	1990–1992	695	6–14	4.6	14.9	26.5	Seifert et al. (2000)
Germany	National (GerEs IV)	2003–2006	462	6–14	<0.6	1.6	–	Schultz et al. (2007)
USA	National (NHANES)	2003–2004	290	6–11	<0.5	5.7	–	CDC (2014)
USA	Arkansas	1980s	197	2–6	14	110	240	Hill et al. (1989)
USA	North Carolina	1997	9	2–5	0.3[e]	–	0.7	Wilson et al. (2003)
USA	North Carolina	2000–2001	128	2–5	0.4	1.9	3.5	Wilson et al. (2007)
USA	Ohio	2001	115	2–5	0.8	3.3	23.8	Morgan et al. (2015)

8 Remedial Action Taken for Management of POP

8.1 *Indian Context*

No relevant information found in Indian context. Research work done in different countries is compiled below.

8.2 *International Context*

PCP during its manufacturing process forms PCDD/Fs, a micro-pollutant which forms a lethal combination with parent compound. Impregnation in soil is ought to be the major reason for their accumulation near production sites, intentional spiking of this toxic POP also shares some part in its environmental presence. Standard remediation techniques using micro-organisms and plants have been tested and tried for PCP removal from soil and sediments (Lopez-Echartea et al., 2016), collected from the proximity of sawmills and other PCP manufacturing units as detailed in Table 7.

Alternatives

Different alternative chemicals are available and widely used in place of PCP and they are both environmentally safe and economical viable. This indicates the technical feasibility in eliminating the use of toxic chemical belonging to chlorophenol family. Therefore, the basic utility of chlorophenols can be met but like any other chemical, they have their own pros and cons as stated below.

Table 7 Biological physicochemical remediation/reduction of PCP

Medium	Experimentation	Inference	References
Soil and sediments	Bioremediation using straw compost and bioaugmentation of remediated soil	56% of spiked PCP reduced within 4 weeks without formation of any partial by-products	Laine and Jørgensen (1996)
Soil and sediments	Pilot-scale composting	90% of chlorophenols were degraded by indigenous microbes in both inoculants medium	Laine and Jørgensen (1997)
Soil and sediments	Farmyard manure	Significant reduction in PCP with formation of tetra and triCP	Jaspers et al. (2002)
Soil and sediments	Composting	72% PCP removal, highest in thermophilic phase	Zeng et al. (2011)
Sawmill soil and impregnated wood	Composting	No harmful metabolites were formed and PCDD/Fs were detected and monitored throughout the process	Laine et al. (1997)
Soil and sediments	Mesophilic, fed-batch composting (FBC) reactor	15–16% reduction of PCDD/Fs with a half-life of 4 months further reduction exhibit complete mineralisation with formation of small amount of metabolic by-product	Narihiro et al. (2010)
Soil and sediments	Effect of nutrient amendments and temperature variation on PCP mineralisation by biopiles	Inorganic nutrient amendment (inorganic chloride) resulted in 76% reduction of PCP, temperature did not have any significant impact	Miller et al. (2004)
Soil and sediments	Effect of inoculation of high organics nurtured microbes in CP reduction	Addition of thin layer of the pine forest humus resulted in 40% increase within 24 days. Control soil did not show any degradation	Sinkkonen et al. (2013)
Physicochemical methods			
Soil and sediments	Partial oxidation by Fenton process	To 99% of TeCDD was converted chlorobenzenes and CPs	Kao and Wu (2000)
Soil and sediments	Evaluation of the efficacy of TeCDD remediation under different oxidation/ reduction conditions and different inoculums	The effectiveness of reductive dechlorination increased in microcosms containing activated sludge as inoculum and sludge cake as carbon source	Kao et al. (2001)
Soil sediments	pCDD/Fs reduction by augmentation of oxygen and nutrient supplement	100% of OCDF and 97.75% of HeCDFs were degraded within 12 weeks of inoculation, with an increase in the concentration of less chlorinated congeners of DD/ Fs, no degradation of OCDD	Chen et al. (2013)
Soil and sediments	OCDF reduction by augmentation of microbial inoculum	82.1% and 87.9% reduction of OCDF and OCDD respectively. Toxicity of the treated soil was also comparatively reduced	Chen et al. (2013)

(continued)

Table 7 (continued)

Medium	Experimentation	Inference	References
Soil and sediments	Compost-amended landfill reactors-based reduction of PCDD/Fs	61% reduction in the presence of paper and pulp industry sludge followed by 36% and 32% reduction by using plant leaves and cattle manure as a secondary treatment step under hypoxic conditions	Chen et al. (2016)
Soil and sediments	Biological reduction of dioxins under sequential bioreactor process 17 weeks anaerobic and 6 weeks aerobic	40% and 60% reduction of TeCDD sequential anaerobic and aerobic treatment, respectively	Binh et al. (2016)
Soil sediments	Dechlorination of 1,2,3,4-TeCDD by bioaugmentation/ biostimulation	Bioaugmentation of sediments from River Kymijoki showed highest removal efficiency of 76–80% after 120 days. Halogenated co-amendments also increased the removal efficiency	Ahn et al. (2008)
Soil and sediments	Dechlorination of 1,2,3,4-TeCDD/F and OCDF by enrichment cultures	Reductive microbial dechlorination ought to be the major removal mechanism	Liu et al. (2014)
Soil and sediments	Phytoremediation by using four local plants—*Trifolium repens* L *Cynodondactylon Agrostis palustris, Huds Zoysia japonica*	*Trifolium repens* L removed 66% DF within 60 days	Wang and Oyaizu (2009)
Soil and sediments	Bio-film-mediated rhizoremediation	White clover seeds augmented with *Comamonas* sp. strain KD7 showed significant reductions in DF (43%), 2,8-DCDF (68%) and DD (58%)	Wang and Oyaizu (2011)

Usage of PCP as a wood pole preserving agent is far more effective in comparison to its alternatives creosote and CCA, which is extensively used in preservation of southern pine and Douglas fir in Western part of the USA.

Distortion in wood cross arms which exert strain on allied electrical wires leading to power outage.

9 Recommendations

India was one of the countries, which opposed the inclusion of PCP and its esters in the list POPs in the seventh COP meeting of the Stockholm Convention, 2015. Therefore, in spite of being the prominent producer and exporter, it has been banned in the agriculture and leather industries but has been used for wood preservation purposes-impregnated wood/particleboards. However, the leather industry in India is faced with a ban in European countries on the use of pentachlorophenol (RIS, 2001). Taking into consideration of its toxic properties, the scientific community in India should assess the usage of this toxic chemical in terms of both environmental sustainability and health impact and give their recommendation to the Government of India.

References

Agramunt, M., Domingo, A., Domingo, J. L., & Corbella, J. (2003). Monitoring internal exposure to metals and organic substances in workers at a hazardous waste incinerator after 3 years of operation. *Toxicology Letters, 146*(1), 83–91. https://doi.org/10.1016/j.toxlet.2003.09.003

Ahn, Y. B., Liu, F., Fennell, D. E., & Haggblom, M. M. (2008). Biostimulation and bio-augmentation to enhance dechlorination of polychlorinated dibenzo-*p*-dioxins in contaminated sediments. *FEMS Microbiology Ecology, 66*, 271–281. https://doi.org/10.1111/j.1574-6941.2008.00557.x

Anderson, H. A., Falk, C., Hanrahan, L., Olson, J., Burse, V. W., Needham, L., Paschal, D., Patterson, D., Hill, R. H., & The Great Lakes Consortium. (1998). Profiles of great lakes critical pollutants: A sentinel analysis of human blood and urine. *Environmental Health Perspectives, 106*(5), 279–289. https://doi.org/10.2307/3434016

Angerer, J., Heinzow, B., Reimann, D. O., Knorz, W., & Lehnert, G. (1992). Internal exposure to organic substances in a municipal waste incinerator. *International Archives of Occupational and Environmental Health, 64*(4), 265–273. https://doi.org/10.1007/BF00378285

ATSDR. (2001). *Toxicological profile for pentachlorophenol*. United States Department of Health and Human Services, Public Health Service, Agency for Toxic Substances and Disease Registry. Available from http://www.atsdr.cdc.gov/toxprofiles/tp51.pdf

Atuma, S. S., & Okor, D. I. (1985). Gas chromatographic determination of pentachlorophenol in human blood and urine. *Bulletin of Environmental Contamination and Toxicology, 35*(3), 406–410. https://doi.org/10.1007/BF01636530

Bader, M., Zimmer, H., & Triebig, G. (2007). Urinary pentachlorophenol in painters and bricklayers in a four-years time interval after the PCP prohibition ordinance in Germany. *Industrial Health, 45*(2), 338–342. https://doi.org/10.2486/indhealth.45.338

Bauchinger, M., Dresp, J., Schmid, E., & Hauf, R. (1982). Chromosome changes in lymphocytes after occupational exposure to pentachlorophenol (PCP). *Mutation Research, 102*(1), 83–88. https://doi.org/10.1016/0165-1218(82)90148-3

Begley, J., Reichert, E. L., Rashad, M. N., Klemmer, H. W., & Siemsen, A. W. (1977). Association between renal function tests and pentachlorophenol exposure. *Clinical Toxicology, 11*(1), 97–106. https://doi.org/10.3109/15563657708989823. PMID: 872545.

Berkowitz, G. S., Obel, J., Deych, E., Lapinski, R., Godbold, J., Liu, Z., et al. (2003). Exposure to indoor pesticides during pregnancy in a multiethnic, urban cohort. *Environmental Health Perspectives, 111*(1), 79–84. https://doi.org/10.1289/ehp.5619

Binh, N. D., Imsapsangworn, C., Kim Oanh, N. T., Parkpian, P., Karstensen, K., Giao, P. H., & DeLaune, R. D. (2016). Sequential anaerobic-aerobic biodegradation of 2,3,7,8-TCDD contaminated soil in the presence of CMC-coated nZVI and surfactant. *Environmental Technology, 37*, 388–398. https://doi.org/10.1080/09593330.2015.1070918

CAREX Canada. (2009). *Pentachlorophenol. Profile*. CAREX Canada. Available from http://www.carexcanada.ca/en/pentachlorophenol/

Casarett, L. J., Bevenue, A., Yauger, W. L., Jr., & Whalen, S. A. (1969). Observations on pentachlorophenol in human blood and urine. *American Industrial Hygiene Association Journal, 30*(4), 360–366. https://doi.org/10.1080/00028896909343138

Castorina, R., Bradman, A., Fenster, L., Barr, D. B., Bravo, R., Vedar, M. G., et al. (2010). Comparison of current-use pesticide and other toxicant urinary metabolite levels among pregnant women in the CHAMACOS cohort and NHANES. *Environmental Health Perspectives, 118*(6), 856–863. https://doi.org/10.1289/ehp.0901568

CDC (Centers for Disease Control and Prevention). (2014, August). *Fourth national report on human exposure to environmental chemicals, updated tables*. Available online http://www.cdc.gov/exposure report/pdf/fourth report_updatedtables_aug2014.pdf. Accessed 11 Oct 2023.

Chen, W.-Y., Wu, J.-H., Lin, S.-C., & Chang, J.-E. (2016). Bioremediation of polychlorinated-*p*-dioxins/dibenzofurans contaminated soil using simulated compost-amended landfill reactors under hypoxic conditions. *Journal of Hazardous Materials, 312*, 159–168.

Chen, W. Y., Wu, J. H., Lin, Y. Y., Huang, H. J., & Chang, J. E. (2013). Bioremediation potential of soil contaminated with highly substituted polychlorinated dibenzo-*p*-dioxins and dibenzofurans: Microcosm study and microbial community analysis. *Journal of Hazardous Materials, 261*, 351–361. https://doi.org/10.1016/j.jhazmat.2013.07.039

Cline, R. E., Hill, R. H., Jr., Phillips, D. L., & Needham, L. L. (1989). Pentachlorophenol measurements in body fluids of people in log homes and workplaces. *Archives of Environmental Contamination and Toxicology, 18*(4), 475–481. https://doi.org/10.1007/BF01055012

Colosio, C., Maroni, M., Barcellini, W., Meroni, P., Alcini, D., Colombi, A., et al. (1993). Toxicological and immune findings in workers exposed to pentachlorophenol (PCP). *Archives of Environmental Health, 48*(2), 81–88. https://doi.org/10.1080/00039896.1993.9938399

Cooper, G. S., & Jones, S. (2008). Pentachlorophenol and cancer risk: Focusing the lens on specific chlorophenols and contaminants. *Environmental Health Perspectives, 116*, 1001–1008.

Domingo, J. L., Schuhmacher, M., Agramunt, M. C., Müller, L., & Neugebauer, F. (2001). Levels of metals and organic substances in blood and urine of workers at a new hazardous waste incinerator. *International Archives of Occupational and Environmental Health, 74*(4), 263–269. https://doi.org/10.1007/s004200000217

Eisler, R. (1989). Pentachlorophenol hazards to fish, wildlife, and invertebrates: A synoptic review. *U.S. Fish and. Wildlife Service Biological Report, 85*(1.17), 72 pp.

Embree, V., Enarson, D. A., Chan-Yeung, M., DyBuncio, A., Dennis, R., & Leach, J. (1984). Occupational exposure to chlorophenates: Toxicology and respiratory effects. *Journal of Toxicology. Clinical Toxicology, 22*(4), 317–329. https://doi.org/10.3109/15563658408992563

EPA. (2010). Toxicological review of pentachlorophenol (CAS No. 87-86-5). In *In support of summary information on the Integrated Risk Information System (IRIS)*. United States Environmental Protection Agency.

Fang, Y., Gao, X., Zha, J., Ning, B., Li, X., Gao, Z., & Chao, F. (2010). Identification of differential hepatic proteins in rare minnow (Gobiocypris rarus) exposed to pentachlorophenol (PCP) by proteomic analysis. *Toxicology Letters, 199(1), 69–79.*

Ferreira, A. J., Vieira, D. N., Marques, E. P., & Pedro, I. S. (1997). Occupational exposure to pentachlorophenol: The Portuguese situation. *Annals of the New York Academy of Sciences, 837*(1), 291–299. https://doi.org/10.1111/j.1749-6632.1997.tb56881.x. PMID: 9472347.

Fiege, H., Voges, H.-M., Hamamoto, T., Umemura, S., Iwata, T., Miki, H., Fujita, Y., Buysch, H.-J., Garbe, D., & Paulus, W. (2000). Phenol derivatives. In *Ullmann's encyclopedia of industrial chemistry*. Wiley-VCH. https://doi.org/10.1002/14356007.a19_313

Gómez-Catalán, J., To-Figueras, J, Planas, J., Rodamilans, M., & Corbella, J. (1987). Pentachlorophenol and hexachlorobenzene in serum and urine of the population of Barcelona. *Human Toxicology, 6*(5), 397–400. https://doi.org/10.1177/096032718700600509

Guo, J., Wu, C., Lv, S., Lu, D., Feng, C., Qi, X., et al. (2016). Associations of prenatal exposure to five chlorophenols with adverse birth outcomes. *Environmental Pollution, 214*, 478–484.

Guvenius, D. M., Aronsson, A., Ekman-Ordeberg, G., Bergman, Å., & Norén, K. (2003). Human prenatal and postnatal exposure to polybrominated diphenyl ethers, polychlorinated biphenyls, polychlorobiphenylols, and pentachlorophenol. *Environmental Health Perspectives, 111*(9), 1235–1241.

Heudorf, U., Letzel, S., Peters, M., & Angerer, J. (2000). PCP in the blood plasma: Current exposure of the population in Germany, based on data obtained in 1998. *International Journal of Hygiene and Environmental Health, 203*(2), 135–139. https://doi.org/10.1078/S1438-4639(04)70018-8

Hill, R. H., Jr., Head, S. L., Baker, S., Gregg, M., Shealy, D. B., Bailey, S. L., et al. (1995). Pesticide residues in urine of adults living in the United States: Reference range concentrations. *Environmental Research, 71*(2), 99–108. https://doi.org/10.1006/enrs.1995.1071

Hill, R. H., Jr., To T, Holler, J. S., Fast, D. M., Smith, S. J., Needham, L. L., et al. (1989). Residues of chlorinated phenols and phenoxy acid herbicides in the urine of Arkansas children. *Archives of Environmental Contamination and Toxicology, 18*(4), 469–474. https://doi.org/10.1007/BF01055011

IARC. (1991). Occupational exposures in insecticide application, and some pesticides. *IARC Monographs on the Evaluation of Carcinogenic Risks to Humans, 53*, 5–586. Available from http://publications.iarc.fr/71

IARC. (1999). Re-evaluation of some organic chemicals, hydrazine and hydrogen peroxide. *IARC Monographs on the Evaluation of Carcinogenic Risks to Humans, 71*, 1–315.

IARC. (2019). Pentachlorophenol and Some Related Compounds. IARC Monographs on the Evaluation of Carcinogenic Risks to Humans Volume 117, IARC, 2019, ISBN-1

INERIS. (2011). *Pentachlorophenol*. Villeurbanne, France Institut national de l'environnement industriel et des risques (INERIS). Available from https://www.ineris.fr/substances/fr/substance/getDocument/2818

Innes, J. R. M., Ulland, B. M., Valerio, M. G., Petrucelli, L., Fishbein, L., Hart, E. R., ... & Peters, J. (1969). Bioassay of pesticides and industrial chemicals for tumorigenicity in mice: a preliminary note. *Journal of the National Cancer Institute, 42(6), 1101–1114.*

IPCS. (1987). *Environmental health criteria no. 71*. International Programme on Chemical Safety, WHO, Geneva.

Jaspers, C. J., Ewbank, G., McCarthy, A. J., & Penninckx, M. J. (2002). Successive rapid reductive dehalogenation and mineralization of pentachlorophenol by the indigenous microflora of farmyard manure compost. *Journal of Applied Microbiology, 92*, 127–133. https://doi.org/10.1046/j.1365-2672.2002.01520.x

Jones, R. D., Winter, D. P., & Cooper, A. J. (1986). Absorption study of pentachlorophenol in persons working with wood preservatives. *Human Toxicology, 5*(3), 189–194. https://doi.org/10.1177/096032718600500307

Jorens, P. G., Janssens, J. J., van Tichelen, W. I., van Paesschen, W., de Deyn, P. P., & Schepens, P. J. (1991). Pentachlorophenol concentrations in human cerebrospinal fluid. *Neurotoxicology, 12*(1), 1–7.

Kao, C. M., Chen, S. C., Liu, J. K., & Wu, M. J. (2001). Evaluation of TCDD biodegradability under different redox conditions. *Chemosphere, 44*, 1447–1454. https://doi.org/10.1016/S0045-6535(00)00464-1

Kao, C. M., & Wu, M. J. (2000). Enhanced TCDD degradation by Fenton's reagent preoxidation. *Journal of Hazardous Materials, 74*, 197–211. https://doi.org/10.1016/S0304-3894(00)00161-8

Klemmer, H. W., Wong, L., Sato, M. M., Reichert, E. L., Korsak, R. J., & Rashad, M. N. (1980). Clinical findings in workers exposed to pentachlorophenol. *Archives of Environmental Contamination and Toxicology, 9*(6), 715–725. https://doi.org/10.1007/BF01055546

Kumari, K., Swamy, S., & Singh, A. (2022). Global monitoring plan on persistent organic pollutants (POPs). In *Persistent organic pollutants: Gaps in management and associated challenges* (1st ed., pp. 227–252). CRC Press.

Kutz, F. W., Cook, B. T., Carter-Pokras, O. D., Brody, D., & Murphy, R. S. (1992). Selected pesticide residues and metabolites in urine from a survey of the U.S. general population. *Journal of Toxicology and Environmental Health, 37*(2), 277–291. https://doi.org/10.1080/15287399209531670

Laine, M. M., Ahtiainen, J., Wågman, N., Öberg, L. G., & Jørgensen, K. S. (1997). Fate and toxicity of chlorophenols, polychlorinated dibenzo-*p*-dioxins, and dibenzofurans during composting of contaminated sawmill soil. *Environmental Science & Technology, 31*, 3244–3250. https://doi.org/10.1021/es970233z

Laine, M. M., & Jørgensen, K. S. (1997). Effective and safe composting of chlorophenol-contaminated soil in pilot scale. *Environmental Science & Technology, 31*, 371–378. https://doi.org/10.1021/es960176u

Laine, M. M., & Jørgensen, K. S. (1996). Straw compost and bioremediated soil as inocula for the bioremediation of chlorophenol-contaminated soil. *Applied and Environmental Microbiology, 62*, 1507–1513.

Liu, H., Park, J.-W., & Häggblom, M. M. (2014). Enriching for microbial reductive dechlorination of polychlorinated dibenzo-*p*-dioxins and dibenzofurans. *Environmental Pollution, 184*, 222–230.

Lopez-Echartea, E., Macek, T., Demnerova, K., & Uhlik, O. (2016). Bacterial biotransformation of pentachlorophenol and micropollutants formed during its production process. *International Journal of Environmental Research and Public Health, 13*(11), 1146.

Maheshwari, N., Khan, A. A., Ali, A., & Mahmood, R. (2022). Oral administration of pentachlorophenol impairs antioxidant system, inhibits enzymes of brush border membrane, causes DNA damage and histological changes in rat intestine. *Toxicol Res (Camb), 11(4):616–627.* https://doi.org/10.1093/toxres/tfac035. PMID: 36051662

Maheshwari, N., Khan, A. A., Mahmood, R. (2023). Pentachlorophenol causes redox imbalance, inhibition of brush border membrane and metabolic enzymes, DNA damage and histological alterations in rat kidney. *Pesticide Biochemistry and Physiology, 190, 105318, ISSN 0048-3575,* https://doi.org/10.1016/j.pestbp.2022.105318

Mari, M., Nadal, M., Schuhmacher, M., & Domingo, J. L. (2013). Body burden monitoring of dioxins and other organic substances in workers at a hazardous waste incinerator. *International Journal of Hygiene and Environmental Health, 216*(6), 728–734. https://doi.org/10.1016/j.ijheh.2013.01.003

Mari, M., Schuhmacher, M., & Domingo, J. L. (2009). Levels of metals and organic substances in workers at a hazardous waste incinerator: A follow-up study. *International Archives of Occupational and Environmental Health, 82*(4), 519–528. https://doi.org/10.1007/s00420-008-0350-0

McQueen, C. (2017). *Comprehensive toxicology*. Elsevier.

Mikoczy, Z., Schütz, A., & Hagmar, L. (1994). Cancer incidence and mortality among Swedish leather tanners. *Occupational and Environmental Medicine, 51*(8), 530–535.

Mikoczy, Z., & Hagmar, L. (2005). Cancer incidence in the Swedish leather tanning industry: updated findings 1958–99. *Occupational and Environmental Medicine, 62(7), 461.*

Miller, M. N., Stratton, G. W., & Murray, G. (2004). Effects of nutrient amendments and temperature on the biodegradation of pentachlorophenol contaminated soil. *Water, Air, and Soil Pollution, 151*, 87–101. https://doi.org/10.1023/B:WATE.0000009903.22105.30

Marsha, M., Jones, P., & Sobus, J. (2015). Short-term variability and predictors of urinary pentachlorophenol levels in Ohio Preschool Children. *International Journal of Environmental Research and Public Health, 12*, 800–815. https://doi.org/10.3390/ijerph120100800

Morgan, M. K. (2015). Predictors of urinary levels of 2,4-dichlorophenoxyacetic acid, 3,5,6-trichloro-2-pyridinol, 3-phenoxybenzoic acid, and pentachlorophenol in 121 adults in Ohio. *International Journal of Hygiene and Environmental Health, 218*(5), 479–488.

Narihiro, T., Kaiya, S., Futamata, H., & Hiraishi, A. (2010). Removal of polychlorinated dioxins by semi-aerobic fed-batch composting with biostimulation of "*Dehalococcoides*". *Journal of Bioscience and Bioengineering, 109*, 249–256.

NPIRS. (2021). *Active ingredient information* (PC Code: 63001). National Pesticide Information.

NTP. (1999). NTP toxicology and carcinogenesis studies of pentachlorophenol (CAS No. 87-86-5) in F344/N rates (feed studies). *National Toxicology Program Technical Report Series, 483*, 1–182.

OSPAR. (2004). *Pentachlorophenol. Hazardous substances series. OSPAR Commission 2001 (2004 Update)*. Committee on the Convention for the Protection of the Marine Environment of the North-East Atlantic.
Pandey, A. K. (2015). Histological changes in liver and kidney of cat fish, *Heteropneustes fossilis,* exposed topentachlorophenol (PCP). *Plant Archives, 15*(2), 1117–1120.
Pekari, K., Luotamo, M., Järvisalo, J., Lindroos, L., & Aitio, A. (1991). Urinary excretion of chlorinated phenols in saw-mill workers. *International Archives of Occupational and Environmental Health, 63*(1), 57–62. https://doi.org/10.1007/BF00406199. PMID:1856025.
RIS. (2001). Research and information system for the non-aligned and other developing countries. In *Product standards and trade in environmentally sensitive goods: A study of South Asian experience* (pp. 1–19). https://www.ris.org.in/sites/default/files/Publication/dp22_pap.pdf
Rylander, C., Lund, E., Frøyland, L., & Sandanger, T. M. (2012). Predictors of PCP, OH-PCBs, PCBs and chlorinated pesticides in a general female Norwegian population. *Environment International, 43*, 13–20. https://doi.org/10.1016/j.envint.2012.02.008
Santosh Kr, K., Satya Eswari, J., Rajput, V. D., Kumar. S, & Kumar, A. (2017). Modeling of simultaneous application of vibrio sp. (SK1) and biochar amendment for removal of pentachlorophenol in Soil. *Environmental Engineering Science, 551–561.* http://doi.org/10.1089/ees.2016.0456
Schuhmacher, M., Domingo, J. L., Agramunt, M. C., Bocio, A., & Müller, L. (2002). Biological monitoring of metals and organic substances in hazardous-waste incineration workers. *International Archives of Occupational and Environmental Health, 75*(7), 500–506. https://doi.org/10.1007/s00420-002-0340-6
Schulz, C., Conrad, A., Becker, K., Kolossa-Gehring, M., Seiwert, M., & Seifert, B. (2007). Twenty years of the German Environmental Survey (GerES): Human biomonitoring–temporal and spatial (West Germany/East Germany) differences in population exposure. *International Journal of Hygiene and Environmental Health, 210*(3–4), 271–297. https://doi.org/10.1016/j.ijheh.2007.01.034. Accessed 11 Oct 2023.
Seifert, B., Becker, K., Helm, D., Krause, C., Schulz, C., & Seiwert, M. (2000). The German Environmental Survey 1990/1992 (GerES II): Reference concentrations of selected environmental pollutants in blood, urine, hair, house dust, drinking water and indoor air. *Journal of Exposure Analysis and Environmental Epidemiology, 10*, 552–565.
Seniori-Costantini, A. S., Paci, E., Miligi, L., Buiatti, E., Martelli, C., & Lenzi, S. (1989). Cancer mortality among workers in the Tuscan tanning industry. *British Journal of Industrial Medicine, 46*(6), 384–388.
Shankar, A., Kongot, M., Saini, V. K., & Kumar, A. (2020). Removal of pentachlorophenol pesticide from aqueous solutions using modified chitosan. *Arabian Journal of Chemistry, 13(1), 1821-1830, ISSN 1878–5352.* https://doi.org/10.1016/j.arabjc.2018.01.016
Sinkkonen, A., Kauppi, S., Simpanen, S., Rantalainen, A. L., Strommer, R., & Romantschuk, M. (2013). Layer of organic pine forest soil on top of chlorophenol-contaminated mineral soil enhances contaminant degradation. *Environmental Science and Pollution Research, 20*, 1737–1745. https://doi.org/10.1007/s11356-012-1047-1
Tewari, P. C., Zulifiqar, S., Andrabi, A., Chaudhary, C. B., Shukla, S., Manohar, R., & Avadh, L. (2011). Screening of pentachlorophenol (PCP) degrading bacterial strains isolated from the tannery effluent sludge of Kanpur, India.
Toxic Link. (2018). *Persistent organic pollutants (POPs) in India: Country situation report* (pp. 4–68). https://ipen.org/sites/default/files/documents/POPs%20Country%20Situation%20report.pdf. Assessed 25 Aug 2023.
U.S. Public Health Service/Environmental Protection Agency. (1989, December). *The toxicological profile for pentachlorophenol*. USPHS & USEPA, Washington.
UNEP/POPS/POPRC.9/13/Add.3. Risk profile on pentachlorophenol and its salts and esters 2013. www.pops.int
United Nations. (2010). *Exploration of management options for pentachlorophenol (PCP).* Paper for the 8th meeting of the UNECE CLRTAP Task Force on Persistent Organic Pollutants, Montreal, 18–20 May 2010.

USEPA. (2006). *Consumer factsheet on: Pentachlorophenol.* United States Environmental Protection Agency, 28 November 2006.

Wang, Y., & Oyaizu, H. (2011). Enhanced remediation of dioxins-spiked soil by a plant-microbe system using a dibenzofuran-degrading *Comamonas* sp. and *Trifolium repens* L. *Chemosphere, 85*, 1109–1114. https://doi.org/10.1016/j.chemosphere.2011.07.028

Wang, Y., & Oyaizu, H. (2009). Evaluation of the phytoremediation potential of four plant species for dibenzofuran-contaminated soil. *Journal of Hazardous Materials, 168*, 760–764. https://doi.org/10.1016/j.jhazmat.2009.02.082

WHO. (1987). *Pentachlorophenol* (Environmental health criteria 71). World Health Organization, International Programme on Chemical Safety.

World Health Organization. (1994). Assessing human health risks of chemicals: Derivation of guidance values for health-based exposure limits. World Health Organization.

Williams, P. L. (1982). Pentachlorophenol: An assessment of the occupational hazard. *American Industrial Hygiene Association Journal, 43*, 799–810.

Wilson, N. K., Chuang, J. C., Morgan, M. K., Lordo, R. A., & Sheldon, L. S. (2007). An observational study of the potential exposures of preschool children to pentachlorophenol, bisphenol-A, and nonylphenol at home and daycare. *Environmental Research, 103*(1), 9–20. https://doi.org/10.1016/j.envres.2006.04.006

Wilson, N. K., Chuang, J. C., Lyu, C. W., Menton, R., & Morgan, M. K. (2003). Aggregate exposures of nine preschool children to persistent organic pollutants at day care and at home. *Journal of Exposure Analysis and Environmental Epidemiology, 13*, 187–202.

Zeng, G., Yu, Z., Chen, Y., Zhang, J., Li, H., Yu, M., & Zhao, M. (2011). Response of compost maturity and microbial community composition to pentachlorophenol (PCP)-contaminated soil during composting. *Bioresource Technology,* ***102***, 5905–5911.

Zheng, W., Yu, H., Wang, X., & Qu, W. (2012). Systematic review of pentachlorophenol occurrence in the environment and in humans in China: Not a negligible health risk due to the re-emergence of schistosomiasis. *Environment International, 42*, 105–116.

Perfluorooctane Sulphonic Acid (PFOS), Its Salts, and Perfluorooctane Sulphonyl Fluoride (PFOS-F)

Kanchan Kumari and Md Shahbaz

1 Introduction

Perfluorooctane sulfonate (PFOS, C8F17SO3) is a representative compound of perfluorinated chemicals, wherein all hydrogen atoms attached with carbon are substituted by fluorine atoms, and it terminates with a sulfonic group (Sun et al., 2017). PFOS is known for its exceptional chemical and physical characteristics, like oil and water resistance, heat resistance, and stability. Consequently, the compound has found various usage like additive in industrial, commercial, and consumer goods. Because of its widespread use, PFOS can be detected in environmental media and organisms worldwide (Martin et al., 2004a, b; Zareitalabad et al., 2013).

However, PFOS is considered a persistent organic pollutant with the ability to accumulate in living organisms. The US Environmental Protection Agency (USEPA) delegated PFOS as a latent contaminant in potable water in 2009 due to its bio accumulative nature (Lau et al., 2007; Pontius, 2019; Yamamoto et al., 2007). Researches have demonstrated that this organic compound exhibits toxic effects on flora and fauna (Austin et al., 2003; Li et al., 2020). For instance, high concentrations of PFOS can hamper the antioxidant defense mechanism of plants and destruct root tissues. While exposure to this POP can affect the hypothalamus of animals, such as rodent (Austin et al., 2003; Li et al., 2020). Therefore, there is a need for the development of effective technologies that can decompose PFOS under mild conditions.

Perfluorooctane sulfonate is an entirely fluorinated anion often utilized as a salt or integrated into more substantial macromolecules. It belongs in the family of perfluoroalkyl sulfonate substances and is known for its extreme persistence,

K. Kumari (✉) · M. Shahbaz
CSIR-National Environmental Engineering Research Institute (NEERI),
Kolkata Zonal Center, Kolkata, West Bengal, India
e-mail: k_kumari@neeri.res.in

K. Kumari, *Pollutants of Global Concern*, Emerging Contaminants and Associated Treatment Technologies, https://doi.org/10.1007/978-3-031-50996-4_20

substantial bioaccumulation, and bio magnification properties. Unlike many other persistent organic pollutants (POPs), PFOS does not primarily accumulate in adipocytes rather it is attached to the liver and blood proteins. Because of its capacity for long-range dispersion and alignment with the toxicity standards outlined in the Stockholm Convention, PFOS is categorized as a profoundly worrisome pollutant. Perfluorohexane sulfonate (PFHxS), perfluoroctanoic acid (PFOA) and PFOS are the three subgroups of Per- and Polyfluoroalkyl Substances (PFASs) currently listed under the Stockholm Convention as industrial POPs. PFOA and PFOS are the two PFAS that have been the most extensively produced and hence most studied among all. Bentel et al. (2019) show hydrated electrons effectively defluorinate PFOS, but highlight its persistence through challenging degradation, urging improved PFAS management. Electrochemical treatment degraded over 50% of PFOA and 33% of PFOS in saturated soil, but also mobilized both perfluorochemicals (PFCs) for potential removal (Hou et al., 2022).

PFOS is generated both deliberately and inadvertently through the degradation of interconnected human-made substances. Its deliberate utilization is extensive and encompasses uses in electric and electronic components, fire-suppressant foam, photographic imaging, hydraulic fluids, and textiles. Some countries continue to produce PFOS despite its environmental concerns (European Union, 2006).

2 Technical Details of PFOS, Its Salts, and PFOSF

The technical details and various physicochemical properties of PFOS have been described in tabular form in Table 1.

3 Status of Usage Within the Country

Production and Stockpiles
At present, there is a lack of production or stockpile data accessible for PFOS.

Regulation
As of now, India lacks dedicated regulations and policies regarding perfluorooctane sulfonic acid. In 2009, the Stockholm Convention incorporated perfluorooctane sulphonic acid (PFOS) (Kumari, 2021), its salts, and perfluorooctane sulphonyl fluoride (PFOSF) into Annex B. They are identified persistent organic pollutants (POPs) used as pesticides to kill insects. As of now, there is no specific regulation for them in India.

Various critical effect studies have assessed the toxicological endpoints and reference doses for specific substances. Luebker et al. (2005) found that reduced pup body weight occurred at 20 ng/kg-day, with an uncertainty factor of 30. This study pertained to lactating women who had a rate of consuming water 0.054 L/kg-day, resulting an RSC (Relative Source Contribution) of 20%. Similarly, in child

Table 1 Technical details and physicochemical properties of perfluorooctane sulphonic acid (PFOS)

Details	Description (perfluorooctane sulfonic acid, its salts, and perfluorooctane sulfonyl fluoride)
Structure	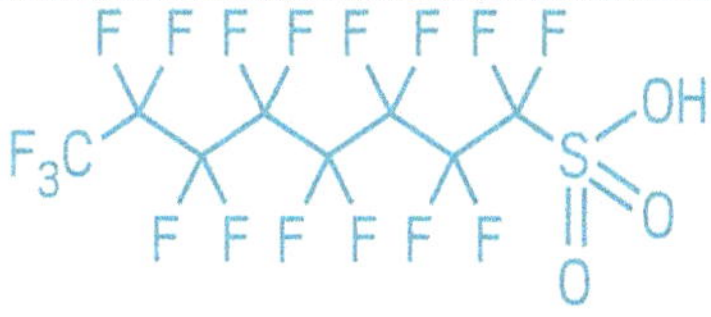 PFOS
IUPAC name	1,1,2,2,3,3,4,4,5,5,6,6,7,7,8,8,8-heptadecafluorooctane-1-sulfonic acid
Molecular formula	$C_8F_{17}SO_3H$
CAS no.	1763-23-1
Molecular weight	500.13 g/mol
Color/form	Liquid/off-white to gray
Odor	–
Boiling point	249 °C
Melting point	No Data
Flash point	11 °C
Solubility	3.2 × 10–3 mg/L at 25 °C
Density	–
Vapor pressure	2.0 × 10–3 mm hg at 25 °C
Octanol/water partition coefficient	Log Kow = 4.49
Octanol/air partition coefficient	–
Henry's law constant	Not measurable
Stability/shelf life	Remains unchanged within the suggested storage parameters
Decomposition	Upon undergoing decomposition through heating, it releases harmful vapors containing sulphur oxides and fluorine
Corrosivity	–

(0–6 years) residential populations, Luebker et al. identified a reduced pup body weight effect at 20 ng/kg-day, applying uncertainty factors of 10 for interspecies and 3 for intraspecies variation.

The water ingestion rate for these children was 0.78 L/day with a body weight of 15 kg, leading to an RSC of 100%. Seacat et al. (2002) found thyroid effects at 80 ng/kg-day, with uncertainty factors of 30. This research centered on adults who consumed 2 L of water daily and had a body weight of 70 kg resulting in an RSC of 20. Another study by Luebker et al. (2005) showed reduced pup body weight at 5.1 ng/kg/day, using uncertainty factors of 3 for interspecies, 10 for intraspecies, and 3 for database variation. Dong et al. (2011) observed immune response effects at 1.8 ng/kg-day, accounting for uncertainty factors of 3 for interspecies and 10 for sensitive subpopulations.

Zeng et al. (2011) found Hippocampus synapse structure effects at 20 ng/kg/day, utilizing an uncertainty factor of 10 for LOAEL to NOAEL and 10 for intraspecies variation. This study focused around child (0–6 years) residential populations, with a water intake rate of 0.64 L/day and a body weight of 15 kg, resulting in an RSC of 100.

4 Uses and Sources of POPs and Regulation

PFOS finds extensive use in a range of applications, such as serving as a surfactant in fire-extinguishing foams, alkaline cleaning agents, metal plating solutions, floor polish formulations, and circuit boards's etching acids. It also serves as an acid catalyst for photoresists and is the active ingredient in ant bait traps.

Additionally, PFOS and its precursors find usage as protective surface coatings for items like carpets, fabrics, and food packaging. They are also employed in specialty chemicals such as fire-fighting foams, hydraulic fluids, mining and oil well surfactants, as well as consumer products (Fig. 1). The study conducted by Jin et al. (2011) highlighted the exposure risk of perfluoroalkyl acids (PFAAs) particularly perfluorohexane sulfonate (PFHxS) to Firefighters using fire-retardants.

The production of PFOS began in the United States in 1948. Nevertheless, in the year 2000, the prominent manufacturer, 3M Corporation, opted to halt the production of numerous renowned Scotchgard stain-repellent items due to PFOS's environmental persistence and its prevalence across diverse species, including humans (Blumenthal et al., 2022). Sun et al. (2016) found that conventional and advanced

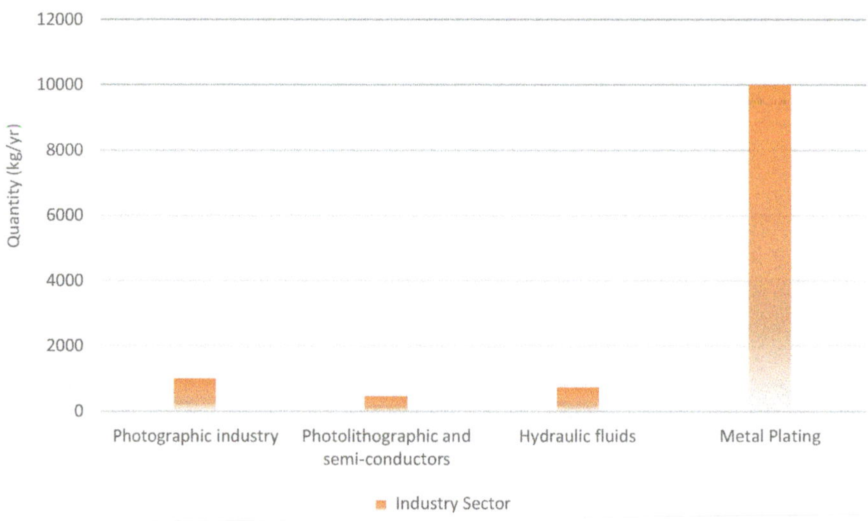

Fig. 1 Status of present industrial demand for PFOS-related substances within EU. (RPA and BRE, 2004)

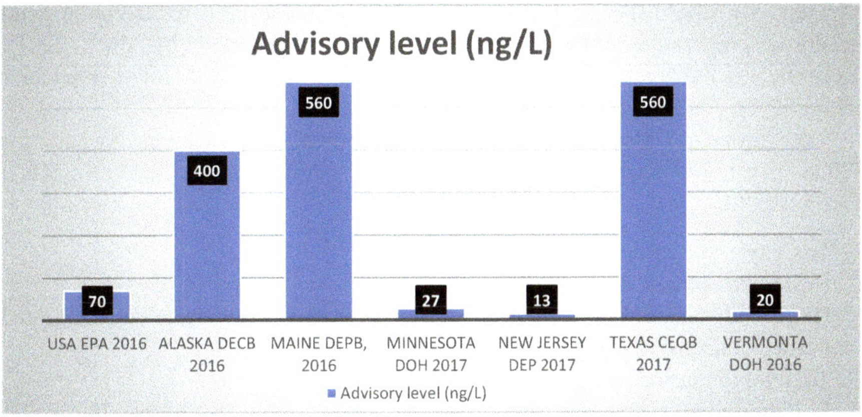

Fig. 2 PFOS potable water guideline levels from various concerned organizations. (Source: Cordner et al., 2019)
CEQ denotes Commission on Environmental Quality, DEC; DEP; DEQ and DOH denote Department of Environmental Conservation, Environmental Protection, Environmental Quality, and Health respectively
"A" denoted individual PFOA and PFOS and cumulative PFOA and PFOS
"B" Texas, Maine and Alaska, as per EPA's HA for public and/or private potable water

water treatment processes at a drinking water plant were ineffective in removing legacy PFASs and emerging Perfluoropolyether carboxylic acids (PFECAs). Figure 2 outlines PFOS's potable water guideline levels from various concerned organizations.

Persistence

The stability of PFOS (perfluorooctane sulfonic acid) is extensively documented, as it withstands hydrolysis, photolysis, and biodegradation across diverse environmental conditions as outlined by OECD (2002) guidelines. Hydrolysis investigations following the OPPTS protocol 835.2210 of US-EPA, carried out under varying pH levels and a temperature of 50 °C, demonstrated no deterioration of PFOS, establishing a half-life exceeding 41 years.

Likewise, photolysis experiments conducted using the OPPTS protocol 835.5270 of US-EPA detected no signs of direct or indirect photolysis under the tested conditions, indicating an indirect photolytic half-life surpassing 3.7 years at 25 °C.

Numerous tests evaluating aerobic and anaerobic biodegradation in different environments failed to reveal any biodegradation of PFOS. The CATABOL system including expert assessments forecasted that more than 99% of the 171-PFC substances studied would break down into highly non-degradable per fluorinated acids. Among these, 109 substances were projected to convert into Perfluorosulfonic acids (PFSA), including PFOS, while 61 substances were anticipated to transform into perfluoroalkyl carboxylic acids (PFCAs) (Dimitrov et al., 2004).

The only established approach for PFOS degradation is incineration with excessive-temperature under suitable criteria, as reported by 3M (2001). Nevertheless, the potential for degradation using incineration with low-temperature has yet to be confirmed.

The European Union (EU) classification proposed by the UK and Sweden for PFOS (perfluorooctane sulfonic acid) includes the following attributes (European Union, 2006):

- Toxicity
- Restricted signs of a carcinogenic effect
- Hazard of causing significant health issues through long exposure when ingested
- Potential hazard to developing fetus
- Toxicity to marine and freshwater animals
- Possibility of causing chronic negative impacts in marine and freshwater environments

Canadian adults potentially ingest around 73 ng/day of these PFOS-related compounds from the foods reported in a study (Tittlemier et al., 2006). In sub-chronic repeated dose studies, PFOS has demonstrated toxicity to mammals even at low concentrations (Fig. 3). Additionally, in rodent reproductional toxicity studies, death of offspring occurred soon after birth. The toxicity of PFOS has also been reported in aquatic animals, with *Chironomus tentans* and shrimp (Mysid) being particularly reactive species.

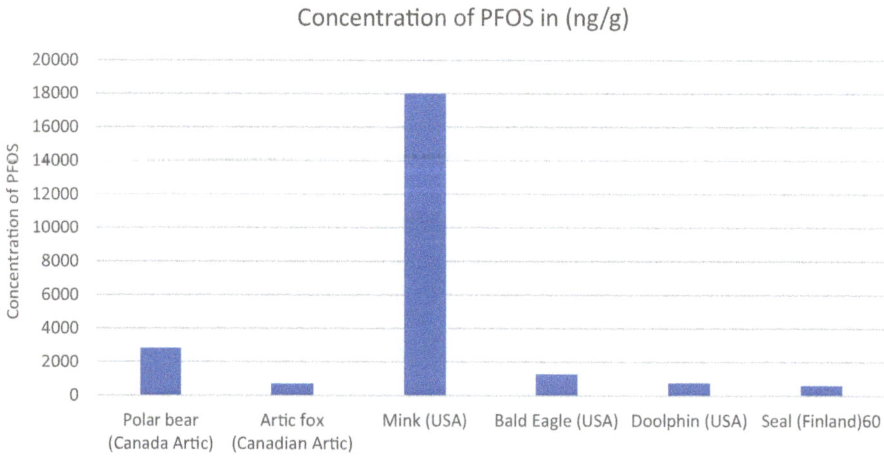

Fig. 3 Bioaccumulation of PFOS in animals. (Source: Giesy & Kannan, 2001; 3M, 2001; Kannan et al., 2004; Martin et al., 2004a)

5 In-House (Countrywide) Scientific Work Done on the Chemical

In a study conducted by Tao et al. in 2008, it was observed that breast milk contains traces of PFAS substances. The research revealed significant concentrations of perfluorooctane sulfonic acid, perfluorooctanoic acid, perfluorohexane sulfonic acid, and perfluorobutane sulfonate in female from Kolkata, Chidambaram, and Chennai. Generally, Indian human milk exhibited PFOS levels of 46 ppt, surpassing the health advisory benchmark issued by US State of Vermont's for potable water of 20 ppt meant for combined PFOA, PFOS, PFHxS, PFHpA, and PFNA. Notably, the greatest exposure of perfluorooctanoic acid in Indian human milk was 16 times higher than the potable water threshold.

The case of India, undergoing rapid industrialization and shifting lifestyles, serves as an intriguing subject for the investigation of PFAS emission and exposure across diverse environmental and socioeconomic contexts. The research conducted by BM Sharma et al. in 2016 delves into PFAS concentrations within river and groundwater, used as potable water in the region, along various points of the Ganges River. The study particularly focuses on PFOS and PFOA emissions. The river showed regular presence of 15 PFAS compounds, with PFHxA (ranging from 0.4 to 4.7 ng/l) and PFBS (<MQL – 10.2 ng/l) demonstrating the highest concentrations among PFCAs and PFSAs. The rampancy of smaller-chain PFAS compounds indicates noticeable effects from substituting PFOA and PFOS in Indian environmental samples. The co-variation between structural patterns of C5-C7 PFCAs and PFOS suggests similar emission drivers. Emissions of PFOS and PFOA into the river displayed substantial variation along the studied areas (ranging from 0.20 to 190 g/d for PFOS and 0.03 to 150 g/d for PFOA). Significantly, the emission trends of PFOS seemed to be associated with the urban resident count in the sub-catchment, rather than the overall population. Per capita emissions were less compared to those in developed countries. Groundwater concentrations and trends closely resembled those seen in surface water, suggesting aquifer contamination from the intake of river water through contaminated water. Crucially, the daily exposure to PFAS through potable water remained under safe thresholds for non-cancer oral risks across all age groups (Sharma et al., 2016). The analysis detected three polyfluoroalkyl substances (PFAS) in both groundwater and the River Ganges. These substances consisted of perfluorooctane sulfonate (PFOS, with a count of 6), perfluorononanoic acid (PFNA, with a count of 6), and perfluorodecanoic acid (PFDA, with a count of 1). Concentrations ranged from <0.001 to 0.033 mg L^{-1} in groundwater and from 0.003 to 0.025 mg L^{-1} in the River Ganges. The highest recorded PFOS concentrations were below the USEPA health advisory level for potable water (0.070 mg L^{-1} for individual analytes, EPA, 2017). However, it's important to highlight that certain PFAS compounds, like PFOA, were not encompassed by the comprehensive screening method utilized in this study (Lapworth et al., 2018).

Perfluorinated compounds, particularly perfluorooctanoic acid (PFOA) and perfluorooctane sulfonate (PFOS), constitute a growing category of environmentally persistent organic pollutants found widely across the environment. The accumulation of PFCs in surface water is mainly attributed to the discharge of contaminated water and untreated effluents. Nonetheless, there is limited information regarding PFC contamination in developing nations, notably in India. Addressing this gap, a 2016 study by Sunantha and Vasudevan undertook a quantitative analysis of PFOA and PFOS levels in rivers (Noyyal, Cauvery) and lakes around Chennai using Ultra-Fast Liquid Chromatography. Concentrations of PFOA ranged from 4 to 93 ng/l, while PFOS ranged from 3 to 29 ng/L. Remarkably, PFOS was not detectable in the Cauvery River. The River Cauvery consistently showed a PFOA concentration of 5 ng/l at all sites. This study offers valuable insights for future monitoring endeavors centered on PFOA and PFOS concentrations in surface water.

6 Relevance of the International Scientific Data as per Indian Conditions

PFAS diversity in waters in Sri Lanka mirrored global trends, with PFOA and PFOS dominating at 42–82 ng/L (Guruge et al., 2007). Sunantha and Vasudevan (2015) conducted a review delving into the exposure and potential human health effects of PFAS. The study highlighted that both PFOA and PFOS possess the ability to breach the placental barrier, resulting in inevitable exposure of the developing fetus. Importantly, exposure during fetal development, infancy, and childhood is particularly worrisome due to the heightened vulnerability of individuals during these stages of human growth. Among PFAS compounds, PFOS, PFHxS, PFOA, and FOSA emerged as the predominant ones in blood samples. The review underscored that India recorded the lowest PFOS blood levels, measuring below 3 ppb (Fig. 4). Intriguingly, India along with South Korea stood out as the countries where levels of PFOA surpassed those of PFOS. Furthermore, the study provided insights into the presence of PFOS and PFOA in human blood, revealing that India showcased the minimum PFOS grades but comparable PFOA to countries like Canada, the United States, Italy, and Spain.

In 2009, Yeung et al. conducted an investigation into the levels and compositions of perfluorinated compounds (PFCs) across various waterbodies (including rivers, lakes, coastal seas, and untreated sewage) comprising 42 samples.

The research also encompassed biota like shrimp (2 samples), fish (28 samples), and the Ganges River dolphin (*Platanista gangetica*; 15 samples). PFOS emerged as the prevailing PFC in most analyzed samples, including water (0.04–3.91 ng/l) and biota (0.248–27.9 ng/g ww), except for untreated sewage. Long-chain perfluoro carboxylates (PFCAs) ranging from C11 to C18 were absent in water samples (<0.2 ng/l), while PFDA (0.061–0.923 ng/g ww) and PFUnDA (0.072–0.998 ng/g ww) were detected in biological samples.

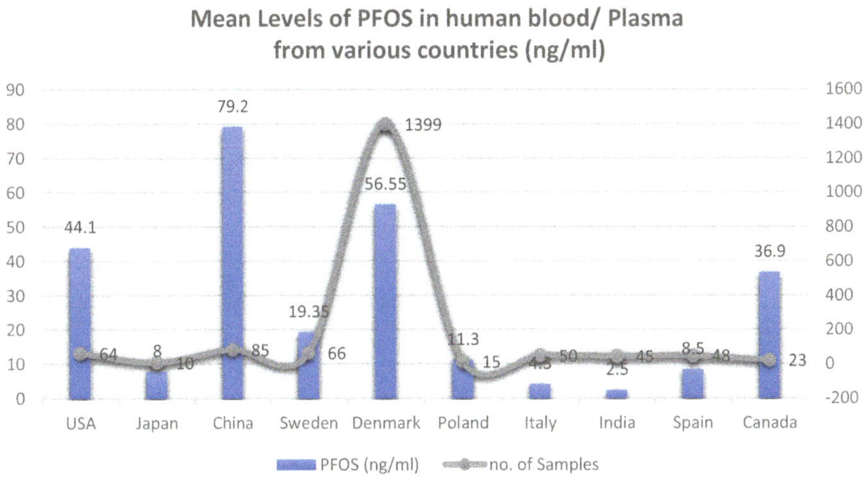

Fig. 4 PFOS and PFOA levels in human blood/plasma in different countries. (Source: Hansen et al., 2001; Harada et al., 2004; Yeung et al., 2004; Kato et al., 2009; Olsen et al., 2012; Kannan et al., 2004; Taniyasu et al., 2005)

Notably, the average PFOS content in the dolphin's liver of River Ganga was 27.9 ng/g. The study estimated bio concentration and bio magnification factors of PFCs within the Ganges River basin's food web. Among untreated sewage samples, the highest PFOA concentration was 23.1 ng/l. Generally, the PFC concentrations in water as well as biological samples from India were less than those documented in different countries. In Indian waters, PFOS dominated the PFC profile, followed by PFOA – a contrast to countries like Korea, Japan, and the United States, where PFOA took precedence. The study conducted in India, also estimated the flux of PFOS, PFOA, and PFNA from the River Ganga to the Bay of Bengal, ranging in the several hundreds of kg per year.

European data were gathered from 14 rivers across various European countries (McLachlan et al., 2007). On average, river water samples from India exhibited less concentration of PFOS compared to those reported in other counties (Nakayama et al., 2007; Murakami et al., 2008). Notably, the highest PFOS and PFOA concentrations in India were recorded in the Cooum River in Chennai (PFOS: 3.91 ng L^{-1}; PFOA: 23.1 ng L^{-1}), whereas the lowest concentrations were found in waters from Goa (PFOS/PFOA: <0.083 – <1 ng L^{-1}). Figure 5 represents concentration of PFOS (ng/L) found in various rivers of the world.

In a study by Xu et al. in 2019, 98 cord samples from Hangzhou, China, were analyzed for 11 PFASs. All target compounds were found to cross the placenta, with median PFOS concentrations of 4.07 ng per mL, 1.05 ng/mL for PFOA, and 0.731 ng/mL for 6:2 Cl-PFESA. The study revealed that advanced age, elevated pre-pregnancy BMI, and being a first-time mother were linked to higher cord PFAS concentrations. Interestingly, after accounting for potential confounders, PFOS exhibited a negative association with birth weight and Ponderal index. PFOS

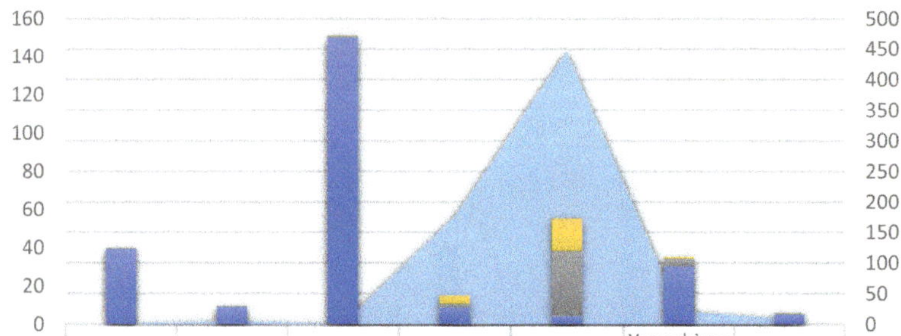

	*Ganga India)	*Yamuna (India)	*Yangtze (China)	*Cape Fear (USA)	Tamagawa (Japan)	Yoneshirogawa (Japan)	Kelani (Sri Lanka)
PFOS (ng/L)	0.363	1.81	0.36	56.3	143	6.9	2.7
PFNA (ng/L)	0.0589	0.0673	0.16	13.2	53	2.3	0.3
PFOA (ng/L)	0.0616	0.0947	2.55	6.84	107	13	1.7
Discharge(m3/s)	124	29.5	470	27.33	13.4	94.8	16.2

Fig. 5 Concentration of PFOS (ng/L) in rivers from several countries. (Source: Yeung et al., 2009a, b; Sarin et al., 1989; Chen et al., 2001; Murakami et al., 2008; McLachlan et al., 2007)

exposure, even at occupationally and environmentally relevant low concentrations, induces ROS (reactive oxygen species) production in human endothelial cells, potentially contributing to endothelial dysfunction and adverse health effects (Qian et al., 2010).

7 Efforts Taken for POPs Management in India

Currently, there is limited relevant information available regarding the persistent organic pollutants (POPs) management in India with respect to this group. The shortage of data on chemical usage across sectors poses a challenge to controlling their use and effectively managing waste. Inventorization of stockpiles for PFAS containing firefighting materials is the need of hour to combat the pollution linked with this POPs chemical. To address this issue, specific regulations should be put in place to stop these POP chemicals production, usage, imports, and exports. Additionally, there is a need for more scientific studies to realize the actual scenario prevailing with respect to its presence in air, food, and water and related health impacts.

In a study by Anderson et al. in 2019, fluorous-phase Ion-Selective Electrodes (ISEs) were utilized to assess equilibrium constants concerning the binding of PFOS to poly (dimethylamine-co-epichloro hydrin) and poly (di-allyl dimethyl ammonium) in simulated groundwater and soil suspensions. This innovative method merged a 1:1 binding model with the ISE response model to interpret potentiometric data regarding surfactant binding to the charged repeat unit of these poly ions. The result permitted for the simple prediction and fitting of experimental potentiometric data in a single step.

In brief, the study determined equilibrium constants for PFOS binding to the poly quaternion polymers poly (di-allyl dimethyl ammonium) chloride and poly (dimethylamine-co-epichlorohydrin), both in the lack and availability of soil, using fluorous-phase ISEs. These sensors used a semi-fluorinated Imidazolium cation as an anion-exchanger specially developed for this purpose. Despite the presence of soil decreasing the affinity of the polyquaternium polymers for PFOS, the study found that PFOS was still bound with affinities of 9.3 ± 0.6 × 104 and 7.3 ± 1.2 × 104 M^{-1} for poly (diallyl dimethylammonium) chloride and poly (dimethylamine-co-epichlorohydrin), respectively. Utilizing these quantitative findings, predictions were made regarding the concentrations of polyquaternium polymers needed to decrease the levels of unbound PFOS. Particularly, concentrations as low as 0.16 g/L of poly (diallyl dimethylammonium) chloride were effective in reducing PFOS concentrations below the EPA health advisory level of 0.1 nM.

Replacement of PFOS
Although substitutes to PFOS exist for specific, these options are not universally accessible, particularly in developing countries where a phased implementation is still required. Technically viable substitutes to PFOS have not yet been developed for particular utilization such as photo imaging, semiconductors, and aviation hydraulic fluids.

8 Recommendations

In India, there is a notable scarcity of information concerning the utilization of this chemical across various sectors, and the collection of such data is essential for regulating usage and managing waste. To prevent PFAS pollution and the resulting costly remediation processes, it is vital for India to rapidly organize an inventory of firefighting foam stocks and promptly transition from PFAS-containing foams to fluorine-free alternatives, if any. Precise controls should be set up to prohibit the manufacturing, usage, import, and export of PFAS. Furthermore, there is a pressing need for more comprehensive scientific investigations into the presence and health implications of PFAS in air, food, and water.

References

3M. (2001). *Analytical laboratory report, determination of the presence and concentration of PFOS, PFOSA, PFOSAA, EtFOSE-OH, M556 and PFOSEA in serum and liver samples of Crl:CD(SD) IGS BR rats exposed to N-ethyl perfluorooctanesulfonamido ethanol.* 3M Environmental Laboratory Report No. Tox-001, Laboratory Request No. U2103, 3M Reference No. T-6316.1.

Anderson, E. L., Mousavi, M. P. S., Aly, Y. H., Chen, X. V., Simcik, M. F., & Bühlmann, P. (2019). Remediation of Perfluorooctylsulfonate contamination by in situ sequestration: Direct monitoring of PFOS binding to polyquaternium polymers. *ACS Omega, 4*(1), 1068–1076. https://doi.org/10.1021/acsomega.8b03275

Austin, M. E., Kasturi, B. S., Barber, M., Kannan, K., MohanKumar, P. S., & MohanKumar, S. M. J. (2003). Neuroendocrine effects of perfluorooctane sulfonate in rats. *Environmental Health Perspectives, 111*, 1485–1489. https://doi.org/10.1289/ehp.6128

Bentel, M. J., Yu, Y., Xu, L., Li, Z., Wong, B. M., Men, Y., & Liu, J. (2019). Defluorination of per- and polyfluoroalkyl substances (PFASs) with hydrated electrons: Structural dependence and implications to PFAS remediation and management. *Environmental Science & Technology, 53*(7), 3718–3728. https://doi.org/10.1021/acs.est.8b06648. Epub 2019 Mar 15.

Blumenthal, J., Diamond, M. L., Hoffmann, M., & Wang, Z. (2022). Time to break the "lock-in" impediments to chemicals management. *Environmental Science & Technology, 56*(7), 3863–3870. https://doi.org/10.1021/acs.est.1c06615. Epub 2022 Mar 19.

Chen, Z. Y., Li, J. F., Shen, H. T., & Wang, Z. H. (2001). Yangtze River of China: Historical analysis of discharge variability and sediment flux. *Geomorphology, 41*, 77–91.

Cordner, A., De La Rosa, V. Y., Schaider, L. A., et al. (2019). Guideline levels for PFOA and PFOS in drinking water: The role of scientific uncertainty, risk assessment decisions, and social factors. *Journal of Exposure Science & Environmental Epidemiology, 29*, 157–171. https://doi.org/10.1038/s41370-018-0099-9

Dimitrov, S., Kamenska, V., Walker, J. D., Windle, W., Purdy, R., Lewis, M., & Mekenyan, O. (2004). Predicting the biodegradation products of perfluorinated chemicals using CATABOL. *SAR and QSAR in Environmental Research, 15*, 69–82. https://doi.org/10.1080/1062936032000169688

Dong, G. H., Liu, M. M., Wang, D., Zheng, L., Liang, Z. F., & Jin, Y. H. (2011). Sub-chronic effect of perfluorooctanesulfonate (PFOS) on the balance of type 1 and type 2 cytokine in adult C57BL6 mice. *Archives of Toxicology, 85*(10), 1235–1244.

Eurpean Union. (2006). *Restrictions on the marketing and use of perfluorooctane sulfonate*. European Parliament legislative resolution on the proposal for a directive of the European Parliament and of the Council relating to restrictions on the marketing and use of perfluorooctane sulfonates (amendment of Council Directive 76/769/EEC) (COM (2005)0618 – C6 – 0418/2005 – 2005/0244(COD)).

Giesy, J. P., & Kannan, K. (2001). Global distribution of perfluorooctane sulfonate in wildlife. *Environmental Science & Technology, 35*, 1339–1342.

Guruge, K., Taniyasu, S., Yamashita, N., & Manage, P. (2007). Occurrence of perfluorinated acids and fluorotelomers in waters from Sri Lanka. *Marine Pollution Bulletin, 54*, 1667–1672. https://doi.org/10.1016/j.marpolbul.2007.05.021

Hansen, K. J., Clemen, L. A., Ellefson, M. E., & Johnson, H. O. (2001). Compound-specific, quantitative characterization of organic fluorochemicals in biological matrices. *Environmental Science & Technology, 35*, 766–770.

Harada, K., Saito, N., Inoue, K., Yoshinaga, T., Watanabe, T., Sasaki, S., Kamiyama, S., & Koizumi, A. (2004). The influence of time, sex and geographic factors on levels of perfluoro octane sulfonate and perfluorooctanoate in human serum over the last 25 years. *Journal of Occupational Health, 46*, 141–147.

Hou, J., Li, G., Liu, M., Liang, C., Yao, Y., Fallgren, P. H., & Jin, S. (2022). Electrochemical destruction and mobilization of perfluorooctanoic acid (PFOA) and perfluorooctane sulfonate (PFOS) in saturated soil. *Chemosphere, 287*(Part 3), 132205. ISSN 0045-6535.

Jin, C., Sun, Y., Islam, A., Qian, Y., & Ducatman, A. (2011). Perfluoroalkyl acids including perfluorooctane sulfonate and perfluorohexane sulfonate in firefighters. *Journal of Occupational and Environmental Medicine, 53*(3), 324–328. https://doi.org/10.1097/JOM.0b013e31820d1314

Kannan, K., Corsolini, S., Falandysz, J., Fillmann, G., Kumar, K. S., Loganathan, B. G., Mohd, M. A., Olivero, J., Wouwe, N. V., Yang, J. H., & Aldous, K. M. (2004). Perfluoro octane sulfonate and related fluorochemicals in human blood from several countries. *Environmental Science & Technology, 38*, 4489–4495.

Kato, K., Calafat, A. M., Wong, L. Y., Wanigatunga, A. A., Caudill, S. P., & Needham, L. L. (2009). Polyfluoroalkyl compounds in pooled sera from children participating in the National Health and Nutrition Examination Survey 2001–2002. *Environmental Science & Technology, 43*, 2641–2647.

Kumari, K. (Ed.). (2021). *Persistent organic pollutants: Gaps in management and associated challenges* (1st ed.). CRC Press. https://doi.org/10.1201/9781003046806

Lapworth, D. J., Das, P., Shaw, A., Mukherjee, A., Civil, W., Petersen, J. O., Gooddy, D. C., Wakefield, O., Finlayson, A., Krishan, G., Sengupta, P., & MacDonald, A. M. (2018). Deep urban groundwater vulnerability in India revealed through the use of emerging organic contaminants and residence time tracers. *Environmental Pollution, 240*, 938–949.

Lau, C., Anitole, K., Hodes, C., & Lai, D. (2007). Perfluoroalkyl acids: A review of monitoring and toxicological findings. *Toxicological Sciences, 99*, 366–394. https://doi.org/10.1093/toxsci/kfm128

Li, R., Tang, T. H., Qiao, W. C., & Huang, J. (2020). Toxic effect of perfluorooctane 651 sulfonate on plants in vertical-flow constructed wetlands. *Journal of Environmental Sciences (China), 92*, 176–186.

Luebker, D. J., York, R. G., Hansen, K. J., Moore, J. A., & Butenhoff, J. L. (2005). Neonatal mortality from in utero exposure to perfluorooctanesulfonate (PFOS) in Sprague–Dawley rats: Dose–response, and biochemical and pharamacokinetic parameters. *Toxicology, 215*(1–2), 149–169.

Martin, J. W., Smithwick, M. M., Braune, B., Hoekstra, P. F., Muir, D. C. G., & Mabury, S. A. (2004a). Identification of long-chain perfluorinated acids in biota from the Canadian Arctic. *Environmental Science & Technology, 38*(2), 373–380.

Martin, J. W., Whittle, D. M., Muir, D. C. G., & Mabury, S. A. (2004b). Perfluoroalkyl contaminants in a food web from Lake Ontario. *Environmental Science & Technology, 38*, 5379–5385.

Mclachlan, M. S., Holmstrom, K. E., Reth, M., & Berger, U. (2007). Riverine discharge of perfluorinated carboxylates from the European continent. *Environmental Science & Technology, 41*, 7260–7265.

Murakami, M., Imamura, E., Shinohara, H., Kiri, K., Muramatsu, Y., Harada, A., & Takada, H. (2008). Occurrence and sources of perfluorinated surfactants in rivers in Japan. *Environmental Science & Technology, 42*(17), 6566–6572. https://doi.org/10.1021/es800353f

Nakayama, S., Strynar, M. J., Helfant, L., Egeghy, P., Ye, X. B., & Lindstrom, A. B. (2007). Perfluorinated compounds in the Cape Fear Drainage Basin in North Carolina. *Environmental Science & Technology, 41*, 5271–5276.

OECD. (2002, November 21). *Hazard assessment of perfluorooctane sulfonate (PFOS) and its salts*. ENV/JM/RD (2002)17/FINAL, Paris, 362 pp.

Olsen, G. W., Lange, C. C., Ellefson, M. E., Mair, D. C., Church, T. R., Goldberg, C. L., Herron, R. M., Medhdizadehkashi, Z., Nobiletti, J. B., Reagen, W. K., & Zobel, L. R. (2012). Temporal trends of perfluoroalkyl concentrations in American Red Cross adult blood donors, 2000–2010. *Environmental Science & Technology, 46*(11), 6330–6338.

Pontius, F. (2019). Regulation of perfluorooctanoic acid (PFOA) and perfluorooctane sulfonic acid (PFOS) in drinking water: A comprehensive review. *Water, 11*(676), 2003. https://doi.org/10.3390/w11102003

Qian, Y., Ducatman, A., Ward, R., Leonard, S., Bukowski, V., Lan Guo, N., Shi, X., Vallyathan, V., & Castranova, V. (2010). Perfluorooctane sulfonate (PFOS) induces reactive oxygen species (ROS) production in human microvascular endothelial cells: Role in endothelial permeability. *Journal of Toxicology and Environmental Health. Part A, 73*(12), 819–836. https://doi.org/10.1080/15287391003689317

RPA. (2004). *RPA & BRE, risk & policy analysts limited in association with BRE environment, perfluorooctane sufonate – Risk reduction strategy and analysis of advantages and drawbacks*. Final report prepared for Department for Environment, Food and Rural Affairs and the Environment Agency for England and Wales.

Sarin, M. M., Krishnaswami, S., Dilli, K., Somayajulu, B. L. K., & Moore, W. S. (1989). Major ion chemistry of the Ganga-Brahmaputra river system – Weathering processes and fluxes to the Bay of Bengal. *Geochimica et Cosmochimica Acta, 53*, 997–1009.

Seacat, A., Thomford, P., Hansen, K., Olsen, G. W., Case, M., & Butenhoff, J. L. (2002). Subchronic toxicity studies on perfluorooctanesulfonate potassium salt in cynomolgus monkeys. *Toxicological Sciences, 68*, 249–264.

Sharma, B. M., Bharat, G. K., Tayal, S., Larssen, T., Bečanová, J., Karásková, P., Whitehead, P. G., Futter, M. N., Butterfield, D., & Nizzetto, L. (2016). Perfluoroalkyl substances (PFAS) in river and ground/drinking water of the Ganges River basin: Emissions and implications for human exposure. *Environmental Pollution, 208*(Pt B), 704–713. https://doi.org/10.1016/j.envpol.2015.10.050

Sun, M., Arevalo, E., Strynar, M., Lindstrom, A., Richardson, M., Kearns, B., Pickett, A., Smith, C., & Knappe, D. (2016). Legacy and emerging perfluoroalkyl substances are important potable water contaminants in the Cape Fear River Watershed of North Carolina. *Environmental Science & Technology Letters, 3*(12), 415–419.

Sun, Z. Y., Zhang, C. J., Chen, P., Zhou, Q., & Hoffmann, M. R. (2017). Impact of humic 710 acid on the photo reductive degradation of perfluorooctane sulfonate (PFOS) by 711 UV/iodide process. *Water Research, 127*, 50–58. https://doi.org/10.1016/j.watres.2017.10.010

Sunantha, G., & Vasudevan, N. (2015). Impacts of perfluorinated compounds on human health. *Bulletin of Environment, Pharmacology and Life Sciences, 4*, 183–191.

Sunantha, G., & Vasudevan, N. (2016). Assessment of perfluorooctanoic acid and perfluorooctane sulfonate in surface water – Tamil Nadu, India. *Marine Pollution Bulletin, 109*(1), 612–618.

Taniyasu, S., Yamashita, N., Wijeratna, S., Mohotti, K. M., & Seneviratne, H. R. (2005). Perfluorinated organic compounds in human blood serum and seminal plasma: A study of urban and rural tea worker populations in Sri Lanka. *Journal of Environmental Monitoring, 7*(4), 371–377.

Tao, L., Ma, J., Kunisue, T., Libelo, E. L., Tanabe, S., & Kannan, K. (2008). Perfluorinated compounds in human breast milk from several Asian countries, and in infant formula and dairy milk from the United States. *Environmental Science & Technology, 42*(22), 8597–8602. https://doi.org/10.1021/es801875v

Tittlemier, S. A., Pepper, K., & Edwards, L. (2006). Concentrations of perfluorooctane sulfonamides in Canadian total diet study composite food samples collected between 1992 and 2004. *Agricultural and Food Chemistry, 54*(5), 8385–8389.

Xu, C., Yin, S., Liu, Y., Chen, F., Zhong, Z., Li, F., et al. (2019). Prenatal exposure to chlorinated polyfluoroalkyl ether sulfonic acids and perfluoroalkyl acids: Potential role of maternal determinants and associations with birth outcomes. *Journal of Hazardous Materials, 380*, 120867.

Yamamoto, T., Noma, Y., Sakai, S. I., & Shibata, Y. (2007). Photodegradation of perfluorooctane sulfonate by UV irradiation in water and alkaline 2-propanol. *Environmental Science & Technology, 41*, 5660–5665. https://doi.org/10.1021/es0706504

Yeung, L. W. Y., So, M. K., Jiang, G. B., Taniyasu, S., Yamashita, N., Song, M. Y., Wu, Y. N., Li, J. G., Giesy, J. P., Guruge, K. S., & Lam, P. K. S. (2004). Perfluorooctane sulfonate and related fluorochemicals in human blood samples from China. *Environmental Science & Technology, 40*, 715–720.

Yeung, L. W. Y., Yamashita, N., Taniyasu, S., Lam, P. K. S., Sinha, R. K., Borole, D. V., & Kannan, K. (2009a). A survey of perfluorinated compounds in surface water and biota including dolphins from the Ganges River and in other waterbodies in India. *Chemosphere, 76*(1), 55–62. ISSN 0045-6535.

Yeung, L. W. Y., Miyake, Y., Wang, Y., Taniyasu, S., Yamashita, N., & Lam, P. K. S. (2009b). Total fluorine, extractable organic fluorine, perfluorooctane sulfonate and other related fluorochemicals in liver samples of Indo-Pacific humpback dolphins (Sousa chinensis) and finless porpoises (Neophocaena phocaenoides) from South China. *Environmental Pollution, 157*, 17–23.

Zareitalabad, P., Siemens, J., Hamer, M., & Amelung, W. (2013). Perfluorooctanoic acid (PFOA) and perfluorooctanesulfonic acid (PFOS) in surface waters, sediments, soils and wastewater – A review on concentrations and distribution coefficients. *Chemosphere, 91*, 725–732. https://doi.org/10.1016/j.chemosphere.2013.02.024

Zeng, H. L., Li, Y.-Y., Zhang, L., Wang, Y., Chen, J., Xia, W., Lin, Y., et al. (2011). Prenatal exposure to perfluorooctanesulfonate in rat resulted in long-lasting changes of expression of synapsins and synaptophysin. *Synapse, 65*, 225–233.

Polychlorinated Naphthalenes (PCNs)

Kanchan Kumari, Shaheen Akhtar, and Piyush Mohapatra

1 Introduction

Counted among the legacy contaminants as labelled by the Stockholm Convention in 2015 (annex A and C) (Kumari, 2021), polychlorinated naphthalenes (PCNs) (Stockholm Convention, 2001) are the most studied in the group of polycyclic aromatic hydrocarbons (Cl-PAHs). Commercially, 75 chlorinated naphthalene congeners and by-products make up PCNs and are often described by the total fraction of chlorine. PCNs' key qualities include chemical inertness, low combustibility, insulation against electricity, recalcitrance, ability to withstand biodegradation, and biocidal functionality. PCNs are structurally similar to PCBs and thus its commercial and industrial applications strongly coincide with that of PCBs (Safe & Phil, 1990).

Cable insulation, engine oil additives, dye carriers, capacitor fluids, refractive index testing oils, flame proofing, wood preservation, and other lubricant usage are among the applications for PCNs. While mono CNs have fungicidal and insecticidal properties and are used as a raw material in dyes and wood preservatives, mono and di CN mixtures have applications as instrument seals and chemical resistant gauge fluids, high boiling specialty solvents, ingredients in motor tune up compounds, and engine crankcase additives (UNEP/POPS/POPRC.8/16/Add.1). PCNs were widely utilized as wood preservatives in the United States of America throughout the 1940s and 1950s, but they are no longer used for this purpose. The purposeful manufacture of PCN is thought to have halted globally, however PCN may be created accidently

K. Kumari (✉) · S. Akhtar
CSIR-National Environmental Engineering Research Institute (NEERI), Kolkata Zonal Center, Kolkata, West Bengal, India
e-mail: k_kumari@neeri.res.in

P. Mohapatra
Toxic Link, New Delhi, India

K. Kumari, *Pollutants of Global Concern*, Emerging Contaminants and Associated Treatment Technologies, https://doi.org/10.1007/978-3-031-50996-4_21

during high-temperature industrial operations in the presence of chlorine (Yang et al., 2019). Some PCNs can be degraded by sunlight or at a slower pace by specific microbes; they also tend to bioaccumulate (tetra- to hepta CNs) and persist in the environment, harming human health in the form of liver illnesses because of prolonged exposure (Domingo, 2004).

2 Technical Details of PCNs

The physicochemical properties of PCN are presented in tabular form given in Table 1.

Table 1 Physicochemical properties of PCN

Details	Description (Polychlorinated naphthalenes/ octachloronaphthalene)
Structure	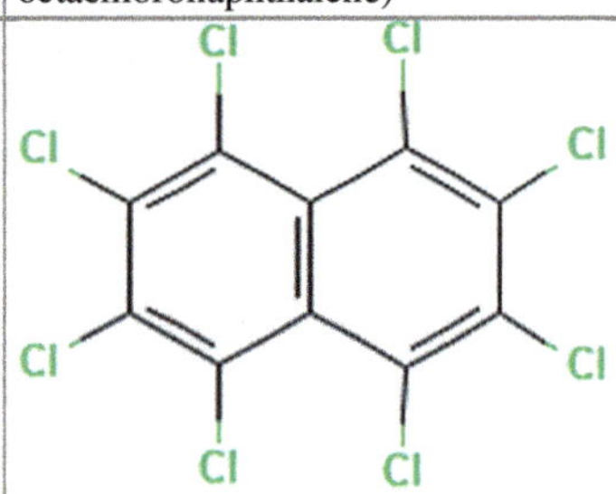
IUPAC name	1,2,3,4,5,6,7,8-octachloronaphthalene
Molecular formula	C10Cl8
CAS no.	2234-13-1
Molecular weight	403.7 g/mol
Colour/form	Crystals from cyclohexane
Odour	Aromatic odour
Boiling point	824 °F at 760 mm Hg
Melting point	387–388 °F
Solubility	Insoluble
Density	2.00 g/cu cm at 25 °C
Vapour pressure	Less than 1 mm Hg at 68 °F
Octanol/water partition coefficient	8.5
Octanol-air partition coefficient	11.05
Decomposition	Decomposes on heating emitting toxic fumes

3 PCN in Stockholm Convention

In the seventh POPRC meeting held in Geneva on October 10–14, 2011, the European Union and its Member States submitted a proposal to list chlorinated naphthalene under Annex A, B, and/or C to the Convention (UNEP/POPS/POPRC.7-4) along with a detailed dossier to support the proposal (UNEP/POPS/POPRC.7/INF/3). The Committee adopted and included Polychlorinated Naphthalene as POPs in Annex A and C to the Convention on May 4–15, 2015, at the seventh meeting of the Conference of the Parties in Geneva. However, the production of polyfluorinated naphthalenes such as octafluoronaphthalene gained special exemptions for PCNs (UNEP/POPS/POPRC.5; Toxics Link, 2015).

4 Usage in India

PCNs stockpile information in India is limited and thus usage details are not readily available (Gaonkar et al., 2021). According to BIS standard IS 15495:2004, it is prohibited in printing ink for food packaging. Polychlorinated naphthalenes are not regulated by any specific policy, rules, or legislation in India (Toxics Link, 2015).

The total export of PCNs from India were 30.52 thousand kgs, 3.47 thousand kgs, 11.08 thousand kgs, and 3.3 thousand kgs in the year of 2015, 2016, 2017, and 2018 respectively to countries such as Canada, China, Germany, Haiti, Belgium, France, Germany, Japan, and Mexico. China among the rest is the main importer of PCN from India where 112.95 thousand kgs was the highest amount to be imported till date (Toxics Link, 2018).

5 Indian Studies

According to studies, average atmospheric PCN concentrations in India and Pakistan were found to be 29 pg/m^3 and 7.7 pg/m^3, respectively, with tri-CNs and tetra-CNs being the main homologues in the air in India. Due to technological combinations such as tri-CN-dominated Halowax, Chennai has the greatest percentage of tri-CNs, while in Punjab, the proportions of tri-CNs were low, but the PCNcom/PCNs ratios were high. A significant fraction of the atmospheric PCNs were produced by industrial thermal emissions in Kolkata, while industrial and technical sources influenced Mumbai having a compositional pattern that indicated tri- and penta-CN-enriched Halowax emission (Xu et al., 2014).

6 Global Studies

The global scenario with PCN production and usage has not yet been widely defined though studies have provided information especially based on its toxicity. Lin et al. (2013) reported the spatial distribution and sources of atmospheric PCNs at 90 sites in North China indicating contamination with a 48 pg m^{-3} of median concentration (range: 10–2460 pg m^{-3}) for 29 PCNs. Sources of these PCNs were traced to thermal processes with a compositional profile of 90% PCNs originating from these. With an estimated TEQ range from 0.97 to 687 fg TEQ m^{-3}, the electronic waste recycling facility was the most at risk for dioxin-like PCNs.

According to recent research on PCN occurrence, the congeners with a high potential for toxicity have been identified in biota, food, human tissues, and several environmental compartments as well (Fernandes et al., 2017) (Table 2). Samples of seal pups showed dominance of hexa-chlorinated congeners in the liver tissues,

Table 2 PCN occurrences in biota and environmental matrices

Study site	Sample details	PCN concentrations	References
North America	Biopsies taken from harbour seal pups	190 ng/kg–1960 ng/kg lipid weight	Ross et al. (2013)
UK marine water stretch (north to Norwegian seas and south to the Algarve)	200 fish samples—sardines, sea bass, herring, sprats, mackerel, turbot, and grey mullet	0.7–265 ng/kg ww	Fernandes et al. (2018)
Lake Ontario	Tissue of fish species—Bluegill, channel catfish, pumpkinseed, northern pike, and brown bullhead	132 ng/kg for individual congeners	Lega et al. (2017)
Lake Erie St. Lawrence River Detroit River St. Lawrence River	Whole fish and herring gull eggs Sediment	7660 & 3020 pg/g ww 34–2370 pg/g ww 264,000 pg/g 440–19,300 pg/g	McGoldrick et al. (2018)
River Chenab, Pakistan	Sediment	8.94–414 ng/g	Mahmood et al. (2014)
Greenland Fjords, Norway	Water	0.0014–0.41 ng/l	Ishaaq et al. (2009)
Ghana	Air	27–95 pg/m^3	Hogargh et al. (2012)
Canada Dunai Island, Russia	Air	0.69 pg/m^3 0.82 pg/m^3	Helm et al. (2004)
Taoyuan City, Taiwan	Air	18.5–153 pg/m^3	Dat et al. (2018)
Qinghai Plateau, China	Air	21.9–75.1 pg/m^3	Hu et al. (2019)
Mace Head, Ireland	Air	1.5–55 pg/m^3	Lee et al. (2005)

while a sum of 23 congeners showed decline since 1984 (Ross et al., 2013). Though studies throughout a time period of >20 years has quantified the occurrence of PCN in the declining phase, the PCN burden has however not reached a null point. Marine species have largely shown PCN contaminations in some of the major fishing regions of the world (Fernandes et al., 2018; Zhihua et al., 2019). Studies on the spatial and temporal trends (1979–2013) and sediments (2011–2013) for PCN concentrations in the Great Lakes and St. Lawrence River revealed different patterns of PCNs in gulls and fish. This was related to routes of exposure, variable metabolic capacity, and resuspension of PCN polluted sediments. In gull eggs, hepta-PCNs predominated, whereas fish had larger amounts of lower chlorinated PCNs (McGoldrick et al., 2018).

Industrial sources have greatly contributed to the PCN pollution of urban regions where waste incineration and burning heat treatment processes have been identified as the main cause. Guo et al. (2008) reported PCN level in city sludge (1.48–28.21 ng/g) where PCN-TEQs content was in detected to be 0.11–2.45 pg/g. The distribution characteristics of PCN congeners showed similarity, i.e. two chlorides and three chloro-naphthalenes deducing that the sources have similar resemblance. Moreover, PCNs present in commercial PCB formulations have also been identified as potential source of environmental PCNs. Studies conducted on Chinese technical formulation of PCB3 of insulating oil of electric capacitor detected PCN levels of 1307.5 μg/g and were significantly higher than capacitor insulating oil formulations of other countries. Hexa- and hepta-CN contributions were prevalent and were present in similar amounts in the homolog profile as in Aroclor 1262 and Clophen A40. Dioxin-like PCN congeners had a Toxic Equivalent Quantity (TEQ) concentration of 0.47 g/g, with CN-73 and CN-66/67 being the main contributors. With the exception of the Russian Sovol formulation, the TEQ concentration from PCN was higher than that in the majority of other PCB formulations (Huang et al., 2015) (Fig. 1).

The possibility of presence of PCN in commercial PCBs indicates PCN contamination in organochlorine chemicals as various studies detected PCN and other unintentional POPs in chlorinated paraffins and high-volume organochlorines (Fernandes et al., 2017).

As a result, additional PCN monitoring in contemporary organochlorine compounds should be done in preparation for the upcoming PCN inventory under the Stockholm Convention.

7 Adverse Effects of PCNs

Bioaccumulation and toxicity of PCNs in birds and fishes have been majorly studied so far. Congeners of PCN such as di, tri, tetra, and pentachloro naphthalenes have shown experimental BCF values in fish greater than 5000 (UNEP/POPS/POPSRC.8/16/Add.1). Elevated PCN levels in the Arctic as a result of air masses from Europe may have designated PCNs as a persistent pollution, identifying it as a

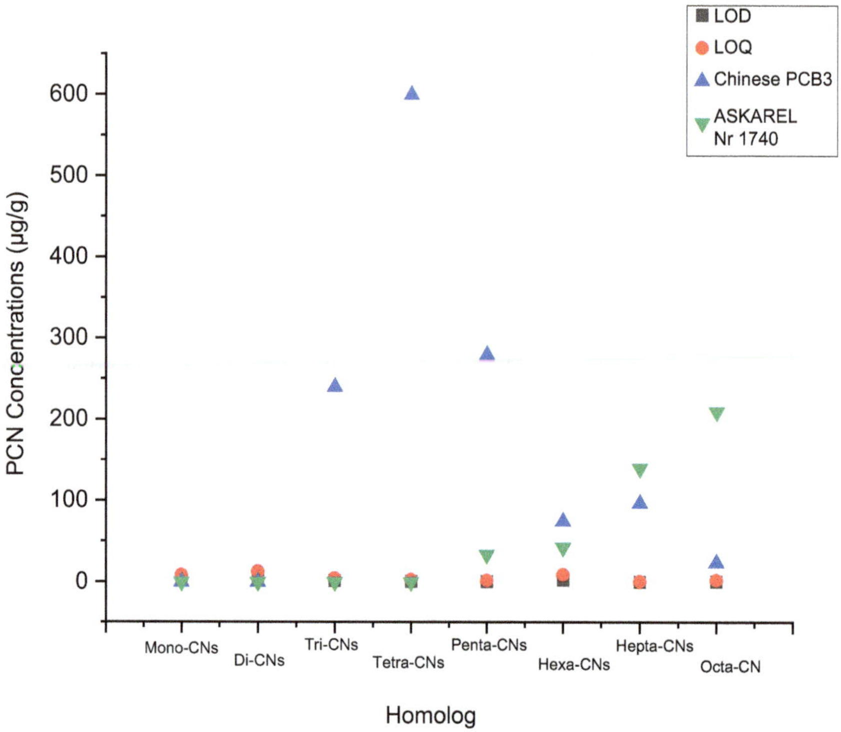

Fig. 1 Concentrations of PCNs homologs (Huang et al., 2015). **LOD-limit of detection, LOQ-limit of quantification

widespread worldwide environmental contaminant. PCN concentrations in the Arctic and Antarctic marine food webs ranged from 1.5 pg/g in krill to 2550 pg/g in South polar skua (w/w), with lower chlorinated PCNs being the predominating congener (Corsolini et al., 2002). Moreover, PCNs can exhibit toxicity such as endocrine disruption in higher organisms (Li et al., 2022).

Human exposure to PCN is mainly considered to be via food intake where studies have shown that cheese and butter samples had PCN concentrations of 5.6–103 pg g^{-1} (w/w) and 5.0–199 pg g^{-1} (w/w) respectively. The major congener, CN73, accounted for more than 40% of the PCN TEQ (toxic equivalence) value, and its mass concentrations were higher than those of the higher chlorinated congeners. Additionally, rather significant quantities of the congeners CN66/67, CN69, and CN63 were discovered. According to Li et al. (2015), PCNs made up about 5% of the overall polychlorinated compound (PCN, PCDD, PCDF, and PCB) TEQ values, whereas PCBs made up 53.8% of the total TEQ, and PCDD and PCDF made up 11.6% and 29.7%, respectively. Also, due to their PCB-like properties, long-term exposure to PCNS have attributed to various teratogenic, carcinogenic, and chronic liver diseases (Fernandes et al., 2022). Another study in China estimated PCN concentrations in human milk, their putative sources, and the health dangers they may

Table 3 Levels of polychlorinated naphthalenes in human samples (pg/g) with predominant congeners including 4, 12, 22, 38, MoCN-OCN and TeCN-OCN

Country	Sample type	Range/mean	References
China	Serum	14,300–50,700	Jin et al. (2019)
China	Serum	1.29×10^4–4.28×10^5 8.29×10^2–6.45×10^5	Wang et al. (2022)
China	Milk	211–2497	Li et al. (2020)
Germany	Adipose tissue	1700–8500	Kunisue et al. (2009)
Germany	Blood	101–1406	Fromme et al. (2015)
Ireland	Milk	59–168	Pratt et al. (2013)
Italy	Adipose tissue	500–14,000	Schiavone et al. (2010)
Pakistan	Serum	0.15–401	Waheed et al. (2020)
South Korea	Blood	2170	Park et al. (2010)
USA	Plasma	345–1540	Horii et al. (2010)
USA	Adipose tissue	21–2500	Kunisue et al. (2009)

pose to nursing infants where concentrations ranged from 211.07 to 2497.43 pg/g lipid with the highly toxic hexa-chlorinated naphthalenes (Hexa-CN66/67) in prevalence (Li et al., 2020).

Studies conducted in Pakistan on PCN exposure to populations through non-dietary sources from neglected e-waste hubs revealed some significant points where PCN were detected in dust and human serum. Combustion based activities emitted higher percentage of PCNs with dominant CN congeners being tri, tetra, and penta—in indoor air, dust, and human serum, while PCNs 66/67 and 73 congeners substantially contributed to the overall TEQ concentrations from the e-waste hubs. The occurrence of PCN in human serums of the e-waste workers indicates a potential health risk (Waheed et al., 2020). Table 3 summarizes various studies wherein levels of polychlorinated naphthalenes have been reported for the analysis carried out in human samples.

8 Management and Remedial Measures

As very limited information is available in the Indian context concerning the production of PCNs except that in the 1970s and 1980s, PCN production has been stopped and substituted by other suitable chemicals which in turn have not been well identified and described. Appropriate standardized remedial measures and regulations could not thus be implemented as far as PCNs are concerned. Reports on the decline of PCN production have been documented since the advent and usage of PCBs and plastic substitutes has taken place (UNEP/POPS/POPRC.9/13/Add.1).

However, to tackle this issue several individual studies have been worked upon to formulate techniques of PCN removal from the contaminated matrices. One such example is that of desulfurization of polychlorinated naphthalene emissions from a sintering factory to remove the polychlorinated naphthalenes. The PCN concentrations at the desulfurization systems demonstrated excellent removal for PCNs (27,888–153,672 pg/m^3 and 11,988–42,245 pg/m^3 at the inlets and outputs, respectively), where PCN removal efficiency increased as chlorination levels increased (Wang et al., 2016).

9 Conclusion

Listed in Annex A and C of Stockholm Convention's POPs listing, PCNs are targeted for a decrease in unintentional release and a gradual elimination. As per the BIS standard IS 15495:2004, PCNs are mainly banned in India from use in printing ink for food packaging, however usage and production details are not readily available and are limited (Toxics Link, 2018). Distribution of PCNs occur in different environmental matrices majorly via evaporation and combustion where the latter is attributed to be the highest contributor of PCNs into the environment. Although their occurrences are lower than that of PCBs, the bioaccumulation of PCN in various organisms especially in fishes has been reported widely (Agunbiade et al., 2020) with a few congeners exhibiting dioxin-like toxicity.

For the public to work thoughtfully with sustainable clean-up efforts, scientific data on the detrimental consequences of PCNs must be produced, as well as awareness that is founded on that evidence. Major sources of emissions should be identified and remedial actions should be taken to reduce the emissions. Moreover, focus on handling of generated wastes can greatly help to mitigate PCNs with a continuous monitoring of this pollutant in order to establish designated PCN inventories.

References

Agunbiade, I. V., Adeniji, A. O., Okoh, A. I., & Okoh, O. O. (2020). A review on occurrence and analytical procedures for the evaluation of polychlorinated naphthalenes in human and environmental matrices. *Environmental Pollutants and Bioavailability, 32*(1), 154–174.

Corsolini, S., Kannan, K., Imagawa, T., Focardi, S., & Giesy, J. P. (2002). Polychloronaphthalenes and other dioxin-like compounds in Arctic and Antarctic marine food webs. *Environmental Science & Technology, 36*(16), 3490–3496.

Dat, N. D., Chang, K. S., & Chang, M. B. (2018). Characteristics of atmospheric polychlorinated naphthalenes (PCNs) collected at different sites in Northern Taiwan. *Environmental Pollution, 237*, 186–195.

Domingo, J. L. (2004). Polychlorinated naphthalenes in animal aquatic species and human exposure through the diet: A review. *Journal of Chromatography A, 1054*(1–2), 327–334.

Fernandes, A., Rose, M., & Falandysz, J. (2017). Polychlorinated naphthalenes (PCNs) in food and humans. *Environment International, 104*, 1–13.

Fernandes, A. R., Mortimer, D., Holmes, M., Rose, M., Zhihua, L., Huang, X., et al. (2018). Occurrence and spatial distribution of chemical contaminants in edible fish species collected from UK and proximate marine waters. *Environment International, 114*, 219–230.

Fernandes, A. R., Kilanowicz, A., Stragierowicz, J., Klimczak, M., & Falandysz, J. (2022). The toxicological profile of polychlorinated naphthalenes (PCNs). *Science of the Total Environment, 837*, 155764.

Fromme, H., Cequier, E., Kim, J. T., Hanssen, L., Hilger, B., Thomsen, C., et al. (2015). Persistent and emerging pollutants in the blood of German adults: Occurrence of dechloranes, polychlorinated naphthalenes, and siloxanes. *Environment International, 85*, 292–298.

Gaonkar, O., Roy-Basu, A., Bharat, G., & Chakraborty, P. (2021). Current scenario of new persistent organic pollutants in India: Review on quantification techniques. *Persistent Organic Pollutants*, 123–154.

Guo, L., Zhang, B., Xiao, K., Zhang, Q., & Zheng, M. (2008). Levels and distributions of polychlorinated naphthalenes in sewage sludge of urban wastewater treatment plants. *Chinese Science Bulletin, 53*(4), 508–513.

Helm, P. A., Bidleman, T. F., Li, H. H., & Fellin, P. (2004). Seasonal and spatial variation of polychlorinated naphthalenes and non−/mono-ortho-substituted polychlorinated biphenyls in arctic air. *Environmental Science & Technology, 38*(21), 5514–5521.

Hogarh, J. N., Seike, N., Kobara, Y., & Masunaga, S. (2012). Atmospheric polychlorinated naphthalenes in Ghana. *Environmental Science & Technology, 46*(5), 2600–2606.

Horii, Y., Jiang, Q., Hanari, N., Lam, P. K., Yamashita, N., Jansing, R., et al. (2010). Polychlorinated dibenzo-p-dioxins, dibenzofurans, biphenyls, and naphthalenes in plasma of workers deployed at the world trade center after the collapse. *Environmental Science & Technology, 44*(13), 5188–5194.

Hu, J., Wu, J., Xu, C., Zha, X., Hua, Y., Yang, L., & Jin, J. (2019). Preliminary investigation of polychlorinated dibenzo-p-dioxin and dibenzofuran, polychlorinated naphthalene, and dioxin-like polychlorinated biphenyl concentrations in ambient air in an industrial park at the Northeastern Edge of the Tibet–Qinghai Plateau, China. *Science of the Total Environment, 648*, 935–942.

Huang, J., Yu, G., Yamauchi, M., Matsumura, T., Yamazaki, N., & Weber, R. (2015). Congener-specific analysis of polychlorinated naphthalenes (PCNs) in the major Chinese technical PCB formulation from a stored Chinese electrical capacitor. *Environmental Science and Pollution Research, 22*, 14471–14477.

Ishaq, R., Persson, N. J., Zebuhr, Y., Broman, D., & Naes, K. (2009). PCNs, PCDD/Fs, and non-ortho PCBs, in water and bottom sediments from the industrialized Norwegian Grenlandsfjords. *Environmental Science & Technology, 43*(10), 3442–3447.

Jin, J., Wang, S., Hu, J., Wu, J., Li, M., Wang, Y., & Jin, J. (2019). Polychlorinated naphthalenes in human serum samples from an industrial city in eastern China: Levels, sources, and sex differences. *Ecotoxicology and Environmental Safety, 177*, 86–92.

Kumari, K. (Ed.). (2021). *Persistent organic pollutants: Gaps in management and associated challenges* (1st ed.). CRC Press. https://doi.org/10.1201/9781003046806

Kunisue, T., Johnson-Restrepo, B., Hilker, D. R., Aldous, K. M., & Kannan, K. (2009). Polychlorinated naphthalenes in human adipose tissue from New York, USA. *Environmental Pollution, 157*, 910–915.

Lee, R. G., Thomas, G. O., & Jones, K. C. (2005). Detailed study of factors controlling atmospheric concentrations of PCNs. *Environmental Science & Technology, 39*(13), 4729–4738.

Lega, R., Megson, D., Hartley, C., Crozier, P., Macpherson, K., Kolic, T., et al. (2017). Congener specific determination of polychlorinated naphthalenes in sediment and biota by gas chromatography high resolution mass spectrometry. *Journal of Chromatography A, 1479*, 169–176.

Li, L., Sun, S., Dong, S., Jiang, X., Liu, G., & Zheng, M. (2015). Polychlorinated naphthalene concentrations and profiles in cheese and butter, and comparisons with polychlorinated dibenzo-p-dioxin, polychlorinated dibenzofuran and polychlorinated biphenyl concentrations. *International Journal of Environmental Analytical Chemistry, 95*(3), 203–216.

Li, C., Zhang, L., Li, J., Min, Y., Yang, L., Zheng, M., et al. (2020). Polychlorinated naphthalenes in human milk: Health risk assessment to nursing infants and source analysis. *Environment International, 136*, 105436.

Li, X., Gu, W., Zhang, B., Xin, X., Kang, Q., Yang, M., et al. (2022). Insights into toxicity of polychlorinated naphthalenes to multiple human endocrine receptors: Mechanism and health risk analysis. *Environment International, 165*, 107291.
Lin, Y., Zhao, Y., Qiu, X., Ma, J., Yang, Q., Shao, M., & Zhu, T. (2013). Spatial distribution of polychlorinated naphthalenes in the atmosphere across north China based on gridded field observations. *Environmental Pollution, 180*, 27–33.
Mahmood, A., Malik, R. N., Li, J., & Zhang, G. (2014). Congener specific analysis, spatial distribution and screening-level risk assessment of polychlorinated naphthalenes in water and sediments from two tributaries of the river Chenab, Pakistan. *Science of the Total Environment, 485*, 693–700.
Mcgoldrick, D. J., Pelletier, M., De Solla, S. R., Marvin, C. H., & Martin, P. A. (2018). Legacy of legacies: Chlorinated naphthalenes in lake trout, walleye, herring gull eggs and sediments from the Laurentian great lakes indicate possible resuspension during contaminated sediment remediation. *Science of the Total Environment, 634*, 1424–1434.
Park, H., Kang, J. H., Baek, S. Y., & Chang, Y. S. (2010). Relative importance of polychlorinated naphthalenes compared to dioxins, and polychlorinated biphenyls in human serum from Korea: Contribution to TEQs and potential sources. *Environmental Pollution, 158*(5), 1420–1427.
Pratt, I., Anderson, W., Crowley, D., Daly, S., Evans, R., Fernandes, A., et al. (2013). Brominated and fluorinated organic pollutants in the breast milk of first-time Irish mothers: Is there a relationship to levels in food? *Food Additives & Contaminants: Part A, 30*(10), 1788–1798.
Ross, P. S., Noël, M., Lambourn, D., Dangerfield, N., Calambokidis, J., & Jeffries, S. (2013). Declining concentrations of persistent PCBs, PBDEs, PCDEs, and PCNs in harbor seals (*Phoca vitulina*) from the Salish sea. *Progress in Oceanography, 115*, 160–170.
Safe, S., & Phil, D. (1990). Polychlorinated biphenyls (PCBs), dibenzo-p-dioxins (PCDDs), dibenzofurans (PCDFs), and related compounds: Environmental and mechanistic considerations which support the development of toxic equivalency factors (TEFs). *Critical Reviews in Toxicology, 21*(1), 51–88. https://doi.org/10.3109/10408449009089873
Schiavone, A., Kannan, K., Horii, Y., Focardi, S., & Corsolini, S. (2010). Polybrominated diphenyl ethers, polychlorinated naphthalenes and polycyclic musks in human fat from Italy: Comparison to polychlorinated biphenyls and organochlorine pesticides. *Environmental Pollution, 158*(2), 599–606.
Stockholm Convention. (2001). *Listing of Pops in Stockholm Convention*. Retrieved June 2, 2023. Retrieved on February 1, 2023, from Http://Chm.Pops.Int/Theconvention/Thepops/Listingofpops/Tabid/2509/Default.Aspx
Toxics Link. (2015). Retrieved on January 10, 2023, from Https://Toxicslink.Org/Wp-Content/Uploads/2022/08/Factsheet-50-Pops-Justified-Pentachlorophenol.Pdf
Toxics Link. (2018). Retrieved on January 14, 2023, from Https://Ipen.Org/Sites/Default/Files/Documents/Pops%20country%20situation%20report.Pdf
Unep/Pops/Poprc.5 Persistent Organic Pollutants Review Committee, Fifth Meeting, Geneva, 12–16 October 2009. Retrieved on January 10, 2023, from Https://Chm.Pops.Int/Portals/0/Download.Aspx?D=Unep-Pops-Poprc.5-10-Add.1.English.Pdf
Unep/Pops/Poprc.7/Inf/3, Persistent Organic Pollutants Review Committee, Seventh Meeting, Geneva, 10–14 October 2011. Retrieved on January 10, 2023, from Https://Chm.Pops.Int/Default.Aspx?Tabid=2267
Unep/Pops/Poprc.7-4, Persistent Organic Pollutants Review Committee, Seventh Meeting, Geneva, 10–14 October 2011. Retrieved on January 10, 2023, from Https://Chm.Pops.Int/Default.Aspx?Tabid=2267
Unep/Pops/Poprc.9/13/Add.1, Persistent Organic Pollutants Review Committee, Fifth Meeting, Geneva, 12–16 October 2009. Retrieved on January 14, 2023, from Https://Chm.Pops.Int/Convention/Popsreviewcommittee/Latestmeeting/Poprc9/Poprc9documents/Tabid/3281/Default.Aspx

Unep/Pops/Popsrc.8/16/Add.1, Persistent Organic Pollutants Review Committee, Eighth Meeting, Geneva, 15–19 October 2012. Retrieved on January 14, 2023, from Https://Www.Env.Go.Jp/Council/05hoken/Y051-155b/900420900.Pdf

Waheed, S., Khan, M. U., Sweetman, A. J., Jones, K. C., Moon, H. B., & Malik, R. N. (2020). Exposure of polychlorinated naphthalenes (PCNs) to Pakistani populations via non-dietary sources from neglected e-waste hubs: A problem of high health concern. *Environmental Pollution, 259*, 113838.

Wang, M., Liu, W., Hou, M., Li, Q., Han, Y., Liu, G., et al. (2016). Removal of polychlorinated naphthalenes by desulfurization and emissions of polychlorinated naphthalenes from sintering plant. *Scientific Reports, 6*(1), 26444.

Wang, S., Jin, J., Guo, C., Li, Z., Xu, T., Wen, X., et al. (2022). Polychlorinated naphthalene concentrations in human serum caused by unintentional production and emissions, and potential effects of polychlorinated naphthalenes on thyroid hormones. *Science of the Total Environment, 806*, 150546.

Xu, Y., Li, J., Chakraborty, P., Syed, J. H., Malik, R. N., Wang, Y., et al. (2014). Atmospheric polychlorinated naphthalenes (PCNs) in India and Pakistan. *Science of the Total Environment, 466*, 1030–1036.

Yang, L., Zheng, M., Zhao, Y., Yang, Y., Li, C., & Liu, G. (2019). Unintentional persistent organic pollutants in cement kilns co-processing solid wastes. *Ecotoxicology and Environmental Safety, 182*, 109373. https://doi.org/10.1016/J.Ecoenv.2019.109373

Zhihua, L., Gong, Y., Holmes, M., Pan, X., Xu, Y., Zou, X., & Fernandes, A. R. (2019). Geospatial visualisation of food contaminant distributions: Polychlorinated naphthalenes (PCNs), potentially toxic elements (PTEs) and aflatoxins. *Chemosphere, 230*, 559–566.

Endosulfan

Kanchan Kumari and Deepak Marathe

1 Introduction

There exist two distinct isomeric configurations of endosulfan, denoted as α-endosulfan and β-endosulfan, correspondingly. Both of these types possess identical chemical structures. Both individuals have symptoms indicative of the existence of some type of biological activity (Wan et al., 2005). The compound referred to as technical endosulfan is a mixture of two isomers, accompanied by minimal amounts of impurities (Sutherland et al., 2004). The compound in question has been assigned the number 115-29-7 by the Chemical Abstracts Service (Landeros et al., 2018). Endosulfan has a significant degree of persistence in several environmental compartments, such as the atmosphere, sediments, and water, among other relevant compartments (Sathishkumar et al., 2021). Furthermore, endosulfan has characteristics commonly linked with bioaccumulation and possesses the ability to disperse across significant distances, perhaps undergoing transportation. Endosulfan has been well acknowledged for its adverse impact on human health and has been empirically shown to cause deleterious effects on a diverse array of aquatic and terrestrial organisms (Soto et al., 1994; Naqvi & Vaishnavi, 1993; Lee et al., 2021). Thus, it was added in Annex A of the Stockholm Convention in 2011 (Kumari, 2021). The chemical endosulfan's current global usage is predominantly concentrated in the 60 nations that have implemented complete bans on endosulfan or are

K. Kumari (✉)
CSIR-National Environmental Engineering Research Institute (NEERI), Kolkata Zonal Center, Kolkata, West Bengal, India
e-mail: k_kumari@neeri.res.in

D. Marathe
CSIR-National Environmental Engineering Research Institute (NEERI), Nagpur, India

K. Kumari, *Pollutants of Global Concern*, Emerging Contaminants and Associated Treatment Technologies, https://doi.org/10.1007/978-3-031-50996-4_22

in the process of gradually eliminating its use, accounting for around 55% of its overall consumption worldwide (Mohapatra et al., 2023).

Endosulfan has been utilised as an insecticide since the 1950s with the objective of managing agricultural pests, tsetse flies, and ectoparasites affecting livestock (Ogbomida et al., 2023; Sethi et al., 2022; Mancini et al., 2006). Furthermore, apart from its function as a wood conservator, endosulfan also functions as an insecticide. In addition to its various applications, endosulfan is utilised for the purpose of wood preservation, hence preventing its degradation (Oliveira-Filho et al., 2009; Singh et al., 2011). Endosulfan, an insecticide with broad-spectrum efficacy (Espinosa et al., 2002), was utilised in agricultural practices for pest management on various crops such as soyabean, cotton, rice, sorghum, and coffee (Janssen, 2011). Endosulfan is an insecticidal compound with broad-spectrum efficacy.

2 Technical Details of Endosulfan

Endosulfan is a cyclic sulphite ester that is 1,5,5a,6,9,9a-hexahydro-6,9-methano-2,4,3-benzodioxathiepine 3-oxide substituted by chloro groups at positions 6, 7, 8, 9, and 10. The various physicochemical characteristics of endosulfan are represented in Table 1.

Table 1 Technical details and physicochemical properties of endosulfan (Weber et al., 2010; Kumar & Philip, 2017)

Details	Description (endosulfan)
Structure	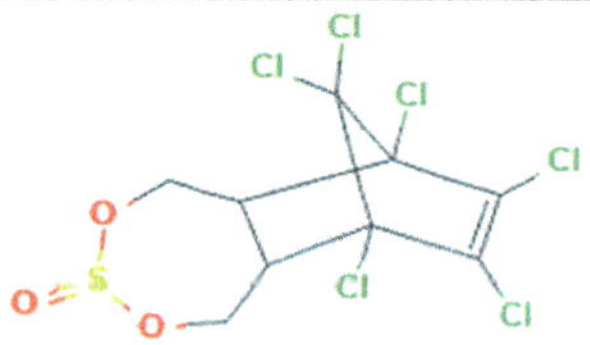
Molecular formula	$C_9H_6Cl_6O_3S$
CAS no.	115-29-7
IUPAC name	1,9,10,11,12,12-hexachloro-4,6-dioxa-5λ4-thiatricyclo[7.2.1.02,8]dodec-10-ene 5-oxide
Molecular weight	406.9 g/mol
Colour/form	Brown crystals
Odour	Similar to terpene
Density	1.745 at 68 °F
Vapour pressure	1e−05 mm Hg at 77 °F
Octanol/water partition coefficient	3.83 (LogP)
Henry's Law Constant	2.42e−05 atm-m^3/mole
Boiling point	106 °C at 0.7 mm Hg
Melting point	406–410 °F
Solubility	Insoluble
Stability/shelf life	Stable to sunlight
Corrosivity	Corrosive to iron

3 Status of Usage Within the Country

Production

According to Sharma et al.'s (2014) research, between 1920 and 2012, India produced and consumed a combined quantity of 4500 tonnes of endosulfan, which established the nation as a key contributor to the worldwide production and usage of this chemical. Because of a ban on the product that was issued by the Supreme Court of India, manufacture of it has been halted in India (Toxics Link, 2018).

Stockpiles

Currently, there is a lack of recorded data about the stockpile status of endosulfan inside the nation (Toxics Link, 2018).

Export and Import of Endosulfan

During the fiscal year 2017–2018, the particular commodity under examination was only sold to Nepal and Zambia, with a cumulative export volume of 0.13 thousand kg. With the exception of these two nations, there is a dearth of recorded evidence suggesting India's engagement in the import or export of this particular item with any other country. The numerical number provided is 18. As per the Writ petition (civil) number 213 of 2011 lodged by the Democratic Youth Federation of India before the Supreme Court of India, the term "regulation" pertains to a set of rules and guidelines established by governing entities to supervise and administer specific activities within a procedural or systemic framework. The substance commonly known as endosulfan poses a substantial threat to the conservation of human life, hence requiring the Court to fulfil its responsibility in protecting human existence, as outlined in Article 21 of the Constitution of India. Moreover, there has been a growing acknowledgement of the significant negative outcomes associated with its usage, particularly in the geographical area of Kerala. As a result, a thorough restriction has been implemented in order to regulate and limit the intake of this particular chemical. The interim judgement delivered by the Supreme Court of India on May 13, 2011, put a comprehensive restriction on the manufacture, distri bution, use, and exports of endosulfan throughout the whole country. The issuance of this declaration was driven by the negative health effects linked to the chemical component. Following a prescribed duration, the Supreme Court issued a verdict that authorised the conveyance of a certain chemical substance, which had been produced prior to the complete enforcement of the prohibition. Despite the issuance of an interim order in 2011 that put limitations on the production, exportation, and importation of endosulfan, the import and export data released by the Government of India in a recent publication reveals a notable surge in the export of endosulfan over the past several years. The revelation of this material has raised apprehensions over the implementation of the verdict issued by the Supreme Court. In light of the divergence in the direction provided by the renowned Supreme Court, there is an urgent need for more investigation into this particular topic (Toxics Link, 2018).

WHO Drinking Water Guidelines (WHO, 2004)

An acceptable daily intake (ADI) of 0.006 milligrams per kg of body weight was developed based on the assessment of the no observed adverse effect limit (NOAEL) of 0.6 milligrams per kg of body weight per day. The establishment of the NOAEL was based on a 2-year study investigating the effects of feeding on rat toxicity. The Acceptable Daily Intake (ADI) was calculated using a safety factor of 100. The Acceptable Daily Intake (ADI) is supported by NOAEL values derived from a diverse variety of scientific findings, which exhibit similarity to the ADI values. One of the studies conducted entailed administering a nutritional intervention to mice for a duration of 78 weeks. The NOAEL (No Observed Adverse Effect Level) value determined in this study was found to be 0.58 mg/kg of body weight per day. During a 1-year feeding trial on dogs, a NOAEL (No Observed Adverse Effect Level) value of 0.77 mg/kg of body weight per day was established. Furthermore, a comprehensive investigation into the issue of developmental toxicity in rats yielded a noteworthy finding on the usefulness of the No-Observed-Adverse-Effect Level (NOAEL). The research investigation identified a significant association between maternal toxicity and a NOAEL (No Observed Adverse Effect Level) value of 0.66 mg/kg of body weight per day. The aforementioned numerical value was derived from a research endeavour focused on the study of developmental toxicity. The determination of the health-based value of 20 micrograms per litre may be achieved by the utilisation of the recommended daily intake (ADI) guideline, which suggests a consumption of 0.006 mg per kg of body weight. Based on the premise that an individual with a body weight of 60 kg drinks a daily quantity of two litres of potable water, this amount corresponds to 10% of the Acceptable Daily Intake (ADI) for water consumption, as per the context of daily activities. Nevertheless, the concentrations of endosulfan often detected in drinking water are far lower than the thresholds associated with the toxicological impacts of this compound. Based on the findings of this analysis, it may be inferred that the establishment of a recommended threshold concentration for endosulfan in drinking water is deemed unnecessary.

EPA Criteria (USEPA, 1980)

Aquatic Life:

As per the established guidelines, the threshold for safeguarding freshwater aquatic organisms from the effects of endosulfan is determined by a 24-hour average concentration of 0.056 g/L, with a maximum allowable concentration of 0.22 g/L at any given moment. The aforementioned barrier has been set. Based on the guidelines, it is important to maintain a concentration of endosulfan in seawater below 0.0087 g/L, on average, throughout a 24-hour period to safeguard the well-being of aquatic organisms. Furthermore, it is highly recommended that the concentration does not exceed 0.034 g/L at any given moment.

Human Health (USEPA, 1980)

In order to mitigate the possible health hazards posed by endosulfan exposure through the ingestion of polluted aquatic species and water, a precise ambient water threshold has been instituted, set at a concentration level of 74 micrograms per litre. In order to mitigate potential risks to human health arising from the eating of

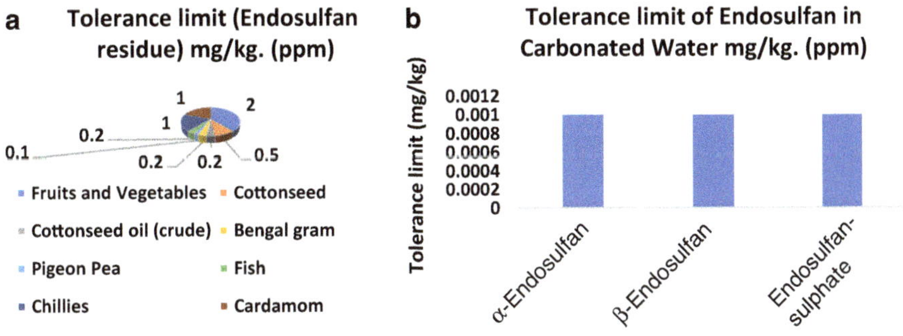

Fig. 1 (**a**) Tolerance limit of endosulfan in different food products and (**b**) carbonated water (mg/kg) (Toxics Link, 2018). (Values are reported as total of endosulfan A and B and endosulfan-sulphate)

endosulfan-contaminated aquatic creatures, a defined ambient water threshold has been implemented, set at a concentration of 159 micrograms per litre of water. The tolerance limit of endosulfan prescribed for different food items and carbonated water have been presented in Fig. 1.

Producers, Trade Names, Uses, and Effects of Endosulfan

Endosulfan was originally produced and marketed by several companies, including Bayer CropScience and Makhteshim Agan Industries (now part of ADAMA Agricultural Solutions) (Janssen, 2011; Bussian et al., 2015). Many countries have banned or severely restricted the production and use of endosulfan due to its harmful effects on health and the environment (Yadav et al., 2015; Wesseling et al., 2005; Quijano, 2000). Therefore, its production and trade have significantly decreased. Endosulfan was sold under various trade names, including Thiodan, Thionex, and Phaser (Janssen, 2011). These names might vary depending on the manufacturer and the region (Janssen, 2011). Endosulfan was used primarily as an insecticide and acaricide in agriculture to control a wide range of pests, including insects that damage crops such as cotton, fruits, vegetables, and ornamental plants (Ware & Whitacre, 2004). It was also used on other crops like tea, coffee, and tobacco. Additionally, it was used for pest control in non-agricultural settings like households and buildings (Salvo et al., 2012). Table 2 comprises of various uses and Sources of POPs and Regulation in India. Figure 2 represents the various uses of endosulfan in fruits and vegetables.

Residential: None
Public Health: None

Effects of Endosulfan to Health and Environment

While endosulfan was effective at controlling pests, it was associated with several serious health and environmental concerns, leading to its ban or restriction in many countries (Enhui et al., 2016). Some of the effects of endosulfan are represented in Fig. 3.

Table 2 Uses and sources of POPs and regulation in India (SC, 2001)

Identified POP	Uses and sources	Annex	Year of listing by SC	Regulation in India
Endosulfan	Pesticides and used as crop-pest complexes	A	2011	Banned by Supreme Court on May 13, 2011

Source: https://chm.pops.int/TheConvention/ThePOPs/The12InitialPOPs/tabid/296/Default.aspx

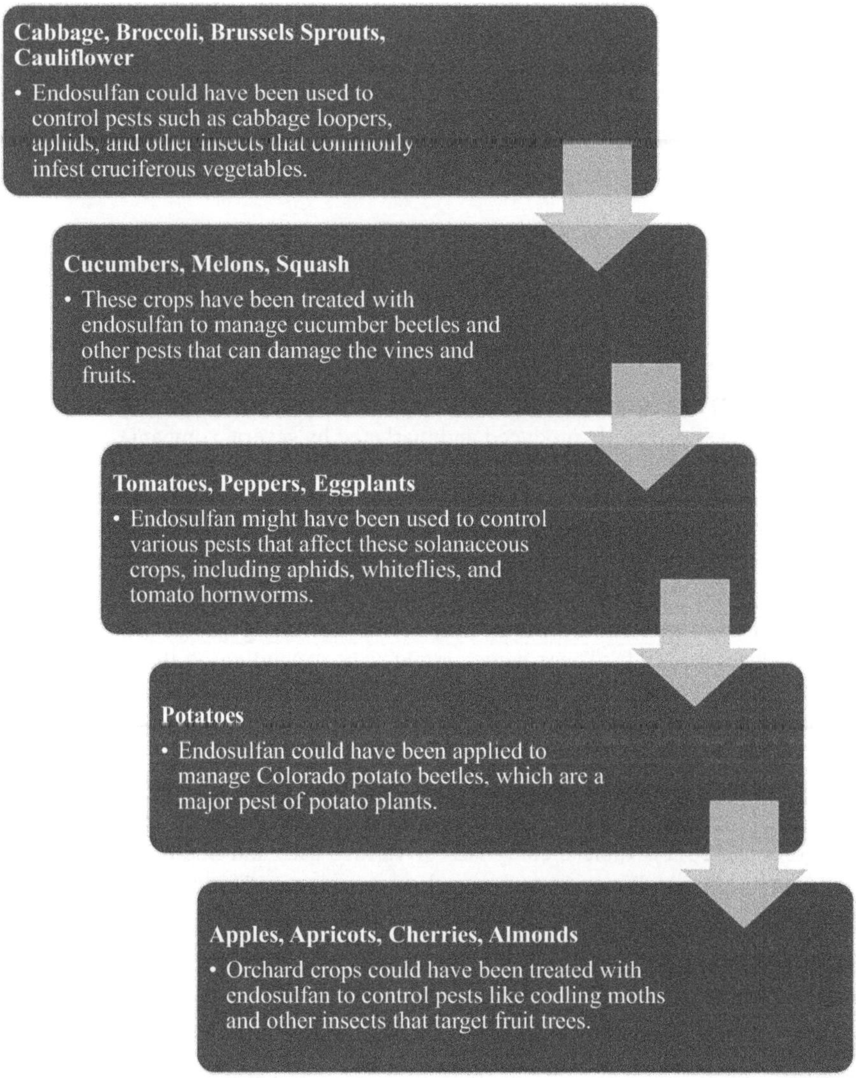

Fig. 2 Summary of uses of endosulfan

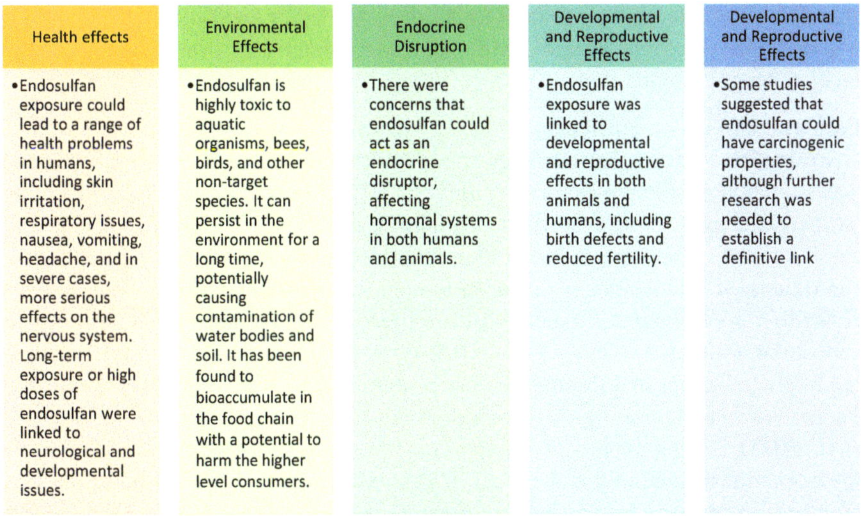

Fig. 3 Impact of endosulfan on humans and environment

ATSDR

The Agency for Toxic Substances and Disease Registry (ATSDR) has established an oral Minimal Risk Level (MRL) of 0.007 mg/kg/day for endosulfan with a short-term exposure (Taylor et al., 2013). The aforementioned statistic was obtained from a study conducted in 1981 by MacKenzie et al. The experiment yielded findings indicating that the No Observed Adverse Effect Level (NOAEL) for the manifestation of neurological symptoms in rabbits was established to be 0.7 mg/kg/day. The maximum residue limit (MRL) was calculated by deducting an uncertainty factor of 100 from the no observed adverse effect level (NOAEL) of 0.7 mg/kg/day (Taylor et al., 2013). The inclusion of this uncertainty factor serves to account for the process of extrapolating findings from animal studies to human populations, as well as the inherent variability observed among individuals. The Agency for Toxic Substances and Disease Registry (ATSDR) has established an oral Minimal Risk Level (MRL) of 0.005 mg/kg/day for endosulfan (Taylor et al., 2013). The aforementioned Maximum Residue Limit (MRL) pertains to instances of moderate period of exposure. The numerical value presented in this study was derived from a No Observed Adverse Effect Level (NOAEL) of 0.45 mg/kg/day (Taylor et al., 2013). This NOAEL was established by Banerjee and Hussain in their inquiry conducted in 1986. The major objective of this study was to investigate the immunological effects of endosulfan, with a specific focus on rats as the experimental subjects. The maximum residue limit (MRL) was calculated by deducting an uncertainty factor of 100 from the No Observed Adverse Effect Level (NOAEL) of 0.45 mg/kg/day (Taylor et al., 2013). This component contributed to the generalisation from animal studies to human populations, as well as the variability observed among individuals. The Agency for Toxic Substances and Disease Registry

(ATSDR) has replaced the Minimal Risk Level (MRL) for endosulfan, which had an oral half-life of chronic duration, with a new MRL that has an oral half-life of intermediate length (Taylor et al., 2013).

Potential Effects on Humans

Consumption of food that has been tainted with endosulfan is the most common way in which people are exposed to this toxin (Menezes et al., 2017). Endosulfan has been found in a variety of dietary products, including oils, fats, fruits, and vegetables, among other things. Individuals have also been subjected to minimum concentrations of endosulfan by cutaneous contact to polluted soil or through the consumption of cigarettes generated from tobacco that included residual levels of endosulfan (Chung & Chen, 2011, 2015; Fontcuberta et al., 2008). Endosulfan is not likely to be ingested through the use of public or private well water supplies as a source of water consumption (Dwivedi et al., 2018; Falconer et al., 2006; Agarwal et al., 2015). During the process of applying the pesticide to the crops, it is possible for worker to breathe in the chemical. It is possible to become exposed to dust via breathing in the particles or coming into close contact with the substance (Yadav & Devi, 2017; Howard, 2017; Karn et al., 2022). Endosulfan exposure can also occur when hazardous waste disposal facilities have accidental leaks or discharges of the chemical into the surrounding environment. Direct contact with contaminated soils is the most common way for people who live in close proximity to hazardous waste sites to be exposed to endosulfan (Misra & Pandey, 2005; Young et al., 2004). This is also the most likely route of exposure for individuals who work at or visit these sites. Endosulfan has an effect on the central nervous system, which disrupts its normally occurring physiological processes as a result of this effect. People who have been given high doses of the substance have reported experiencing various symptoms, including hyperactivity, nausea, dizziness, headaches, and even convulsions (Taylor et al., 2013). It is possible for one to die as a result of ingesting poisonous chemicals. According to the findings of studies that were carried out on the effects of endosulfan on animals, there is a possibility that extended exposure to this drug might result in adverse effects on the kidneys, testes, and liver. In addition, there is evidence to suggest that endosulfan may have a detrimental effect on the immune system of the body, which would reduce the body's capability to ward off illnesses (Taylor et al., 2013). Nevertheless, it is still uncertain whether or not these repercussions will occur in human beings. Endosulfan may have carcinogenic effects on humans; however, it is not yet known whether or not these effects will occur. Uncertain conclusions have been drawn from studies that have been carried out on animals (Taylor et al., 2013). Some of the summary of the toxicological endpoints for endosulfan and permissible limit of endosulfan in drinking water levels of comparison for acute dietary exposure is represented in Tables 3 and 4.

Screening and response level Based upon the criteria specified by IS10500:2012 and other sources, the screening and response level of endosulfan can be determined. The specified concentration of endosulfan in soil, surface water, and groundwater are given in Table 5.

Table 3 Summary of the toxicological endpoints for endosulfan (USEPA, 2002)

Exposure scenario	Study	Dose (mg/kg/day)	Endpoint
Acute dietary	Acute neurotoxicity study in rats Acute RfD = 0.015 mg/kg/day aPAD = 0.0015 mg/kg/day	NOAEL = 1.5 UF = 100 FQPA SF = 10	Oral LOAEL = 3 mg/kg/day; based on increased incidence of convulsions seen in female rats within 8 hours after dosing
Chronic dietary	Combined chronic toxicity/ carcinogenicity study in rats Chronic RfD = 0.006 mg/kg/day cPAD = 0.0006 mg/kg/day	NOAEL = 0.6 UF = 100 FQPA SF = 10	LOAEL = 2.9 mg/kg/day based on reduced body weight gain, enlarged kidneys, increased incidences of marked progressive glomerulonephrosis, and blood vessel aneurysms in male rats

Table 4 Endosulfan drinking water levels of comparison for acute dietary exposure (USEPA, 2002)

Population subgroup	Acute PAD (mg/kg/day)	Food exposure (mg/kg/day) @ 99.9th percentile	Ground water EEC (ppb)	Surface water peak EEC (ppb)	DWLOCacute (ppb)	Max. allowable water exposure (mg/kg/day)
Infants <1 yr	0.0015	0.0014	0.012	23.86	<1	0.0001
Children 1–6 yrs	0.0015	0.0023	0.012	23.86	0	0
US population	0.0015	0.0012	0.012	23.86	3	0.0003
Females (13–50 yrs)	0.0015	0.00098	0.012	23.86	16	0.00052

Table 5 Screening and response level of endosulfan in environmental matrix

Soil						Ground water			Surface water
Levels in soil (HW rules, 2008)	Response levels (Dutch intervention levels) 2	Screening levels soil quality guidelines for the protection of environmental and human health 3)				Indian standard for drinking water[a] (max acceptable concentration) mg/l	Guidelines for Canadian drinking water quality mg/l	WHO guidelines for drinking water mg/l	The Environment (Protection 1986 Schedule VI (land for irrigation)
50 mg/kg	4 mg/kg	–	–	–	–	0.0004 mg/l	–	10[a]	10[a]

[a]IS10500:2012 Source: NPRPS (2015)

4 In-House (Countrywide) Scientific Work Done on the Chemical

The concentration of ∑endosulfan, which includes the combined levels of α -endosulfan and β-endosulfan sulphate, ranges from 1.5 to 13 ng/g, with a value of 4 ng/g serving as the average. This data was uncovered over the course of a study that was carried out by Khuman et al. (2020). The suburban section of the Howrah district was found to have the greatest significant concentration of endosulfan, with a value of 13 ng/g. This was determined based on the findings of the study and was stated in the paper. Previous study that was carried out by Chakraborty et al. (2016) revealed that there are considerably larger levels of endosulfan found in the soil of suburban areas that are situated inside the Howrah district. In addition to this, Chakraborty et al. (2010) have demonstrated evidence that substantial concentrations of endosulfan may be present in the air in the same places as reported in the previous research. The mean endosulfan concentrations, was highest in Howrah (4.7 ng/g) than Hooghly (2.9 ng/g) and North 24 Parganas (3.5 ng/g), although Kolkata (5 ng/g) has the lowest concentration.

As shown in Fig. 4, the presence of α-endosulfan is something that may be found in relatively high concentrations in urban as well as suburban transects of Hooghly river in Bengal (Khuman et al., 2020). In addition, more than half of the total quantity of endosulfan that was discovered in the sample was determined to be in the form of β-endosulfan. To the overall endosulfan concentration, both α-endosulfan and endosulfan sulphate provided considerable contributions; the proportions of these contributions to the total endosulfan concentration were 23% and 30%, respectively. The mean amounts of endosulfan sulphate and β-endosulfan found in the suburban transect are much higher than those found in the urban transect (Khuman et al., 2020; Chakraborty et al., 2010, 2016). There are 2.33 times as many α-endosulfan molecules as there are β-endosulfan molecules in endosulfan of

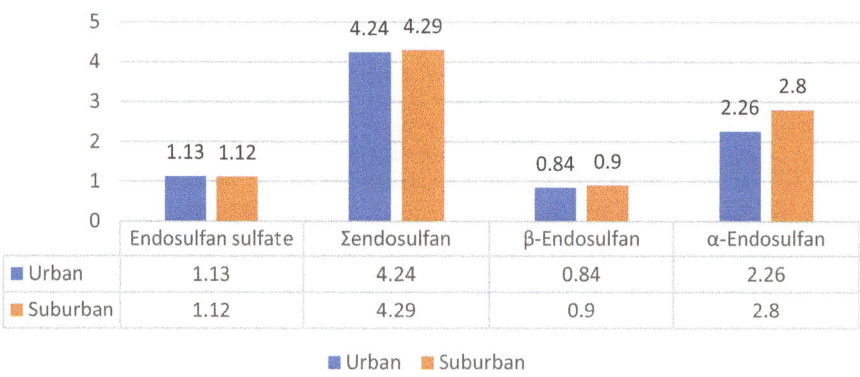

Fig. 4 Presence of endosulfan (ng g^{-1}) along urban as well as suburban transects of Hooghly riverine sediment in Bengal (Khuman et al., 2020)

Table 6 Concentration of individual and toal OCPs in ng g^{-1} in surface water of Brahmaputra and Hooghly river in Bengal (Chakraborty et al., 2010, 2016)

	Brahmaputra (n = 16)		Hooghly (n = 20)	
Compounds	Range	Avg ± SD	Range	Avg ± SD
Total OCP	2-245	47 ± 67	12-154	62 ± 35
ΣEndosulfan	ND-53	3 ± 13	ND-10	2 ± 4
β-Endosulfan	ND-45	3 ± 11	ND-10	2 ± 4
α-Endosulfan	ND-9	1 ± 2	ND	ND

technical grade. It was discovered that the β-endosulfan values in almost half of the urban and suburban regions were greater than 2.33, which is evidence that endosulfan is still being employed in a technical capacity (Khuman et al., 2020). Table 6 represents the concentration of different types of isomers in surface water of Bhrahmaputra and Hooghly rivers, West Bengal (Chakraborty et al., 2010, 2016).

Sruthi et al. (2017) conducted research to investigate the occurrence of several organochlorine pesticides (OCPs) within the rice fields of the Kuttanad agroecosystem (KAE). The researchers also investigated the distribution patterns of organochlorine pesticides (OCPs). The Kuttanad wetland region, located within the geographical confines of the state of Kerala in the Republic of India, serves a crucial function as an essential component of the Vembanad wetland system. The system in question has received formal recognition as a Ramsar site of significant global significance. Based on the findings of Ramesh and Vijayalakshmi (2002), the concentration range of endosulfan in the human population dwelling in the Kasaragod area is between the ranges of 1 and 10 ng/g. The pesticide known as endosulfan has been a subject of considerable debate in the state of Kerala, mostly due to its detrimental effects on the environment. In the subset of endosulfan compounds, it was discovered that the soil samples exhibited the occurrence of -endosulfan in three distinct instances, leading to an average concentration of 4.47 ng/g. On the other hand, the identification of β-endosulfan was confined to a solitary soil specimen, wherein it exhibited a concentration of 5.31 ng/g. The average concentration of α-endosulfan was found to be 4.47 ng/g. The findings of the current investigation demonstrate the detection of endosulfan in the soil specimens obtained from KAE, exhibiting concentrations that vary between approximately 50% and 90% of the levels seen in Kasaragod. The findings of a research investigation on the ratios of isomers suggest a clear prevalence of α-endosulfan compared to β-endosulfan, as evidenced by a ratio of less than one between the two isomers. The findings further suggest that the likely factor influencing the occurrence of endosulfan in the KAE is its recent usage. The presence of α-endosulfan and β-endosulfan, in the soil samples obtained from KAE, Kerala suggests recent usage of this substance in the particular region. The observed range of half-life values, which span from 12 to 39 days, supports this. Figure 5, Tables 7, and 8 represent average concentration of endosulfan (ng/l, ng/g) in water and sediment (post and pre-monsoon) Kuttanad agroecosystem KAE Kerala (Sruthi et al., 2017).

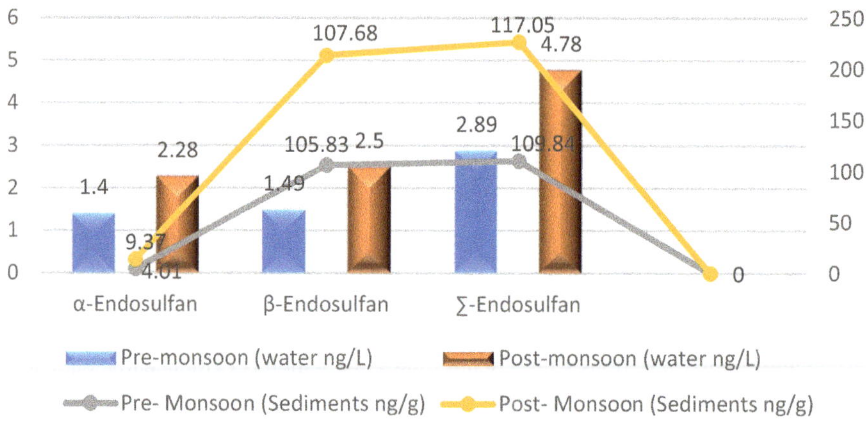

Fig. 5 Average concentration of endosulfan (ng/L, ng/g) in water and sediment (Post- and pre-monsoon) Kuttanad agroecosystem KAE Kerala (Sruthi et al., 2017)

Table 7 Summary of OCP concentration (ng/g) in soil samples of Kuttanad agroecosystem KAE Kerala (Sruthi et al., 2017)

Compound	Observed range	Mean ± SD
α-Endosulfan	0.74–8.9	4.47 ± 4.12
β-Endosulfan	<5.31	–

The research that was carried out by Baqar et al. (2018) looked at the presence of organochlorine pesticides in the tributaries of the River Ravi that are situated in the country of Pakistan. This research was conducted with the sole purpose of determining whether dermal exposure poses any potential dangers to the health of humans. In addition, the purpose of the study was to investigate the ecological dangers that are connected to the use of these pesticides, ascertain the distinctive qualities of the places where they are produced, and investigate the geographical and temporal distribution patterns of these compounds. In the sediment samples, the concentrations of organochlorine pesticides (OCPs) showed a wide range of values, ranging from 13.61 to 1992.18 ng/g dry weight. In water samples, the amounts ranged from 12.89 to 128.16 ng/L. Endosulfan was determined to be the predominant organochlorine pesticide (OCP) that was identified in sediment, whereas p,p'-DDT was shown to be the primary OCP that was found in water. The ecological danger, as determined by the risk quotient (RQ), was discovered to be N1 in every single stream that was surveyed throughout both seasons. This was the case regardless of the time of year. This finding lends credence to the hypothesis that there is an elevated risk associated with the presence of endrin and endosulfan. The amounts of the chemical endosulfan that were found in the sediments that were analysed for this study were comparable to the values that were previously reported for Mocu and Xnoha, Mexico (Hinojosa-Garro et al., 2016). This was the conclusion reached by the researchers

Table 8 Concentration of OCP in Kuttanad agroecosystem KAE Kerala (Sruthi et al., 2017)

OCP	Water (ng/L)				Sediments (ng/g)			
	Pre-monsoon		Post-monsoon		Pre-monsoon		Post-monsoon	
	Mean ± SD	Range	Mean ± SD	Range	Mean ± SD	Range	Mean ± SD	Range
α-Endosulfan	1.4 ± 0.94	0.40–4.02	2.28 ± 1.95	0.42–9.36	4.01 ± 2.47	0.64–9.35	9.37 ± 11.23	1.81–48.52
β-Endosulfan	1.49 ± 0.68	0.49–2.93	2.50 ± 2.21	0.70–11.63	105.83 ± 127.83	1.16–379.91	107.68 ± 152.20	8.70–585.73
∑ - Endosulfan	2.89 ± 1.45	0.95–6.80	4.7 ± 3.43	1.62–14.39	109.84 ± 129.03	2.20–384.97	117.05 ± 153.21	11.72–591.28

who conducted this research. Despite this, the levels identified in this inquiry were determined to be more than those disclosed in any and all previous investigations that were relevant. According to the research carried out by Chakraborty et al. (2016), the concentrations of endosulfan found in the water matrix had similarities to those discovered in the Brahmaputra River and the Hooghly River. In spite of this, when contrasted with other aquatic formations, it revealed a significantly lower level.

5 Relevance of the International Scientific Data as per Indian Conditions

A research conducted by Sruthi et al. (2017). aimed to investigate the occurrence of organochlorine pesticides (OCPs) in rice fields situated within the Kuttanad agro-ecosystem (KAE), Kerala. Additionally, the investigators analysed the distribution patterns of the aforementioned organochlorine pesticides (OCPs). The Kuttanad wetland region is located within the geographical confines of the Indian state of Kerala and plays a crucial role as a component of the broader Vembanad wetland system. The Vembanad wetland system has been officially recognised as a Ramsar site, signifying its significant worldwide value. The concentration of endosulfan recorded in India was found to be considerably lower when compared to levels detected in other nations. The phenomena under observation displayed a significant discrepancy. Table 9 represents the comparison of OCPs concentration (ng/g) in agricultural soil of KAE Kerala with other parts of the world.

The purpose of this study was to investigate the concentration of organochlorine pesticides (OCPs), as well as their geographical distribution and risk assessment, in the agricultural soils of the Aghili plain, which is located in the province of

Table 9 Comparison of OCPs concentration (ng/g) in agricultural soil of KAE Kerala with other parts of the world (Sruthi et al., 2017)

References	αEndosulfan	Area
Mao et al. (2013)	NA	Lishui, China
Murugan et al. (2013)	Nd–4.79	Andaman Islands
Nawab et al. (2003)	NA	Aligarh, UP, India
Gonzalez et al. (2003)	0.08 ± 0.07	Argentina
Kumar et al. (2011)	0.18 ± 0.14	North India
Dan et al. (2012)	NA	Zhangzhou City, China
Jiang et al. (2009)	0.13 ± 0.30	Shanghai, China
Mishra et al. (2012)	NA	Assam, India
Agnihotri et al. (1996)	1–113	Indo-Gangetic Plain 96
Fang et al. (2007)	2.04 ± 3.0	Taihu Lake, China

Table 10 Concentrations of OCPs in agricultural soils from different regions mean (range) (ng/g)

Study area	Central China	Mexico	Uzbekistan	Shanghai, China
α-endosulfan	0.34 (ND–5.88)	–	ND	0.13 (ND–1.77)
β endosulfan	0.94 (ND–23.04)	57	10.47(ND–74.56)	–
Endosulfan sulfate	1.28 (ND–23.04)	–	–	–

Khuzestan in Iran. According to the findings, the agricultural soils displayed contamination from organochlorine pesticides (OCPs), and it was highly unlikely that OCPs had been sprayed during the past few years. The likelihood of this happening is low. Concentrations of OCPs in agricultural soils from different regions mean (range) (ng/g) are represented in Table 10.

6 Remedial Action Taken for Management of POP

The process of microbial degradation of endosulfan was monitored for a duration of 21 days. The identification of the bacterial species *Pseudomonas* and *Bacillus* was accomplished via the use of biochemical tests subsequent to their successful isolation from their respective habitats (Harikumar et al., 2013). Significant degradation of endosulfan isomers was seen on a nutrient-rich agar medium containing numerous bacterial cultures. Within a span of just 14 days, a notable deterioration was seen in the constituents as well as an escalation in the bacterial biomass. After a 2-week incubation period, it was observed that *Bacillus* species had a limited ability to degrade 50 0.08% of the endosulfan. In contrast, *Pseudomonas* species has shown a significantly higher capacity to digest 98.02 0.18% of the endosulfan at a concentration of 119.48 0.53 g/L (Harikumar et al., 2013). Following a 14-day incubation period, it was shown that the uninoculated controls retained over 50% of the substrate, suggesting that only a negligible degree of chemical degradation or volatilisation of endosulfan had occurred. The results of the monitoring conducted on the 16th day revealed that complete degradation of endosulfan occurred through the process of bioremediation using *Pseudomonas* species. In contrast, the uninoculated control group retained over 50% of the endosulfan. The results of a study examining the biodegradation of endosulfan by two distinct bacterial species are reported in Tables 11, 12, and Fig. 6, following a 14-day examination period (Harikumar et al., 2013).

Table 11 Microbial degradation of a-endosulfan (a-ES), b-endosulfan (b-ES) and its toxic metabolite, endosulfan's sulfate (ESS), Endo ether (EE), and Endo lactone (EL) (Kumar & Philip, 2017)

Microorganism	Compound	Initial concentration (mg L^{-1})	Maximum degradation rate (mg L^{-1} d)	Cell density	References
Pseudomonas aeroginosa	ES	50	4.00	2.7–2.8[b]	Narkhede et al. (2015)
Agrobacterium tumefaciens	a-ES	70	15.83	0.7–0.8[c]	Thangadurai and Suresh (2014)
	b-ES	30	7.67	0.7–0.8[c]	
Alcaligenes faecalis JBW4	a-ES	70	12.24	1.9[c]	Kong et al. (2013)
	b-ES	30	5.04	1.9[c]	
Aspergillus sydoni	a-ES	70	3.69	0.07[g]	Goswami et al. (2009)
	b-ES	30	1.62	0.07[g]	
Acromobacter xylosoxidans	a-ES	70	3.95	0.7[c]	Li et al. (2009)
	b-ES	30	1.73	0.7[c]	
Mixed culture (*Stenotrophomonas maltophilia* and *Rhodococcus Erythropolis*)	a-ES	70	3.41	1.2[d]	Kumar and Philip (2007)
	b-ES	30	1.62	1.2[d]	
Mixed bacterial culture (*Bacilus circulans*-I, *Bacillus circulans*-II, and *Staphylococcus*)	a-ES	70	1.07	ND[c]	Kumar and Philip (2006)
	b-ES	30	0.458	ND[c]	
Klebsiella oxytoca KE- 8	a-ES	121	11.14	ND[c]	Kwon et al. (2005)
	b-ES	29	1.71	ND[c]	
	ESS	173	9.36	ND[c]	
Klebsiella pneumoniae KE-1	a-ES	65	6.17	0.9[b]	Kwon et al. (2002)
	b-ES	28	2.55	0.9[b]	
Fusarium ventricosum	a-ES	70	14.2	ND[c]	Siddique et al. (2003)
	b-ES	30	6.6	ND[c]	
Pandoraea sp.	a-ES	70	8.19	0.52[d]	
	b-ES	30	3.18	0.52[d]	
Chlorococcum sp.	a-ES	4.24	0.13	ND[c]	Sethunathan et al. (2004)
	ESS	8.44	0.12	ND[c]	
Scenedesmus sp.	a-ES	4.24	0.14	ND[c]	
	ESS	8.44	0.05	ND[c]	
Anabaena flosaquae	a-ES	10	2.25 (4.0)[a]	50[e]	Lee et al. (2003)
Anabaena sp. PCC7120	a-ES	10	2.25	50[e]	

(continued)

Table 11 (continued)

Microorganism	Compound	Initial concentration (mg L^{-1})	Maximum degradation rate (mg L^{-1} d)	Cell density	References
Bacillus sp. co-culture	a-ES	388	29.7	2×10^6(cfu)[f]	Awasthi et al. (2003)
	b-ES	386	27.7	2×10^6(cfu)[f]	
Mycobacterium sp. ESD	b-ES	125	<30	0.78[g]	Sutherland et al. (2002a)
Enriched culture	ESS	125	25	0.4[g]	Sutherland et al. (2002b)
	a-ES	125	17.8	0.4[g]	
	b-ES	125	12.5	0.4[g]	
Phanerochaete chrysosporium	a-ES	1	0.48	ND[c]	Kullman and Matsumura (1996)

[a]pH controlled culture (pH 7.2)
[b]OD550 nm
[c]Not determined
[d]OD600 nm
[e]Chlorophyll concentration (mg L^{-1})
[f]Initial cell concentration
[g]OD595 nm

Table 12 Percentage removal of endosulfan on bioremediation by isolated bacterial species (Harikumar et al., 2013)

Bacterial species	Initial concentration of endosulfan (μg/L)	Final concentration of endosulfan (μg/L)	Removal of endosulfan (%)
Control	119.48 ± 0.53	110.87 ± 0.07	7.20 ± 0.63
Bacillus	119.48 ± 0.53	61.30 ± 0.10	50.0 ± 0.08
Pseudomonas	77.12 ± 0.49	16.41 ± 0.09	78.73 ± 0.12
Bacillus	77.12 ± 0.49	33.40 ± 0.47	56.62 ± 0.61
Control	77.12 ± 0.49	69.34 ± 0.37	10.08 ± 0.49
Pseudomonas	119.48 ± 0.53	2.37 ± 0.21	98.02 ± 0.18

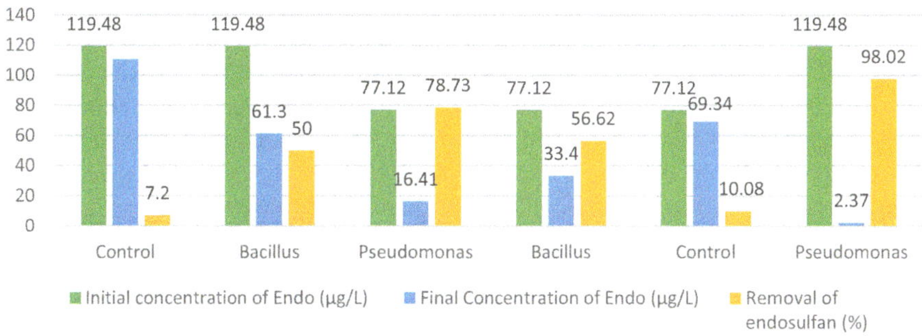

Fig. 6 Percentage removal of endosulfan on bioremediation by isolated bacterial species (Harikumar et al., 2013)

The purpose of this study was to assess the extent to which 20 different species of fungi might degrade the chlorinated organic insecticides endosulfan and chlorpyrifos. The initial stage of the fungal screening procedure involved the utilisation of soil extract agar to assess the organisms' tolerance levels towards different concentrations of a certain pesticide. Subsequently, an examination was conducted to investigate the degradation of individual pesticides in a soil extract broth, utilising a concentration of 25 mg/L. Gas chromatography played a crucial role in assessing the degree of pesticide degradation. In addition to the aforementioned factors, this study also assessed other parameters, namely pH and mycelial weight. The findings indicate that chlorpyrifos exhibited much greater growth inhibition in all investigated fungus as compared to endosulfan. The aforementioned result is derived from an assessment of the growth inhibition percentage exhibited by the test fungus, as well as an examination of the EC50 values in the subsequent investigation. The fungus *Trametes hirsuta* exhibited complete degradation of both the endosulfan isomers, achieving a degradation efficiency of 100%. On the other hand, the fungus *Cladosporium cladosporioides* demonstrated the highest degradation rate for chlorpyrifos among the three fungi studied. With the exclusion of *Phanerochaete chrysosporium*, *Trichoderma harzianum*, and *Trichoderma virens*, all the fungi that were subjected to experimentation have shown a notable enhancement in their capacity to break down endosulfan when compared to chlorpyrifos. In contrast, it has been observed that these three fungal species have a higher capacity for the degradation of chlorpyrifos compared to endosulfan. Furthermore, the study revealed that all of the fungi examined had a greater level of effectiveness in the breakdown of α-endosulfan as compared to β-endosulfan. The research findings indicate that endosulfan sulphate emerged as the predominant degradation product generated by the various fungi examined in this study. Fungi exhibiting a greater extent of endosulfan degradation also exhibited a higher production of endosulfan sulphate at a same magnitude. Although *T. hirsuta* and *Trametes* versicolor have demonstrated more efficacy in the degradation of endosulfan, it has been observed that the use of either of these organisms leads to a considerable reduction in the presence of endosulfan sulphate (Bisht et al., 2019).

The pesticide solution was subjected to gamma ray irradiation at a concentration of 100 parts per million (ppm). The decomposition of acetonitrile can be attributed to the presence of endosulfan inside the molecule. The degradation kinetics exhibit a first-order behaviour, characterised by an exponential decay pattern that elucidates the temporal evolution of the dose-concentration relationship. Consequently, the presence of acetonitrile and water in the solution facilitates an expedited progression of degradation. The incomplete degradation of endosulfan has been seen even when exposed to higher absorbed doses of 45 kGy and beyond. Consequently, a study was conducted to assess the interaction between a solution of gold nanoplates and a solution of endosulfan. The metabolites generated from this interaction were then analysed using GC-MS QP2010 PLUS. The findings derived from the gas chromatography-mass spectrometry (GC-MS) analysis indicate that the gold nanoplates synthesised using the radiolytic reduction of a 300 micromolar solution of auclide effectively eliminated endosulfan and demonstrated a rapid degradation mechanism. The study has established that the approach using the interaction of Au nanoplates is more favourable compared to the method employing gamma irradiation (Mitra et al., 2014).

There exist many geographical contexts wherein alternatives to endosulfan, encompassing both chemical and non-chemical options, are accessible. These scenarios encompass both established nations and developing ones. In nations where the use of endosulfan has been either fully prohibited or is now undergoing a gradual phase-out, several alternative options have been proposed.

7 Recommendations

Currently, there is a lack of recorded information on the present stockpile status of endosulfan inside the nation. From 1920 to 2012, India had a significant role in the manufacture and consumption of endosulfan, with a cumulative output of 4500 metric tonnes. Consequently, India has emerged as a significant participant in the global manufacture and consumption of this chemical component, garnering substantial recognition. The presence of endosulfan contamination in the natural environment has been demonstrated via extensive investigations. The use of effective corrective actions is crucial for attaining successful management.

References

Agarwal, A., Prajapati, R., Singh, O. P., Raza, S. K., & Thakur, L. K. (2015). Pesticide residue in water—A challenging task in India. *Environmental Monitoring and Assessment, 187*, 1–21.

Agnihotri, N. P., Kulshrestha, G., Gajbhiye, V. T., Mohapatra, S. P., & Singh, S.B. (1996). Organochlorine insecticide residues in agricultural soils of the Indo-Gangetic plain. *Environmental Monitoring and Assessment, 40*(3), 279–288.

Awasthi, N., Singh, A. K., Jain, R. K., Khangarot, B. S., & Kumar, A. (2003). Degradation and detoxification of endosulfan isomers by a defined co-culture of two Bacillus strains. *Applied Microbiology and Biotechnology, 62*, 279–283.

Baqar, M., Sadef, Y., Ahmad, S. R., Mahmood, A., Li, J., & Zhang, G. (2018). Organochlorine pesticides across the tributaries of River Ravi, Pakistan: Human health risk assessment through dermal exposure, ecological risks, source fingerprints and spatio-temporal distribution. *Science of the Total Environment, 618*, 291–305.

Bisht, J., Harsh, N. S. K., Palni, L. M. S., Agnihotri, V., & Kumar, A. (2019). Bioaugmentation of endosulfan contaminated soil in artificial bed treatment using selected fungal species. *Bioremediation Journal, 23*(3), 196–214.

Bussian, B. M., Pandelova, M., Lehnik-Habrink, P., Aichner, B., Henkelmann, B., & Schramm, K. W. (2015). Persistent endosulfan sulfate is found with highest abundance among endosulfan I, II, and sulfate in German forest soils. *Environmental Pollution, 206*, 661–666.

Chakraborty, P., Zhang, G., Li, J., Xu, Y., Liu, X., Tanabe, S., & Jones, K. C. (2010). Selected organochlorine pesticides in the atmosphere of major Indian cities: Levels, regional versus local variations, and sources. *Environmental Science & Technology, 44*(21), 8038–8043.

Chakraborty, P., Khuman, S. N., Selvaraj, S., Sampath, S., Devi, N. L., Bang, J. J., & Katsoyiannis, A. (2016). Polychlorinated biphenyls and organochlorine pesticides in River Brahmaputra from the outer Himalayan Range and River Hooghly emptying into the Bay of Bengal: Occurrence, sources and ecotoxicological risk assessment. *Environmental Pollution, 219*, 998–1006.

Chung, S. W., & Chen, B. L. (2011). Determination of organochlorine pesticide residues in fatty foods: A critical review on the analytical methods and their testing capabilities. *Journal of Chromatography A, 1218*(33), 5555–5567.

Chung, S. W., & Chen, B. L. (2015). Development of a multiresidue method for the analysis of 33 organochlorine pesticide residues in fatty and high water content foods. *Chromatographia, 78*(7–8), 565–577.

Dan, Y. A., Shi-Hua, Q. I., Zhang, J. Q., Ling-Zhi, T. A., Zhang, J. P., Zhang, Y., Feng, X. U., Xin-Li, X. I., Ying, U., Wei, C. H., Jun-Hua, Y. A. (2012). Residues of organochlorine pesticides (OCPs) in agricultural soil of Zhangzhou city, China. *Pedosphere, 22*(2), 178–189.

Dwivedi, S., Mishra, S., & Tripathi, R. D. (2018). Ganga water pollution: A potential health threat to inhabitants of Ganga basin. *Environment International, 117*, 327–338.

Enhui, Z., Na, C., MengYun, L., Jia, L., Dan, L., Yongsheng, Y., et al. (2016). Isomers and their metabolites of endosulfan induced cytotoxicity and oxidative damage in SH-SY 5 Y cells. *Environmental Toxicology, 31*(4), 496–504.

Espinosa, P. J., Bielza, P., Contreras, J., & Lacasa, A. (2002). Insecticide resistance in field populations of Frankliniella occidentalis (Pergande) in Murcia (south-east Spain). *Pest Management Science, 58*(9), 967–971.

Falconer, I. R., Chapman, H. F., Moore, M. R., & Ranmuthugala, G. (2006). Endocrine-disrupting compounds: A review of their challenge to sustainable and safe water supply and water reuse. *Environmental Toxicology: An International Journal, 21*(2), 181–191.

Fang, W., Jiang, X., Bian, Y. R., Yao, F. X., Gao, H. J., Yu, G. F., Munch, J. C., & Schroll, R. (2007). Organochlorine pesticides in soils under different land usage in the Taihu Lake region, China. *Journal of Environmental Sciences, 19*(5), 584–590.

Fontcuberta, M., Arqués, J. F., Villalbi, J. R., Martínez, M., Centrich, F., Serrahima, E., et al. (2008). Chlorinated organic pesticides in marketed food: Barcelona, 2001–06. *Science of the Total Environment, 389*(1), 52–57.

Gonzalez, M., Miglioranza, K. S., Aizpún de Moreno, J. E., & Moreno, V. J. (2003). Organochlorine pesticide residues in leek (Allium porrum) crops Environ Sci Pollut Res grown on untreated soils from an agricultural environment. *Journal of Agricultural and Food Chemistry, 51*(17), 5024–5029.

Goswami, S., Vig, K., & Singh, D. K. (2009). Biodegradation of a and b endosulfan by Aspergillus Niger. *Chemosphere, 75*, 883–888.

Harikumar, P. S., Jesitha, K., & Sreechithra, M. (2013). Remediation of endosulfan by biotic and abiotic methods. *Journal of Environmental Protection, 4*(5), 418.

Hinojosa-Garro, D., Burgos Chan, A. M., & Rendón-von Osten, J. (2016). Organochlorine pesticides (OCPs) in sediment and fish of two tropical water bodies under different land use. *Bulletin of Environmental Contamination and Toxicology, 97*, 105–111.

Howard, P. (2017). *Handbook of environmental fate and exposure data: For organic chemicals, volume III pesticides*. Routledge.

https://chm.pops.int/TheConvention/ThePOPs/ListingofPOPs/tabid/2509/Default.aspx.

Janssen, M. P. M. (2011). Endosulfan: A closer look at the arguments against a worldwide phase out.

Jiang, Y. F., Wang, X. T., Jia, Y., Wang, F., Wu, M. H., Sheng, G. Y., & Fu, J. M. (2009). Occurrence, distribution and possible sources of organochlorine pesticides in agricultural soil of Shanghai, China. *Journal of Hazardous Materials, 170*, 989–997.

Karn, S. K., Upadhyay, A., & Kumar, A. (2022). Biomonitoring of endosulfan toxicity in human. *Biocell, 46*(7), 1771.

Khuman, S. N., Bharat, G., & Chakraborty, P. (2020). Spatial distribution and sources of pesticidal persistent organic pollutants in the Hooghly riverine sediment. *Environmental Science and Pollution Research, 27*, 4137–4147.

Kong, L., Zhu, S., Zhu, L., Xie, H., Su, K., Yan, T., et al. (2013). Biodegradation of organochlorine pesticide endosulfan by bacterial strain Alcaligenes faecalis JBW4. *Journal Environment Science, 25*(11), 2257–2264.

Kullman, S. W., & Matsumura, F. (1996). Metabolic pathway utilized by Phenerochete chrysosporium for degradation of the cyclodine pesticide endosulfan. *Applied and Environment Microbiology, 62*, 593–600.

Kumar, M., & Philip, L. (2006). Bioremediation of endosulfan contaminated soil and water optimization of operating conditions in laboratory scale reactors. *Journal of Hazardous Materials, 136*(2), 354–364.

Kumar, M., & Philip, L. (2007). Biodegradation of endosulfan contaminated soil in a pilot scale reactor—bioaugmented with mixed bacterial culture. Journal of Environmental Science and Health. *Part B: Pesticides, Food Contaminants, and Agricultural Wastes, 42*(6), 707–715.

Kumar, M., & Philip, L. (2017). Remediation of endosulfan contaminated system by microbes. In *Microbe-induced degradation of pesticides* (pp. 59–81).

Kumar, B., Kumar, S., Gaur, R., Goel, G., Mishra, M., Singh, S. K., Prakash, D., & Sharma, C.S. (2011). Persistent organochlorine pesticides and polychlorinated biphenyls in intensive agricultural soils from North India. *Soil and Water Research, 4*, 190–197.

Kumari, K. (Ed.). (2021). *Persistent organic pollutants: Gaps in management and associated challenges* (1st ed.). CRC Press. https://doi.org/10.1201/9781003046806

Kwon, G., Kim, J., Kim, T., Sohn, H., Koh, S., Shin, K., et al. (2002). Klebsiella pneumoniae KE-1 degrades endosulfan without formation of the toxic metabolite, Endosulfan sulfate. *FEMS Microbiology Letters, 215*, 255–259.

Kwon, J. S., Sohn, H. Y., Shin, K. S., Kim, E., & Seo, B. I. (2005). Biodegradation of the organochlorine insecticide, endosulfan, and the toxic metabolite, endosulfan sulfate, by Klebsiella oxytoca KE 8. *Applied Microbiology and Biotechnology, 67*, 845–850.

Landeros, C. R., Díaz, C. E. B., & Bilyeu, B. (2018). Dissemination of Endosulfan into the environment. In *Persistent organic pollutants*. IntechOpen.

Lee, D. Y., Choi, G. H., Bae, Y. S., Lee, S. W., Kim, S. K., Bae, J. Y., et al. (2021). Fate of endosulfan in ginseng farm and effect of granular biochar treatment on endosulfan accumulation in ginseng. *Environmental Geochemistry and Health, 44*, 1–13.

Lee, S. E., Kim, J. S., Kennedy, I. R., Park, J. W., Kwon, G. S., Koh, S. C., et al. (2003). Biotransformation of an organochlorine insecticide, endosulfan, by Anabaena species. *Journal of Agriculture and Food Chemistry, 51*, 1336–1340.

Li, W., Dai, Y., Xue, B., Li, Y., Peng, X., Zhang, J., et al. (2009). Biodegradation and detoxification of endosulfan in aqueous medium and soil by Achromobacter xylosoxidans strain CS5. *Journal Hazard Materials, 167*, 209–216.

Mancini, F., Termorshuizen, A. J., & Van Bruggen, A. H. (2006). *Impact of integrated pest management farmer field schools on pesticide use and farmers' ecological knowledge on cotton farming in India*. Impact of IPM Farmer Field Schools on the environment, health and livelihoods of cotton growers in Southern India.

Mao, Y. E., Xing-Lun, Y. A., Ming-Ming, S. U., Yong-Rong, B. I., Fang, W. A., Cheng-Gang, G. U., Hai-jiang, W. E., Yang, S. O., Lei, W. A., Xin, J. I., & Jiang, X. (2013). Use of organic solvents to extract organochlorine pesticides (OCPs) from aged contaminated soils. *Pedosphere, 23*(1), 10–19.

Menezes, R. G., Qadir, T. F., Moin, A., Fatima, H., Hussain, S. A., Madadin, M., et al. (2017). Endosulfan poisoning: An overview. *Journal of Forensic and Legal Medicine, 51*, 27–33.

Misra, V., & Pandey, S. D. (2005). Hazardous waste, impact on health and environment for development of better waste management strategies in future in India. *Environment International, 31*(3), 417–431.

Mishra, K., Sharma, R. C., & Kumar, S. (2012). Ecotoxicolgy and environmental safety contamination levels and spatial distribution of organochlorine pesticides in soils from India. *Ecotoxicology and Environmental Safety, 76*, 215–225.

Mitra, D., Francis, S., & Varshney, L. (2014). Remediation of pesticide Endosulfan in solution using gold nanoplates and gamma irradiation a comparative studies using GCMS

QP2010PLUS. In *Proceedings of the DAE-BRNS twelfth biennial Trombay symposium on radiation and photochemistry*.

Mohapatra, P., Nøklebye, E., Arora, T., & Basu, A. R. (2023). International regulatory frameworks and best practices in management of POPs. In *Managing persistent organic pollutants in India: Case studies on Vapi and Surat, Gujarat* (pp. 127–151). Springer Nature Switzerland.

Murugan, A. V., Swarnam, T. P., & Gnanasambandan, S. (2013). Status and effects of pesticide residues in soils under different land uses of Andaman Islands, India. *Environmental Monitoring and Assessment, 185*(10), 8135–8145.

Narkhede, C. P., Patil, A. R., Koli, S., Suryawanshi, R., Wagh, N. D., Salunke, B. K., et al. (2015). Studies on endosulfan degradation by local isolate Pseudomonas aeroginosa. *Biocatalysis Agricultural Biotechnology, 4*(2), 259–265.

Naqvi, S. M., & Vaishnavi, C. (1993). Bioaccumulative potential and toxicity of endosulfan insecticide to non-target animals. *Comparative Biochemistry and Physiology Part C: Comparative Pharmacology, 105*(3), 347–361.

Nawab A, Aleem A, Malik A (2003) Determination of organochlorine pesticides in agricultural soil with special reference to γ-HCH degradation by Pseudomonas strains. *Bioresour Technol 88*:41–46

NPRPS. (2015, Dec). *National Program for Rehabilitation of Polluted Sites in India Guidance document for assessment and remediation of contaminated sites in India.* 1st Edn, pp. 1–564. https://cpcb.nic.in/uploads/hwmd/MoEFCC_guidelines_contaminatedsites.pdf

Ogbomida, E., Emeribe, C., Akukwe, T., Nwobodo, T., Aganmwonyi, I., Eze, E., et al. (2023). Knowledge levels of environmental effects of pesticides usage among smallholder farmers in etsako west lg. a, edo state, Nigeria: Pesticide usage, smallholder farmers, environmental effects, farming operations. *Fudma Journal of Agriculture and Agricultural Technology, 9*(2), 16–28.

Oliveira-Filho, E. C., Grisolia, C. K., & Paumgartten, F. J. R. (2009). Effects of endosulfan and ethanol on the reproduction of the snail Biomphalaria tenagophila: A multigeneration study. *Chemosphere, 75*(3), 398–404.

Quijano, R. F. (2000). Risk assessment in a third-world reality: An endosulfan case history. *International Journal of Occupational and Environmental Health, 6*(4), 312–317.

Ramesh, A., & Vijayalakshmi, A. (2002). Environmental exposure to residues after aerial spraying of endosulfan: Residues in cow milk, fish, water, soil and cashew leaf in Kasargod, Kerala, India. *Pest Management Science: Formerly Pesticide Science, 58*, 1048–1054.

Salvo, L. M., Bainy, A. C., Ventura, E. C., Marques, M. R., Silva, J. R. M., Klemz, C., & Silva de Assis, H. C. (2012). Assessment of the sublethal toxicity of organochlorine pesticide endosulfan in juvenile common carp (Cyprinus carpio). *Journal of Environmental Science and Health, Part A, 47*(11), 1652–1658.

Sathishkumar, P., Mohan, K., Ganesan, A. R., Govarthanan, M., Yusoff, A. R. M., & Gu, F. L. (2021). Persistence, toxicological effect and ecological issues of endosulfan–a review. *Journal of Hazardous Materials, 416*, 125779.

SC, Stockholm convention. (2001).

Sethi, A., Lin, C. Y., Madhavan, I., Davis, M., Alexander, P., Eddleston, M., & Chang, S. S. (2022). Impact of regional bans of highly hazardous pesticides on agricultural yields: The case of Kerala. *Agriculture & Food Security, 11*(1), 1–13.

Sethunathan, N., Megharaj, M., Chen, Z. L., Williams, B. D., Lewis, G., & Naidu, R. (2004). Algal degradation of a known endocrine disrupting insecticide, alpha-endosulfan and its metabolites, endosulfan sulfate, in liquid medium and soil. *Journal of Agriculture and Food Chemistry, 52*, 3030–3035.

Siddique, T., Okeke, B. C., Arshad, A., & Frankenberger, W. T. (2003). Enrichment and Isolation of endosulfan- Degrading Microorganisms. *Journal of Environmental Quality, 32*, 47–54.

Singh, J. K., Nath, A., Kumar, A., Md Ali, M., & Kumar, R. (2011). Study on the effect of endosulfan on testosterone level and seminiferous tubule of testis of mice. *World Journal of Environmental Polluation, 1*, 1–4.

Sharma, B. M., Bharat, K., Tayal, S., Nizzetto, L., & Larssen, T. (2014). The legal framework to manage chemical pollution in India and the lesson from the persistent organic pollutants (POPs). *The Science of the Total Environment, 490C*, 733–747.
Soto, A. M., Chung, K. L., & Sonnenschein, C. (1994). The pesticides endosulfan, toxaphene, and dieldrin have estrogenic effects on human estrogen-sensitive cells. *Environmental Health Perspectives, 102*(4), 380–383.
Sruthi, S. N., Shyleshchandran, M. S., Mathew, S. P., & Ramasamy, E. V. (2017). Contamination from organochlorine pesticides (OCPs) in agricultural soils of Kuttanad agroecosystem in India and related potential health risk. *Environmental Science and Pollution Research, 24*, 969–978.
Sutherland, T. D., Horne, I., Harcourt, R. L., Russell, R. J., & Oakeshott, J. G. (2002a). Isolation and characterization of a Mycobacterium strain that metabolizes the insecticide endosulfan. *Journal of Applied Microbiology, 93*(3), 380–389.
Sutherland, T. D., Weir, K. M., Lacey, M. J., Horne, I., Russel, R. J., & Oakeshott, J. G. (2002b). Enrichment of a microbial culture capable of degrading endosulphate, the toxic metabolite of endosulfan. *Journal of Applied Microbiology, 94*, 541–548
Sutherland, T. D., Horne, I., Weir, K. M., Russell, R. J., & Oakeshott, J. G. (2004). Toxicity and residues of endosulfan isomers. *Reviews of Environmental Contamination and Toxicology, 2004*, 99–113.
Taylor, J., Pohl, H. R., Mumtaz, M., Ruiz, P., Llados, F., Rudisill, C., & Citra, M. J. (2013). *Draft toxicological profile for endosulfan*. https://stacks.cdc.gov/view/cdc/21427
Toxics Link. (2018). *Persistent organic pollutants (POPs) in India: country situation report.*
Thangadurai, P., & Suresh, S. (2014). Biodegradation of endosulfan by soil bacterial cultures. *International Biodeterioration Biodegradation, 94*, 38–47.
United States Environmental Protection Agency (USEPA). (1980). Ambient Water Quality Criteria for Endosulfan, EPA 440/5-80-046, https://www.epa.gov/sites/default/files/2019-03/documents/ambient-wqc-endosulfan-1980.pdf
United States Environmental Protection Agency (USEPA). (2002). Reregistration Eligibility Decision for Endosulfan, EPA 738-R-02-013, https://archive.epa.gov/pesticides/reregistration/web/pdf/endosulfan_red.pdf
Wan, M. T., et al. (2005). Toxicity of α-, β-,(α+ β)-endosulfan and their formulated and degradation products to Daphnia magna, Hyalella azteca, Oncorhynchus mykiss, Oncorhynchus kisutch, and biological implications in streams. *Environmental Toxicology and Chemistry, 24*, 1146–1154.
Ware, G. W., & Whitacre, D. M. (2004). An introduction to insecticides. In *The pesticide book* (Vol. 6).
Weber, J., Halsall, C. J., Muir, D., Teixeira, C., Small, J., Solomon, K., et al. (2010). Endosulfan, a global pesticide: A review of its fate in the environment and occurrence in the Arctic. *Science of the Total Environment, 408*(15), 2966–2984.
Wesseling, C., Corriols, M., & Bravo, V. (2005). Acute pesticide poisoning and pesticide registration in Central America. *Toxicology and Applied Pharmacology, 207*(2), 697–705.
World Health Organization (WHO). (2004). Background document for development of WHO Guidelines for Drinking-water Quality 'Endosulfan in Drinking-water' WHO/SDE/WSH/03.04/92. https://cdn.who.int/media/docs/default-source/wash-documents/wash-chemicals/endosulfan-bd.pdf?sfvrsn=2dd251fd_4.
Yadav, I. C., & Devi, N. L. (2017). Pesticides classification and its impact on human and environment. *Environmental Science and Engineering, 6*, 140–158.
Yadav, I. C., Devi, N. L., Syed, J. H., Cheng, Z., Li, J., Zhang, G., & Jones, K. C. (2015). Current status of persistent organic pesticides residues in air, water, and soil, and their possible effect on neighboring countries: A comprehensive review of India. *Science of the Total Environment, 511*, 123–137.
Young, S., Balluz, L., & Malilay, J. (2004). Natural and technologic hazardous material releases during and after natural disasters: A review. *Science of the Total Environment, 322*(1–3), 3–20.

Decabromodiphenyl Ether (decaBDE)

Kanchan Kumari

1 Introduction

The Deca Bromodiphenyl Ether (decaBDE) is extremely persistent, highly capable of bioaccumulation and food web bio magnification, and capable of long-distance transport. It is a member of the Polybrominated Diphenyl Ether family (PBDEs). DecaBDE is one of the three commercial forms (Penta, Octa, and Deca) of PBDEs that have been extensively used as flame retardants. Due to their compliance with the definition of chemicals deemed to be persistent, toxic, bio-accumulative, and subject to long-distance transport, two of the three vital commercial formulations (Octa and Penta) have been eliminated from the European (1998) and the North American (2004) marketplaces. The single congener of decaBDE, BDE-209 makes up over 90% of the market, however decaBDE, i.e. BDE-209 is banned from 2017 onwards after listing in POPs chemicals by the Stockholm Convention and also phased out from the domestic markets in Canada and the USA at the end of 2013.

More than 80% of the current PBDE production consists of decaBDE, which serves as an additive flame retardant. It finds diverse applications, including use in plastics, polymers, sealants, adhesives, textiles, garments, rubber, coatings, inks, and various materials used in furnishings. Plastics used in wires, cables, pipes, and carpets often contain decaBDE or are composed of materials containing it. While the peak of commercially produced decaBDE usage occurred in the early 2000s, c-decaBDE is still widely employed on a global scale.

K. Kumari (✉)
CSIR-National Environmental Engineering Research Institute (NEERI),
Kolkata Zonal Center, Kolkata, West Bengal, India
e-mail: k_kumari@neeri.res.in

K. Kumari, *Pollutants of Global Concern*, Emerging Contaminants and Associated Treatment Technologies, https://doi.org/10.1007/978-3-031-50996-4_23

2 Technical Details of decaBDE

The physicochemical properties of decaBDE are presented in tabular form given in Table 1.

Table 1 Technical details and physicochemical properties of Deca Bromodiphenyl Ether (commercial mixture, c-decaBDE)

Details	Description (Decabromodiphenyl ether (commercial mixture, c-decaBDE))
Structure	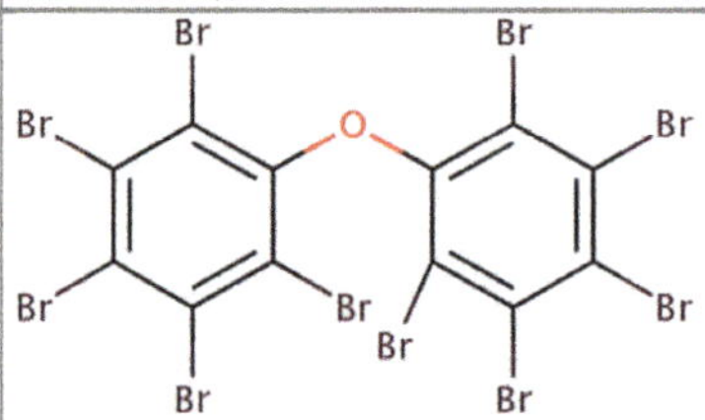
IUPAC name	1,2,3,4,5-pentabromo-6-(2,3,4,5,6-pentabromophenoxy) benzene
Formula (molecular)	$C_{12}Br_{10}O$
CAS no.	1163-19-5
Molecular weight	959.2 g/mol
Colour/form	Yellow prisms from toluene
Odour	Odourless
Boiling point	530.0 °C (ECB, 2000)
Melting point	569.1 to 576.5 °F (ECB, 2000)
Flash point	241 °C (466 °F; 514 K)
Solubility	<1 mg/mL at 68 ° F
Density	3 g/cu cm at 25 °C
Solubility in water	˂0.1 μg/L at 25 °C (ECB, 2000)
Vapour pressure	Less than 1 mm Hg at 68 °F (ECB, 2000)
Octanol/water partition coefficient	Log K_{ow} = 9.97 (ECB, 2000)
Octanol-air partition coefficient	Log K_{oa} = 13.1 (Kelly et al., 2008)
Henry's Law Constant (atm-m^3/mol)	1.62×10^{-6}
Stability/shelf life	Remain stable when stored in recommended conditions
Decomposition	It releases deadly hydrogen bromide fumes when heated to the point of disintegration

3 Status of Works and Usage in India

3.1 Production and Stockpiles

Currently in India, there is no specific information available on the production or stockpiles of decaBDE (Kumari et al., 2022). However, in a review article, Sharkey et al. (2020) stated that HBB (hexabromobiphenyl), HBCDD (hexabromocyclododecane), and other PBDEs are no longer produced, traded, imported, or used in India, and the country has also set PBDE concentration restrictions in specific electrical products (Kumari, 2021).

3.2 Regulation

According to the RoHS Regulation (2016), new electrical and electronic machinery and its components, consumables, and parts may not contain Cadmium, Lead, Mercury, Chromium, Hexavalent, and PBDEs in amounts greater than 0.1% by weight in homogeneous materials. The restrictions or limits of PBDEs in waste of recyclable products are not yet in place in India. An appropriate commercial good or waste screening mechanism has yet to be implemented. The European Commission has published Commission Regulation (EU) 2017/227 amending Annex XVII to REACH Regulation (EC) No 1907/2006 which introduces a restriction on the use of the substance decabromodiphenyl ether (decaBDE). According to the US EPA (Tables 2 and 3), the following screening and removal thresholds for domestic residential soil, commercial soil, and tap water have been reported.

Table 2 US EPA regional screening levels for chemical contaminants (US EPA, 2015), decabromodiphenyl ether, 2,2′,3,3′,4,4′,5,5′,6,6′- (BDE-209)

Resident soil (mg/kg)	4.40e+01
Industrial soil (mg/kg)	5.70e+02
Tap water (ug/L)	1.40e+01
Risk-based SSL (mg/kg)	7.80e+00
Chronic oral reference dose (mg/kg-day)	7.00e−04
Fraction of contaminant absorbed in gastrointestinal tract	1
Fraction of contaminant absorbed dermally from soil	0.1

Table 3 US EPA regional removal management levels for chemical contaminants, decabromodiphenyl ether, 2,2′,3,3′,4,4′,5,5′,6,6′-(BDE-209)

Resident soil (mg/kg)	4.40e+02
Industrial soil (mg/kg)	5.70e+03
Tap water (ug/L)	1.40e+02
Chronic oral reference dose (mg/kg-day)	7.00e−04
Fraction of contaminant absorbed in gastrointestinal tract	1
Fraction of contaminant absorbed dermally from soil	0.1

Table 4 Persistence and bioaccumulation criteria as per CEPA, 1999

	Persistency	Bioaccumulation
Medium	Half-life (the time taken to reduce the concentration of a substance by half, by transformation, in a matrix)	
Air	≥ 48 hrs or 2 days or depends on atmospheric transport from its place of origin to a remote area	BAF ≥ 5000 BCF ≥ 5000 log K_{ow} ≥ 5
Water	≥182 days (≥6 months)	
Sediment	≥365 days (≥12 months)	
Soil	≥182 days (≥6 months)	

- The Federal Sediment Environmental Quality Guidelines (FSeQG) for BDE 209 has been set as 19 (ng g^{-1} dw).
- The chronic oral reference dose (RfD) for decaBDE has been determined by the EPA. (U.S EPA, 2014) as 7×10^{-3} (mg/kg/day).
- The drinking water unit risk for decaBDE-209 is 2.0×10^{-8} (μg/L), and the EPA has assigned an oral cancer slope factor (CSP) of 7×10^{-4} (mg/kg/day)-1 to decaBDE-209 (EPA IRIS, 2008).
- According to EPA risk assessments, the drinking water concentration that represents a 1×10^{-6} cancer risk level for decaBDE-209 is 50 μg/L (EPA IRIS, 2008).
- Several states, including Maine, Maryland, Oregon, and Washington, have prohibited the sale of specific products containing decaBDE. Additionally, pentaBDE and octaBDE have been banned in various states, such as California, Illinois, Hawaii, Maine, Maryland, Minnesota, Michigan, Oregon, New York, Rhode Island, and Washington (EPA, 2009).
- In December 2009, the two US producers of decaBDE and the primary US importer made a commitment to cease the use of the chemical in consumer, transportation, and military applications by the end of 2013 (EPA, 2012).

- As per the Canadian Environment Protection Act (CEPA), 1999 (Table 4), a substance could be considered persistent in case at least one criterion is met in any one matrix.
- The Bio Concentration Factor (BCF) of a substance will be taken into account when the Bio Accumulation Factor (BAF) of a substance is impossible to determine using widely accepted methods; however, if neither the BAF nor the BCF

Table 5 Uses and sources of decaBDE and regulation in India

Identified POP	Uses and sources	Annex	Year of listing by SC	Regulation in India
Decabromodiphenyl ether (commercial mixture, c-decaBDE)	Industrial chemicals and used in vehicles, aircraft, textiles, additives in plastic housings, etc., and polyurethane foam for building insulation	A	2017	The use of decaBDE in electrical and electronic equipment is restricted above a maximum concentration of 0.1% (by weight) in homogenous materials, according to the E-Waste (Management) Rules, 2016

of a substance can be determined using recognized methods, then the log Kow will be taken into account.

- There are no specific regulations in line with the Stockholm Convention's banning of decaBDE implied in India, but the uses of the particular chemical is restricted in electronic products (Table 5).

Persistency of BDE-209 (the Main Constituent of DecaBDE)

The term "persistency" denotes BDE-209's capability to remain in the environment without undergoing significant degradation over an extended period. There are two primary mechanisms through which BDE-209 undergoes transformation in the environment: photodegradation and biodegradation, as noted in reports by ECA in 2006 and 2010. Given BDE-209's limited water solubility (0.1 grams per litre at 25 °C) and the absence of easily hydrolysable functional groups, hydrolysis is unlikely to be a significant environmental degradation pathway, as indicated by ECHA in 2012.

BDE-209 is nearly entirely adsorbed onto airborne particles, providing a protective shield for its molecules. Consequently, photolysis does not significantly break down BDE-209 in the atmosphere. Several research studies have established that BDE-209 displays high persistence in soil, sediment, and the air. The degree of light exposure and delayed biodegradation processes can influence its persistence, as mentioned in ECHA's 2012 report and ECA's 2010 findings.

The type of particle to which BDE-209 becomes attached could potentially impact its degradation rate. For instance, studies on the photolytic degradation of various solid matrices have shown that BDE-209 adsorbed to montmorillonite or kaolinite has half-lives of 36 and 44 days, respectively, while BDE-209 adsorbed onto organic carbon-rich natural sediment exhibits a half-life of 150 days (Ahn et al., 2006).

Estimated removal half-lives for BDE-209 in a photolytic experiment were approximately 35–37 hours in sand, 100 hours in sediment, and 200 hours in soil (Söderström et al., 2004; Tysklind et al., 2001, as cited in ECHA's 2012 report). Additionally, other organic compounds, such as humic substances and sand particles coated with humic acid can reduce photodegradation in natural waters by absorbing light or interacting with BDE-209 through hydrophobic interactions (Leal et al., 2013).

Furthermore, the biodegradation of BDE-209 can depend on the nature of dissolved organic matter, the presence of suspended particulates, and BDE-209's adsorption to solid surfaces (Leal et al. 2013). An additional factor contributing to extended environmental half-lives in natural conditions is increased adsorption to the soil or sediment matrix with ageing, as highlighted by ECHA in 2013. BDE-209, when found in deep-sea sediments, exhibits a high level of persistence due to minimal exposure to sunlight, limiting its potential for photodegradation (as noted by ECHA in 2012 and references therein). Because BDE-209 has low water solubility, assessing its half-lives in water presents challenges. However, recent research has unveiled environmental half-lives in water ranging from a few hours to 660 days when considering the use of solvents and the presence of natural light (Kuivikko et al., 2007; Leal et al., 2013).

Tokarz et al. (2008) reported the longest environmental half-life for BDE-209, spanning from 6 to 50 years in sediment, with an average of approximately 14 years. This data was obtained through a laboratory microcosm experiment conducted over 3.5 years at 22 °C in dark conditions. In a small Swiss city lake, Kohler and Zennegg (2008) observed concentrations of BDE-209 in sediments and how they changed over time. BDE-209 first appeared in sediment layers corresponding to the mid-1970s and steadily increased to 7.4 ng/g dry weight (dw) in 2001, with a doubling time of approximately 9 years. This study, spanning nearly 30 years, did not reveal evidence of long-term transformation processes related to sediment.

Studies conducted on sludge and soil further support the notion of BDE-209's persistence. Liu and Zhu (2011) discovered that soil samples treated with BDE-209 remained non-degradable even after 180 days in the absence of light.

Countrywide Scientific Work Done on the Chemical Several observations of PBDEs in varied matrices have been reported from Sweden, Japan, Holland, the USA, and other countries; however, there are few measurements for India. In order to develop the skills necessary to analyse PBDEs and BFRs, Kumari et al. (2014) reported a preliminary screening of several PBDE congeners in various outdated electronic and consumer products. To detect the presence of PBDE congeners such as BDE-28, BDE-47, BDE-99, BDE-100, BDE-153, BDE-154, BDE-183, and BDE-209, six different samples, including foam from upholstery, a computer motherboard, children's toy composite, electrical wire, and PVC flooring were collected and then analysed. Three congeners (BDE-47, BDE-153, and BDE-209) out of the six samples were found to have PBDEs present, and of which, BDE-209 was present in the highest concentration. The sample with the highest concentration of BDE-209 was the motherboard of the computer, which was followed by a window blind and upholstery foam. Therefore, the preliminary findings of this study show that PBDEs still exist in older commercial products, that can be a significant source of vulnerability for the population. Products like computer motherboard contains varying concentrations of PBDEs, with BDE-47 at 1.173 mg/kg, BDE-153 at 0.259 mg/kg, and notably high levels of BDE-209 at 11,583.8 mg/kg. Toxics Link, 2011, reported the presence of PBDEs and PBB in plastic wastes. While performing the field sampling in Delhi, the pellet samples were randomly taken from the recycling facilities at Mundka, Bawana, Bhopura, and Pooth Gaon, which are located at the centre of the city's plastic recycling. The plastic samples were collected from various units, specialized in plastic pellet production. Following the recycling of the E-Plastics to produce pellets, the plastic samples were gathered. The plastic pellets consisting primarily of ABS (Acrylonitrile Butadiene Styrene) and HIPs (High Impact Polystyrenes), which are frequently utilized in EEE casings, were gathered for screening. Ten samples of HIPS and ABS were tested under this study. In terms of the geographic scope and location of processing, the trading and processing of plastic was extensive, which created a barrier to creating a direct connection between WEEE and pellets. The assembled samples were then sent to a well-equipped authorized laboratory in Delhi to be examined for various PBDE and PBB congeners. Brominated flame retardants were detected in 5 of the 10 samples that were gathered for testing. Although various PBDE congeners were identified in five of the plastic pellet samples, no traces of PBB were observed in any of them. Deca-BDE dominated the pattern in all of the samples for all PBDE congeners that were found. A range of 140 mg/kg to 39,602 mg/kg of PBDEs had been found. All five of the samples that had PBDEs found in them also included decaBDE.

In a research study, grab sediment and sediment cores were examined by Tiwari et al. (2018), from Thane Creek, Mumbai, for the dispersion of 15 PBDEs congeners, and their concentrations were tracked across time and space. A triangle-shaped body of brackish water known as the creek opens to the Arabian Sea. The total PBDE values in grab sediments varied between 15.98 and 132.72 ng/g dry weight. Additionally, findings from sediment cores indicate the widespread use of PBDEs in the vicinity over the past two decades. The most prevalent congener, BDE-209, contributed between 19 and 35% of the total PBDEs. In sediments collected from Thane Creek, commercial pentaBDE, octaBDE, and decaBDE contributed an average of 24 ± 5, 5 ± 1, and 69 ± 7% ($p\backslash 0.001$) to the profile, respectively. Comparing all monitoring locations across Thane Creek had the greatest concentration of PBDEs in the sediment.

4 International Data Versus the Indian Scenario

IPEN (a network for civil society worldwide), Arnika (an environmental association in the Czech Republic), and Toxics Link (an environmental organization of India) conducted a global survey. According to the study, toys for kids and other items made from recycled plastics produced from e-waste contain extremely hazardous flame retardants, as well as Rubik's Cubes, a brain-training puzzle toy typically constructed of recycled plastic, have been reported to have chemical pollutants that can impair the nervous system and affect intellectual development of children. Harmful chemicals such as decaBDE, octaBDE, and HBCD (Hexabromocyclododecane) are utilized in the plastic coverings of electronic equipment. If these toxic chemicals are not eliminated, recycled plastic will transfer these toxins into new products. Ninety percent of the samples tested positive for octaBDE or decaBDE, according to the survey that was carried out in 26 nations, including India and 43% of them contained HBCD. Six samples from India were evaluated for the study, and it was discovered that four of the samples had high quantities of octaBDE and decaBDE, and concerns are raised since one of the samples had a significant percentage of HBCD. These persistent compounds are very much known to damage the reproductive system and cause disruption of hormonal systems, which negatively affects IQ, attention, learning, and memory. According to a research study in China, the toxic effects of BDE-209 found on human embryonic kidney cells (HEK293T) (Li et al., 2014a, Li et al., 2014b). The BDE-209 treated HEK293T cells may change how nucleosomes are organized by drastically altering the expression of histone gene clusters.

4.1 Relevance of the International Scientific Data as per Indian Conditions

PBDEs concentration in sediments from Thane Creek, India was compared with the concentrations in different parts of the world (Fig. 1).

Emissions from certain PBDEs were calculated and in use and waste stock quantity estimated for ∑5BDE (28, 47, 99, 153, 183) were 25 and 13 kilotons, respectively; nevertheless, decaBDE had values of 400 kt and 100 kt, respectively (Abbasi

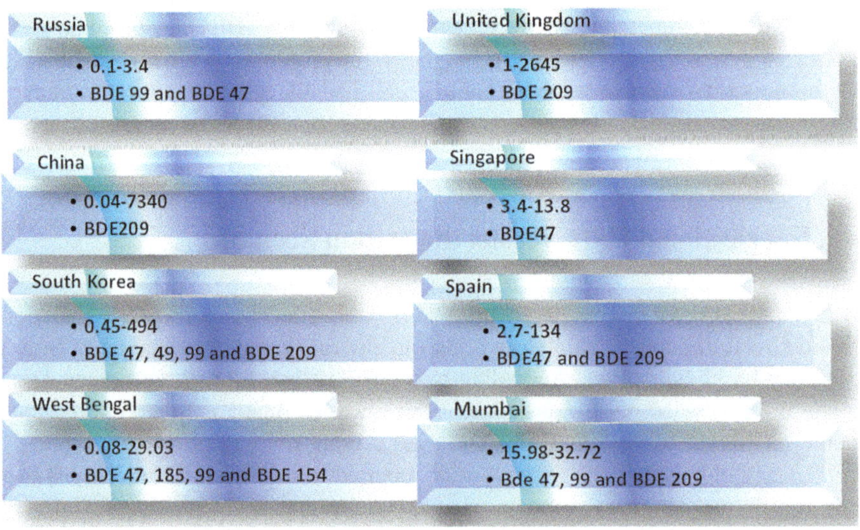

Fig. 1 PBDE levels in sediment (ng/g) and the main congeners from various locations of the world. (Source: Labusanska et al., 2010; Vane et al., 2010; Mai et al., 2005; Wurl & Obbard, 2005; Moon et al., 2007; Eljarrat et al., 2005; Binelli et al., 2007; Wu et al., 2017)

et al., 2019) in the year 2018. According to Yuan et al. (2020), 13% of China's total manufactured decaBDE is exported, while the total production is estimated to be 465 kt, from 1993 to 2018. BFRs have total market shares of 12, 6.40, and 28%, respectively, in North America, Western Europe, and Asia. In 1999, North America was the only region in the globe where 98% of pentaBDE was utilized (Renner, 2000). According to Choo et al. (2019), the usage of PBDEs is rapidly expanding, which poses a serious threat to world health because these emissions are able to enter the human body and can accumulate in environments. Abbasi et al. (2019) also reported that a total of 70 kt of $\sum$5BDE and BDE209 was recycled within products since 1970 and as this recycling practices are promoted under the Circular Economy, an additional 45 kt of PBDEs (mainly BDE209) may reappear in new products. Through voluntary and legislative efforts, technical combinations of octaBDE and pentaBDE were phased out in 2004, while decaBDE was only used sparingly up until 2013. Studies on biomonitoring have revealed that PBDEs are widely present in both the US population and the population worldwide. On the other hand, human serum concentrations may be decreasing with time, according to some research, but it is uncertain if this trend will hold. *All these findings clearly indicate that despite the fact that laws to ban PBDEs from the market have successfully reduced exposure to certain of their congeners, these data suggest that human exposure to these persistent pollutants may plateau and continue to be pervasive in human populations.*

5 Adverse Effects of decaBDE

In aquatic creatures, mammals, and birds, BDE-209 exhibits neurotoxic, reproductive, developmental, endocrine, and toxic effects, along with reports of impacts on growth, mortality, and survival. On the other hand, decaBDE compounds have lower acute toxicity, with oral and dermal 50% lethal dose values ranging from 2 to 5 g/kg body weight (as per WHO, 1994; Norris et al., 1975). Kumari et al. (2022) reported adverse effects of BDE-209 on the optimum metabolism of fish, *Labeo rohita*. In a prolonged research study involving rats at approximately 80 mg/kg body weight, effects such as thyroid hyperplasia, liver enlargement, and kidney hyaline degeneration were observed. Furthermore, foetal resorptions were noted in rats at a dose of 100 mg/kg body weight.

According to a research study, carcinomas and adenomas were observed, but only at extremely high doses (1200 mg/kg body weight/day and above). PBDEs have been demonstrated to have adverse impacts on fish, invertebrates, and algae. Holm et al. (1993) found that PBDEs could induce fatty liver and reduce fish spawning success. BDE-209 and other PBDE exposures combined may pose significant health risks, particularly for children aged 6 months to 3 years, who are more likely to be exposed to PBDEs than any other age group (Kortenkamp et al., 2014).

Villanger and Jenssen (2011), Villanger and Lydersen (2011), Villanger and Gabrielsen (2013) provided evidence that mixtures of organohalogen contaminants containing multiple PBDEs (BDE-28, BDE-47, BDE-99, BDE-100, and BDE 153) could impact the thyroid homeostasis of Arctic marine animals. These findings align with those of Kortenkamp (2014), even though the impact of BDE-209 was not assessed in these studies, and only correlations were reported.

Recent studies on rodents indicate that PBDE exposure may lead to behavioural issues, disruptions in thyroid hormone activity, and liver effects (Eriksson et al., 2001). Gascon (2012) also reported that PBDE exposure occurs during early human development in utero through placental transfer and via maternal milk, with the presence of BDE-209 levels in colostrum associated with poorer mental development scores in children aged 12–18 months.

6 Management and Remedial Measures

According to BioWin projections, it is expected that hexaBDEs to nonaBDEs are resistant to biodegradation, and there is no anticipation of aerobic biodegradation of these compounds. Additionally, in a 28-day closed bottle test, no degradation was observed based on oxygen uptake (OECD 301D). Gerecke et al. (2005) reported the breakdown of nonaBDEs 206 and 207 to OctaBDEs utilizing sewage sludge inoculum in a research report on decaBDE, and other research study by Hale et al. (2012) has verified this degradation. For hexaBDE to nonaBDE, AOPWIN estimates half-lives for reactivity with ambient free radicals of hydroxyl ranging from 30.4 to 161.0 d, respectively. It is anticipated that octaBDE will significantly adsorb to airborne suspended particles and be eliminated through wet and/or dry deposition in the environment. Numerous BDEs' photodecomposition has been analysed in a variety of matrices, including an 80:20 methanol/water mixture (Eriksson et al., 2001). As per Sierra et al. (2006), hexa BDE153 exhibits a short photochemical half-life in aquatic environments, and several common penta- and tetra-brominated diphenyl ether congeners are swiftly photohydrodibrominated from it. Various advanced oxidation processes were critically reviewed for the evaluation of effectiveness of particular POPs and have been represented in Table 6.

To replace c-decaBDE in plastics and textiles, there is a range of non-POP chemical alternatives readily accessible in the market. Furthermore, non-chemical alternatives and technical solutions, such as non-flammable materials and physical barriers, are also available as substitutes.

7 Conclusion

Based on current trends, it is expected that PBDEs will surpass PCBs as the predominant pollutant in various fish and marine animal populations within the next decade. In India, the production and utilization of decaBDE have been subject to regulation under the Stockholm Convention on Persistent Organic Pollutants

Table 6 Evaluating the effectiveness of selective POPs by advance oxidation process (AOP)

POPs	Advanced oxidation process (AOP)	Removal efficiency	Problem	References
BDEs 28, 100, 153, and 183 (10 ng/mL in hexane)	Photolysis-Hg lamp500W	8.5%-BDE 28, 5.4%-BDE 100, 49%-BDE 153 83%-BDE 183 in 4 hrs	Poor debromination with 0.10–2.64 h^{-1} rate constant	Sun et al. (2012)
BDE-209 (5.5 μg/L)	Photocatalysis-Xe lamp (360 nm); powdered TiO_2 (1 mg/mL)	4 hrs-30%	Clustering of bare titania	Li et al. (2014a, b)
BDE-209, (10 mg/L)	Photocatalysis-Hg lamp 100 W (360 nm); Pd doped TiO_2 (0.2 g/L)	40 min-60% 60 min-100%	Photo catalysis rate is significantly higher for BDE concentrations measured in mg/L, which is considerably different from real-world conditions where concentrations are typically in μg/L	Lai et al. (2012)
BDE-209 (75 μg/L)	1% nano-sized TiO_2 enhances the solubility of BDE in visible light irradiation	After 14 days 40–50%	The degradation rate in real-world environmental conditions is exceptionally slow	Daramola et al. (2012)

(POPs), of which India is a signatory. The primary objective of the Stockholm Convention is to curb or limit the production and utilization of specific POPs, including decaBDE, due to their detrimental impacts on both human health and the environment.

Therefore, it is likely that the production of decaBDE in India was either phased out or significantly restricted in compliance with the Stockholm Convention and other relevant environmental regulations. DecaBDE was found to meet the screening demands listed in after a thorough review by the POPs Review Committee. Considering Norway's proposal to include c-decaBDE in the Stockholm Convention on POPs, along with the additional scientific insights presented during its ninth meeting, it was determined that c-decaBDE, a synthetic substance with no known natural origins, extensively used as a flame retardant in various global applications meets the criteria to be listed as POPs. DecaBDE (BDE 209) shows persistency in the environment that bio accumulates and biomagnifies in living beings as well as in food webs.

References

Abbasi, G., Li, L., & Breivik, K. (2019). Global historical stocks and emissions of PBDEs. *Environmental Science & Technology, 53*, 6330–6340. https://doi.org/10.1021/acs.est.8b07032.ABRELPE, 2017. Panorama Dos Resíduos Śolidos No Brasil 2016 https://doi.org/ISSN:2179-657 8303. [accessed Jan 12 2024].

Ahn, M. Y., Filley, T. R., Jafvert, C. T., Nies, L., Hua, I., & Bezares-Cruz, J. (2006). Photodegradation of decabromodiphenyl ether adsorbed onto clay minerals, metal oxides, and sediment. *Environmental Science & Technology, 40*(1), 215–220.

Binelli, A., Sarkar, S. K., Chatterjee, M., Riva, C., Parolini, M., deb Bhattacharya, B., ... & Satpathy, K. K. (2007). Concentration of polybrominated diphenyl ethers (PBDEs) in sediment cores of Sundarban mangrove wetland, northeastern part of Bay of Bengal (India). *Marine Pollution Bulletin, 54*(8), 1220–1229.

CEPA. (1999). Canadian Environmental Protection Act, 1999 Federal Environmental Quality Guidelines Polybrominated Diphenyl Ethers (PBDEs). https://www.ec.gc.ca/ese-ees/05DF7A37-60FF-403F-BB37-0CC697DBD9A3/FEQG_PBDE_EN.pdf

Choo, G., Lee, I.-S., & Oh, J.-E. (2019). Species and habitat-dependent accumulation and biomagnification of brominated flame retardants and PBDE metabolites. *Journal of Hazardous Materials, 371*, 175e182.

Daramola, M., Aransiola, E., & Ojumu, T. (2012). Potential applications of zeolite membranes in reaction coupling separation processes. *Materials, 5*, 2101.

ECB (European Chemicals Bureau). (2000). *European Union risk assessment report diphenyl ether, pentabromo deriv.* Institute for Health and Consumer Protection, Joint Research Centre, European Commission

ECHA European Chemicals Agency. (2012). Support Document Bis(pentabromophenyl) ether [decabromodiphenyl ether] (Member State Committee, 29 November 2012).

ECHA, European Chemicals Agency. (2013). Draft background document for Bis(pentabromophenyl)ether (decabromodiphenyl ether; decaBDE). Document developed in the context of ECHA's fifth Recommendation for the inclusion of substances in Annex XIV. 24 June 2013.

Eljarrat, E., De La Cal, A., Larrazabal, D., Fabrellas, B., Fernandez-Alba, A. R., Borrull, F., ... & Barcelo, D. (2005). Occurrence of polybrominated diphenylethers, polychlorinated dibenzo-p-dioxins, dibenzofurans and biphenyls in coastal sediments from Spain. *Environmental Pollution, 136*(3), 493–501.

EPA. (2008). 2,2′,3,3′,4,4′,5,5′,6,6′ - Decabromodiphenyl ether (BDE-209) (CASRN 1163-19-5). www.epa.gov/IRIS/subst/0035.htm

EPA. (2009). Polybrominated Diphenyl Ethers (PBDEs) Action Plan Summary. www.epa.gov/oppt/existingchemicals/pubs/actionplans/pbdes_ap_2009_1230_final.pdf

EPA. (2012). DecaBDE Phase-out Initiative. www.epa.gov/oppt/existingchemicals/pubs/actionplans/deccadbe.html

Eriksson, P., Jakobsson, E., & Fredriksson, A. (2001). Brominated flame retardants: A novel class of developmental neurotoxicants in our environment? *Environmental Health Perspectives, 109*, 903–908.

Gascon, M., Fort, M., Martinez, D., Carsin, A. E., Forns, J., Grimalt, J. O., Santa Marina, L., Lertxundi, N., Sunyer, J., & Vrijheid, M. (2012). Polybrominated diphenyl ethers (PBDEs) in breast milk and neuropsychological development in infants. *Environmental Health Perspectives, 120*(12), 1760–1765.

Gerecke, A. C., Hartmann, P. C., Heeb, N. V., Kohler, H.-P. E., Giger, W., Schmid, P., Zennegg, M., & Kohler, M. (2005). Anaerobic degradation of decabromodiphenyl ether. *Environmental Science and Technology, 39*, 1078–1083.

Hale, R. C., La Guardia, M. J., Harvey, E., Chen, D., Mainor, T. M., Luellen, D. R., & Hundal, L. S. (2012). Polybrominated diphenyl ethers in U.S. Sewage Sludges and Biosolids: Temporal and geographical trends and uptake by corn following land application. *Environmental Science & Technology, 46*, 2055–2063.

Holm, G., Norrgren, L., Andersson, T., & Thu, N. A. (1993). Effects of exposure to food contaminate with PBDE, PCN or PCB on reproduction, liver morphology and cytochrome P450 activity in the three-spine stickleback, Gasterosteus aculeatus. *Journal of Aquatic Pollution and Toxicology, 27*, 33–50.

Kelly, B. C., Ikonomou, M. G., Blair, J. D., & Gobas, F. A. (2008). Bioaccumulation behaviour of polybrominated diphenyl ethers (PBDEs) in a Canadian Arctic marine food web. *Science of the Total Environment, 401*(1–3), 60–72.

Kohler, M., & Zennegg, M. (2008). Temporal trends, congener patterns, and sources of octa-, nona-, and decabromodiphenyl ethers (PBDE) and hexabromocyclododecanes (HBCD) in Swiss lake sediments. *Environmental Science & Technology, 42*(17), 6378–6384.

Kortenkamp, A., Martin, O., Evans, R., Faust, M., & Backhaus, T. (2014). *Risk of combination effects between decabromodiphenyl ether and other polybrominated diphenyl ethers* (p. 152). Norwegian Environmental Protection Agency. Available at: http://www.miljodirektoratet.no/no/Publikasjoner/2014/Juli-2014/Risk-of-Combination-EffectsBetween-Decabromodiphenyl-Ether-and-Other-Polybrominated-Diphenyl-Ethers/

Kuivikko, M., Kotiaho, T., Hartonen, K., Tanskanen, A., & Vahatalo, A. V. (2007). Modeled direct photolytic decomposition of Polybrominated diphenyl ethers in the Baltic Sea and the Atlantic Ocean. *Environmental Science & Technology, 41*(20), 7016–7021.

Kumari, K. (Ed.). (2021). *Persistent organic pollutants: Gaps in management and associated challenges* (1st ed.). CRC Press. https://doi.org/10.1201/9781003046806

Kumari, K., Sharma, J. K., Kanade, G. S., et al. (2014). Investigation of polybrominated diphenyl ethers in old consumer products in India. *Environmental Monitoring and Assessment, 186*, 3001–3009. https://doi.org/10.1007/s10661-013-3596-2

Kumari, K., Singh, A., Swamy, S., Singhar, R. S., & Thakur, S. (2022). Use of enzymatic biomarkers of *Labeo rohita* to study the effect of polybrominated diphenyl ether (BDE- 209) via dietary exposure in laboratory conditions. *Environmental Monitoring and Assessment, 194*(7), 499. https://doi.org/10.1007/s10661-022-09963-0

Labunska, I., Brigden, K., Santillo, D., Kiselev, A., & Johnston, P. (2010). *Russian Refuse #2: an update on PBDEs and other contaminants detected in St-Petersburg area, Russia.* Greenpeace Research Laboratories Technical Note 04/2010.

Lai, C. K., Man, Y. B., Zheng, J. S., Liang, Y., Tam, N. F. Y., & Wong, M. H. (2012). Characterizing the optimal operation of photocatalytic degradation of BDE-209 by nano-sized TiO2. *Journal of Environmental Sciences, 24*(9), 1670–1678. ISSN 1001-0742. https://doi.org/10.1016/S1001-0742(11)60992-3

Leal, J. F., Esteves, V. I., & Santos, E. B. H. (2013). BDE-209: Kinetic studies and effect of humic substances on photodegradation in water. *Environmental Science & Technology, 47*, 14010–14017.

Li, L., Chang, W., Wang, Y., Ji, H., Chen, C., Ma, W., & Zhao, J. (2014a). Rapid, photocatalytic, and deep debromination of polybrominated diphenyl ethers on Pd-TiO2: Intermediates and pathways. *Chemistry - A European Journal, 20*, 11163–11170.

Li, M., Liu, Z., Gu, L., Yin, R., Li, H., Zhang, X., Cao, T., & Jiang, C. (2014b). Toxic effects of decabromodiphenyl ether (BDE-209) on human embryonic kidney cells. *Frontiers in Genetics, 5*, 118. https://doi.org/10.3389/fgene.2014.00118

Liu, L., & Zhu, W. (2011). Effect of decabromodiphenyl ether (BDE 209) and dibromodiphenyl ether (BDE 15) on soil microbial activity and bacterial community composition. *Journal of Hazardous Materials, 186*(1), 883–890.

Mai, B., Chen, S., Luo, X., Chen, L., Yang, Q., Sheng, G., Peng, P., Fu, J., & Zeng, E. Y. (2005). Distribution of polybrominated diphenyl ethers in sediments of the Pearl River Delta and adjacent South China Sea. *Environmental science & technology, 39*(10), 3521–3527.

Martin, S., Harrad, S., Abdallah, M. A.-E., Drage, D. S., & Berresheim, H. (2020). Phasing-out of legacy brominated flame retardants: The UNEP Stockholm Convention and other legislative action worldwide, 106041, ISSN 0160-4120. *Environment International, 144*. https://doi.org/10.1016/j.envint.2020.106041

Moon, H. B., Kannan, K., Choi, M., & Choi, H. G. (2007). Polybrominated diphenyl ethers (PBDEs) in marine sediments from industrialized bays of Korea. *Marine Pollution Bulletin, 54*(9), 1402–1412.

Norris, J. M., Kociba, R. J., Schwetz, B. A., Rose, J. Q., Humiston, C. G., Jewett, G. L., Gehring, P. J., & Mailhes, J. B. (1975). Toxicology of octabromobiphenyl and decabromodiphenyl oxide. *Environmental Health Perspectives, 11*, 153–161.

Renner, R. (2000). Increasing levels of flame retardants found in North American environment. *Environmental Science & Technology, 34*, 452ae453a.

Sierra, R., Wan, P., & Ikonomou, M. (2006). Photochemistry of a major commercial polybrominated diphenyl ether flame retardant congener: 2,2′,4,4′,5,5′-Hexabromodiphenyl ether (BDE153). *Environment International, 32*(5), 575–585. ISSN 0160-4120. https://doi.org/10.1016/j.envint.2006.01.009

Söderström, G., Sellström, U., De Wit, C. A., & Tysklind, M. (2004). Photolytic debromination of decabromodipehnyly ether (BDE-209). *Environmental Science & Technology, 38*, 127–132.

Sun, C., Zhao, J., Ji, H., Ma, W., & Chen, C. (2012). Photocatalytic debromination of preloaded decabromodiphenyl ether on the TiO2 surface in aqueous system. *Chemosphere, 89*, 420–425.

Tiwari, M., Sahu, S. K., Bhangare, R. C., Ajmal, P. Y., & Pandit, G. G. (2018). Polybrominated diphenyl ethers (PBDEs) in core sediments from creek ecosystem: Occurrence, geochronology, and source contribution. *Environmental Geochemistry and Health, 40*(6), 2587–2601. https://doi.org/10.1007/s10653-018-0125-9

Tokarz Iii, J. A., Ahn, M. Y., Leng, J., Filley, T. R., & Nies, L. (2008). Reductive debromination of polybrominated diphenyl ethers in anaerobic sediment and a biomimetic system. *Environmental Science & Technology, 42*(4), 1157–1164.

Toxics Link. (2011). *Brominated Flame Retardants Spreading the Fire - A Report on the Recycling of Brominated Flame Retardant Contaminated Plastic.* http://toxicslink.org/docs/BFR_Report_2011.pdf.

Tysklind, M., Sellström, U., Söderström, G., & de Wit C. (2001). *Abiotic transformation of polybrominated diphenyl ethers (PBDEs): Photolytic debromination of decabromo diphenyl ether.* The Second International Workshop on Brominated Flame Retardants, BFR 2001, May 14–16, Stockholm.

US EPA. (2014). *Technical Fact Sheet - Polybrominated Diphenyl Ethers (PBDEs) and Polybrominated Biphenyls (PBBs)*, U.S. Environmental Protection Agency, Washington, DC, EPA 505-F-14-006, 2014

US EPA. (2015). Regional Screening Levels (RSLs) - Generic Tables. https://www.nrc.gov/docs/ML1307/ML13078A051.pdf

Vane, C. H., Ma, Y. J., Chen, S. J., & Mai, B. X. (2010). Increasing polybrominated diphenyl ether (PBDE) contamination in sediment cores from the inner Clyde Estuary, UK. *Environmental Geochemistry and Health, 32*, 13–21.

Villanger, G. D., & Gabrielsen, K. M. (2013). Effects of complex organohalogen contaminant mixtures on thyroid homeostasis in hooded seal (Cystophora cristata) mother-pup pairs. *Chemosphere, 92*(7), 828–842.

Villanger, G. D., & Jenssen, B. M. (2011). Exposure to mixtures of organohalogen contaminants and associative interactions with thyroid hormones in East Greenland polar bears (Ursus maritimus). *Environment International, 37*(4), 694–708.

Villanger, G. D., & Lydersen, C. (2011). Disruptive effects of persistent organohalogen contaminants on thyroid function in white whales (Delphinapterus leucas) from Svalbard. *Science of Total Environment, 409*(13), 2511–2524.

WHO. (1994). World Health Organization - brominated diphenyl ethers. Environmental health criteria 162. Int. Prog. Chem. Saf.

Wu, M. H., Xu, B. T., Xu, G. et al., (2017). Occurrence and profiles of polybrominated diphenyl ethers (PBDEs) in riverine sediments of Shanghai: A combinative study with human serum from the locals. *Environmental Geochemistry and Health, 39*, 729–738. https://doi.org/10.1007/s10653-016-9843-z

Wurl, O., & Obbard, J. P. (2005). *Chemosphere, 58*(7), 925–933.

Yuan, C., Li, J., & Tan, Q. (2020). Trends of production, consumption and environmental emissions of Decabromodiphenyl ether in mainland China. *Environmental Pollution, 260*, 114022. ISSN 0269-7491. https://doi.org/10.1016/j.envpol.2020.114022

Short-Chain Chlorinated Paraffins (SCCPs)

Kanchan Kumari and Karthik Raghunathan

1 Introduction

Chlorinated paraffins, also known as CPs, are combinations of polychlorinated n-alkanes that can have varied chain lengths and different chlorination patterns. According to the length of the carbon chain, chlorinated paraffins are classified into one of three subgroups: short-chain (SCCPs), middle-chain (MCCPs), and long-chain chlorinated paraffins (LCCPs, $C>18$) (Nevondo & Okonkwo, 2021). Typically ranging from 30% to 70% chlorination, short-chain chlorinated paraffins (SCCPs) are a heterogeneous blend of polychlorinated n-alkanes with carbon chain lengths of 10–13 carbon atoms. Since the 1930s, SCCPs have been widely used as plasticizers or fire retardants in paints, cohesive, sealants, fat liquors for leather, plastics, lubricants, textiles, rubber, and polymeric materials (Kumari, 2021) because of their excellent physicochemical properties in applications involved in metalworking, polyvinyl chloride processing, and others. It is widely accepted that chlorinated paraffins (CPs) emerged as substitutes for polychlorinated biphenyls (PCBs) in many applications following the prohibition of PCB usage in the mid-1980s (Stejnarova et al., 2005; POPRC, 2007; RPA, 2010; Fiedler, 2010). The current manufacturing volumes of CPs are significantly larger than the production volumes of PCBs at any point in history. Furthermore, the manufacturing volumes of CPs are significantly greater than they have ever been. To a limited extent, information regarding the physicochemical features, emission potentials, and ambient levels of

K. Kumari (✉)
CSIR-National Environmental Engineering Research Institute (NEERI),
Kolkata Zonal Center, Kolkata, West Bengal, India
e-mail: k_kumari@neeri.res.in

K. Raghunathan
CSIR-National Environmental Engineering Research Institute (NEERI), Nagpur, India

K. Kumari, *Pollutants of Global Concern*, Emerging Contaminants and Associated Treatment Technologies, https://doi.org/10.1007/978-3-031-50996-4_24

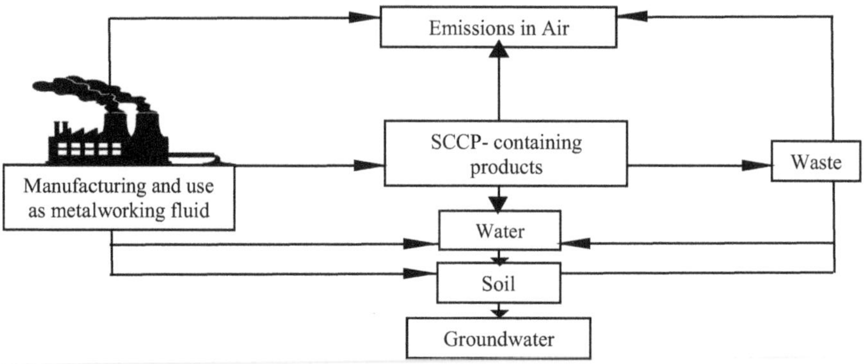

Fig. 1 Pathways of exposure of SSCPs to humans

chlorinated paraffins (CPs) and short-chain chlorinated paraffins (SCCPs) is available. However, this information is dispersed throughout numerous sources, and there is currently a lack of a comprehensive compilation of data on these substances. Nevertheless, over the past 20 years, it has been discovered that SCCPs exhibit toxicity toward aquatic creatures and persist in the environment. In addition, they possess a significant capacity for bioaccumulation and can be transported over great distances even at low concentrations. Subsequently, there has been a progressive imposition of restrictions on the production and use of substances containing these chemicals, with their initial inclusion in various environmental statutes and regulations. One of the main obstacles to evaluating this molecule in the environment is the lack of a solid method for its detection. Research and development (R&D) plays a crucial role in the design of analytical methods and the exploration of alternative products as substitutes for their industrial applications.

According to the EPA (2009), SCCPs released into the atmosphere are likely to exist in the form of vapour and particulate matter, which humans may breathe in. SCCPs that are present in metalworking and metal cutting fluids have the potential to be discharged into aquatic ecosystems, which can act as a non-point source for human exposure through the ingestion of food that has been exposed to SCCPs (EPA, 2009). The exposed soil has the potential to contaminate ground water, which is the ultimate source of human exposure because it is used for drinking (Fig. 1).

SCCPs are hydrolytically stable and have a persistence in air that allows them to be transported over long distances. In addition, they have a significant amount of persistence in air to allow for this. Numerous SCCPs can accumulate within biotic organisms. SCCPs are expected to have major environmental and human health consequences due to their long-range environmental mobility.

2 Technical Details of the SCCPs

The physicochemical properties of SCCPs are presented in tabular form given in Table 1.

Table 1 Technical details and physicochemical properties of the SCCPs (NCBI, 2023)

Details	Description (short-chain chlorinated paraffins (SCCPs)
Structure	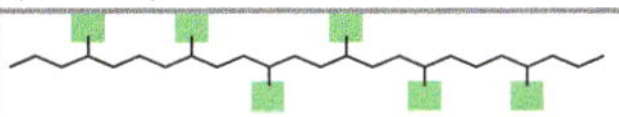
Molecular formula	$C_{24}H_{44}Cl_6$
CAS no.	85535-84-8
Molecular weight	320–500 g/mole
Colour/form	Clear to yellowish liquid
Density	1.2–1.6 g/cm^3
Vapour pressure	1 mg/mL at 75 °F
Octanol/water partition coefficient	log Kow 4.39–8.01
Toxicity	Highly toxic
Boiling point	>200 °C
Flash point	166–202 °C
Solubility	0.15–0.47 mg/l (at 20 °C)
Autoflamibility	Not stated (decomposes above 200 °C)
Explosivity	Not explosive
Oxidizing properties	No oxidizing properties
Toxicity	Highly toxic

3 Status of Work and Usage in India

At this time, there is no regulation or law in place in India regarding SCCPs to govern the uses of these products or their distribution on the market (UNEP SCCPs, 2019a). However, 20 firms in India have a total installed capacity of 110,000 tonnes of CP production per year (POPRC, 2015). Although India is now the second biggest producer of CPs (Nevondo & Okonkwo, 2021), there is no particular information on SCCPs that is accessible.

4 International Data Versus the Indian Scenario

In accordance with the Pollutant Release and Transfer Register Law in Japan, the production of SCCPs has been discontinued (Bayen et al., 2006). In Canada, SCCPs were added to the "Priority Substances List" under the Canadian Environmental Protection Act in 1999 (Bayen et al., 2006; CEPA, 1999), but the Prohibition of Certain Toxic Substances Regulations did not add SCCPs until 2013, when restrictions on use and production began to take effect. It is now illegal to sell or produce SCCPs in Europe, and the European Water Framework Directive (Bayen et al., 2006; Van Mourik et al., 2015, Directive 2000/60/EC) has included them on a list of priority compounds that pose a risk to human health. According to the findings of a

Table 2 Status of usage of SCCPs

Identified POPs	Uses and Sources	Specific exemption/acceptable purposes	Annex	Year of listing by SC
Short-chain chlorinated paraffins (SCCPs)	Industrial chemical; primarily used in metal working; softeners in plastics; fire-retardant additives	*Production:* parties listed in the register are allowed. *Use:* additives in transmission and rubber conveyor belts, lubricant, leather, paints, tubes for outside decoration bulbs, metal processing, adhesives, plasticizers	A	2017

research published in 2011 (Corden et al., 2011), the efficacy of abatement strategies to decrease SCCPs emissions within the European Union was evaluated. In May 2017, the Stockholm Convention on Persistent Organic Pollutants (POPs) included short-chain chlorinated paraffins (SCCPs) in Annex A (Table 2) because of their environmental characteristics, such as persistence, toxicity, long-range transport potential, and bioaccumulation (Table 3) (POPRC, 2017; UNEP, 2017). With a chlorine level >48% by weight and a carbon chain length ranging from C10 to C13, SCCPs was banned (UNEP, 2017). Because SCCPs are released into the environment at various points during the product's life cycle, a strict limit of 1% by weight has been established for the amount of SCCPs that may be found in different mixtures of CP (UNEP, 2019a, b, c).

According to the results of simulation tests, SCCPs that have a low chlorine content (e.g. less than 50% by weight of chlorine) may biodegrade slowly in the environment. In sediment, SCCPs' half-life has been described as more than 1 year (POPRC, 2015). Furthermore, SCCPs pose significant health threats to both humans and animals (POPRC, 2015; Wei et al., 2016). According to Fridén et al. (2011), the most common way for humans to be exposed is through the consumption of infected food and the act of breathing in contaminated dust. The International Agency for Research on Cancer (IARC) has categorized SCCPs as "possible human carcinogens" due to observed dose-dependent elevations in the occurrence of adenomas and carcinomas in the thyroid, liver, and kidney of rodents. According to the International Agency for Research on Cancer (IARC) in 1990, an experiment that was carried out on mice because of CP exposure for 103 weeks by gavage with 0.125 or 250 mg/kg revealed mortality between 112 and 114 weeks of age (IARC, 1990).

In-House (Countrywide) Scientific Work Done on Chemicals

Environmental Fate of SCCPs in the Atmosphere

According to Li et al. (2012), the concentrations of short-chain chlorinated paraffins (SCCPs), which include both the gas and particle phases, were observed to vary throughout East Asia from 0.28 to 517 ng/m^3 (with a mean concentration of 67.6 ng per m^3). SCCPs are known to be carcinogenic to humans. According to the study conducted by Chaemfa et al. (2014), the concentrations of pollutants in South Asia varied from non-detectable (ND) to 47.4 ng/m^3, with a mean value of 8.51 ng/

Table 3 Short-chain chlorinated paraffin: persistence, bioaccumulation, potential for long-range environmental transport, and adverse effects

POPs criteria	Evidence	Data
1(b). Persistence	(i) Half-life >1 year	On the basis of simulation studies, low chlorine containing SCCPs (e.g. <50% wt. chlorine) may take more time to biodegrade half-life in sediment, >1 year; in air, approximately 1.9–7.2 days
	(ii) Potentially persistent(P) and very persistent (vP)	Concluded as potentially persistent (P) and perhaps very persistent (vP)
1(c). Bioaccumulation	(i) BCF/BAF	BCFs in fish: 1173–7816 (radioactivity measurements) and 574–7273 (parent compound analysis) (Madeley & Maddock, 1983) BCF in Lake Trout: 36500 for C10–13 SCCPs (Filyk et al., 2003) log Kow value: 4.39–8.69, indicating a strong potential for bioaccumulation
1(d). Long-range environmental transport	(i) Monitoring data	Found almost in similar concentrations in Inuit women's breast milk sampled from localities near Canada's Hudson Strait (Stern et al., 1997) and breast milk of women in the UK (Thomas et al., 2006). Also reported in Arctic marine mammals
	(ii) Chemical characteristics for long-range environmental transport such as vapour pressure (VP) and Henry's Law Constant (HLC)	The EU assumes that an SCCP with an approximate 50% chlorine concentration will have a vapour pressure of 2.1 102 Pa at 40 °C (range- 2.8×10^{-7} to 2.5 Pa). The estimated atmospheric half-life for SCCPs ranges from 1.9 to 7.2 days. (European Commission, 2000)
1(e). Adverse effects	(i) Animal study	Direct evidence is lacking for humans or ecosystems; however, SCCPs is classified as 'possible human carcinogens' by IARC (adenomas and carcinomas in thyroid, liver, and kidney of rodent in response to incremental dose)
	(ii) Toxicity or eco toxicity data	Low acute toxicity-fishes High chronic toxicity: rainbow trout and sheep head minnow (NOECs <0.040 and 0.28 mg/L respectively) Aquatic invertebrates: lower NOECs (0.012–3.7 mg/L). Rats and mice-NOAELs of 100 and 1000 mg/kg-bwt/day, respectively)

Source: Scientific Committee on Toxicity, Ecotoxicity and The Environment (CSTEE, 2003), (European Commission, 2000)

Table 4 Concentrations of short-chain chlorinated paraffins in gas and particle phase

Sr no	Sample type	Location	Concentration ng/m^3	References
1	Sum of gas and particle phase	East Asia	0.28–5717	Li. et.al (2012)
2	Sum of gas and particle phase (range)	South Asia	ND-47.4	Chaemfa et al. (2014)
3	Using passive air samples (average)	India (2006)	10.2	Chaemfa et.al. (2014)

m^3. In a study carried out by Chaemfa et al. (2014), it was discovered that China demonstrated the greatest levels of SCCPs (ranging from 0.95 to 517 ng/m^3), whereas India revealed lower levels (ranging from non-detectable to 47.4 ng/m^3) of the substance (Table 4).

5 Relevance of the International Scientific Data as per Indian Conditions

Numerous investigations have detected chlorinated paraffins contents in sediments of rivers, lakes, and coastal locations (Table 5). In their study, Zeng et al. (2011) reported the detection of notably increased concentrations of short-chain chlorinated paraffins (SCCPs) in sediment samples collected from lakes located d/s of a sewage treatment plant (STP). The quantities of SCCPs ranged from 1100 to 8700 ng/g. In samples collected from sediments of coastal areas of Barcelona, and wastewater effluent from a nearby emissary, increased levels of PCNs (3.27 vs 6.56. ng/g), SCCPs (1.17 vs. 2.09 μg/g), and PCBs (44.00 vs. 37.74 ng/g) were found, highlighting the emissary's role in coastal pollutant dispersion (Castells et al., 2008). These results suggest that effluents from STPs could potentially contribute significantly to the presence of SCCPs in lake sediments.

Based on the findings of Chen et al. (2011), the sediments collected from the Pearl River Delta had the highest levels of SCCPs, with levels ranging from 320 to 6600 ng per g. Notably, the significantly increased amounts were found almost exclusively in the sediments of ponds that were situated in locations that recycle electronic trash (e-waste). The evidence presented clearly indicates that e-waste recycling activities make a substantial contribution to the occurrence of sedimentary chlorinated paraffins (CPs) in this specific geographical area. In the study conducted by Zhao et al. (2013), it was observed that the marine sediment originating from the East China Sea had the most minimal concentration range, specifically ranging from 9.0 to 37 ng/g dry weight. Analysis of data from the East China Sea (R(2) = 0.332, $p < 0.01$) revealed that TOC, latitude, and longitude are key factors influencing surficial SCCP levels (Zeng et al., 2013).

Table 5 Environmental fate in sediment

Sr. no	Sample type	Location/country	SCCP (ng/g dw)	References
1	Pond, river, tributary, estuary, and core sediment	Pearl River Delta, China	320–6600	Chen et al. (2011)
2	Lake and river sediment	Liaohe River basin, China	40–48	Zeng et al. (2011)
3	Marine sediment	Bohai Sea, China	97–1757	Ma et al. (2014a)
4		Bohai Sea, China	65–541	Ma et al. (2014b)
5		Firthe of Clyde, Scotland	0.4–69	Gao et al. (2012)
6		Bohai and Yellow Sea, China	15–85	Hussy et al. (2012)
7		East China Sea, China	5.8–65	Zeng et al. (2012)
8		East China Sea, China	9.0–37	Zeng et al. (2012)
9	Lake sediment	Gaobeidan Lake, China	1100–8700	Zhao et al. (2013)
10		Canadian Arctic from remote northern lakes	1.6–257	Tomy et al. (1999)
11		Near harbour areas of Lake Ontario	5.9–290	Tomy et al. (2000)
12		Arctic Lake, Canada	7	Tomy et al. (1997)
13		Detroit River, Lake Erie, and Middle Sister Island, western Lake Erie (1995)	245 μg/kg dw	Tomy et al. (1997)
14		Lake Ontario (2001)	5.9–290	Muir et al. (2001)
15		Lake Ontario, Canada, industrialized area	410	Marvin et al. (2003)
16		Besòs River, Spain	250–3040	Parera et al. (2004)
17		Rivers in Germany, France, and Norway	2–75	Hütting and Oehme (2006), Ballschmiter (1994)
18	River sediments	Germany	<5–83 μg/kg dw	Ballschmiter (1994)
20		Japan (2003)	4.9–484.4	Iino et al. (2005)
21		Liaohe River in Northern China	39.8–480.3	Gao et al. (2012)

Compound to be monitored and core matrices (status 2019)

	Air	Human milk	Human blood	Water
SCCP (C10–C13) alkanes	SCCPs	SCCPs	SCCPs	SCCPs

At the moment, there is a need for further advancement in analytical methodologies in order to make recommendations on analytes of SCCPs and metrices. The analytes and metrices that are suggested in the guidance document on the Global Monitoring plans for Persistent Organic pollutants, 2019 (UNEP/POPS/COP.9/INF/36), are air, human milk, human blood, and water. Chlorinated paraffins are difficult to analyse because of their intricate technical makeup and varied content in food and environmental samples.

6 Adverse Effects of SCCPs

Several studies have shown that SCCPs have an effect on the thyroid gland, which led to their inclusion in a database that is used to prioritize suspected endocrine disruptors that are related with thyroid cancers and thyroid hypertrophy (EU, 2000). The global frequency of human thyroid cancer has been steadily increasing in recent decades (Pellegriti et al., 2013). However, it is hypothesized that specific environmental carcinogens may have exercised a targeted effect on the thyroid gland. The authors of the research say that the mechanisms leading to the observed spike in thyroid cancer remain unknown.

Scientific literature has shown the presence of short-chain chlorinated paraffins (SCCPs) in the breast milk of lactating mothers, as reported by Van Mourik et al. (2015) and Qiao et al. (2018). The study conducted by Yang et al. (2018) projected a decline in the consumption of SCCPs as individual's age. The intake for infants aged up to 6 months was found to be only 2.5 μg/kg/day. However, the intake was higher for infants during the first 3 months (7.1 μg/kg/day) and the first month (13.0 μg/kg/day).

SCCPs have been measured on the surface and in the ocean, in sediments, soils, sewer sludge, and in the air. Most of the available data come from North America, East Asia, and Europe. While levels have been observed to decrease in few regions, there has been a documented escalation in levels in environmental samples in alternative locations. Constructing a temporal trend for faraway locations is not feasible. Although there have been substantial breakthroughs in the research of short-chain chlorinated paraffins (SCCPs), the measurement of SCCPs still presents a number of problems.

Through atmospheric deposition, water movement, or migratory patterns of species, short-chain chlorinated paraffins (SCCPs) are transported to Antarctic, Arctic, and subarctic areas, increasing their prevalence outside of their original origins. Multiple homologues of short-chain chlorinated paraffins (SCCPs) are able to survive for an extended period in the environment. Extensive research on food chains and food webs has shown evidence that certain SCCPs accumulate to a large degree in invertebrates, as well as in fish that live in both freshwater and saltwater environments. Research carried out in controlled laboratory conditions has shown that there is a possibility that fish may suffer detrimental effects even at low doses. It has been discovered that some forms of SCCPs collect and go through the process of bio magnification in the Arctic biota, including a variety of mammalian species. The effects of short-chain chlorinated paraffins (SCCPs) on mammalian animals may be seen in a variety of their physiological systems, such as the renal system, the hepatic system, and the thyroid hormone system. Synthetic chlorinated paraffins have surprisingly found in geographically distant areas, for example, the Arctic, where they have been discovered in various environmental compartments, encompassing air, sediment, and biota. It is important to note that SCCPs have also been identified in the Antarctic, more especially in air samples with concentrations reaching as high as 20.8 picograms per cubic meter (pg/m^3). Tomy et al. (2000) found that the mean concentrations of SCCPs were greatest in ringed seals (0.5 μg/g of live weight), followed by walruses (0.4 μg/g of live weight), and then the Arctic blubber of beluga whales (0.2 μg/g of live weight). Further in an another study conducted by Zhou et al., (2018) reported high concentration of SCCPs (10−1300 μg g^{-1} lipid weight) in aquatic organisms with high trophic magnification factor.

Measurements of SCCPs have also been carried out on the sediments found in arctic lakes. The sodium carboxymethyl cellulose polymers, often known as SCCPs, pose a significant threat to the invertebrate life in aquatic environments. Because of the important part played by invertebrates in fresh water and marine ecosystems, there is substantial cause for concern over the potential effects that submerged chemical pollutants (SCCPs) might have on species of invertebrates that live in sediment.

7 Remedial Action Taken for Management of POPs

- *Waste Management*

  It is highly recommended to exercise control over the utilization and disposal of chlorinated paraffins in order to prevent their release into the environment due to the fact that these substances bio-accumulate and are toxic to the species that live in the environment. This is especially important when considering the difficulties that are associated with monitoring the levels of these substances in the environment. When getting rid of waste items that include chlorinated paraffins, resource recovery, destructive incineration, or landfill procedures are often used as disposal options. These methods of disposal are often used, and they are carried out in accordance with the legislation and standards that are in place in the specific locations where special wastes are stored. It is predicted that chlorinated paraffins, owing to their inherent thermal instability, would experience deterioration when exposed to incineration at lower temperatures, hence minimizing the risk of their volatilization in the exhaust emissions of an incinerator. This is because chlorinated paraffins have an inherent thermal instability. According to the International Programme on Chemical Safety (IPCS, 1996), the deposition of CPs in landfills has the potential to leak into water sources. However, since they have a low solubility in water and a strong attraction to solid things, it is anticipated that only trace amounts will make it into water bodies. This is owing to both of these factors.

- *Thermal Degradation*

  It is possible for considerable thermal degradation of short-chain chlorinated paraffins (SCCPs) to occur either via prolonged exposure to temperatures as low as 70 degrees Celsius or during shorter exposure times at temperatures as high as 200 degrees Celsius. The elimination of hydrochloride (HCl), which results in the generation of unsaturated compounds that are then capable of undergoing polymerization and/or further degradation processes, is the basic mechanism that underlies the process of heat breakdown. It is possible to draw parallels between the thermal decomposition of polyvinyl chloride (PVC) and the first stage of the HCl elimination process. There is no study that has been done yet that proves a direct connection between these compounds and chloroparaffins (Bogdal et al., 2015).

8 Recommendations

The Stockholm Convention has a substantial impact on the chemical industry and many industries that use certain hazardous substances, such as flame retardants. Once a hazardous item is included in Annex A, it becomes subject to a global ban. The implementation of initiatives targeted at the elimination or substitution of hazardous compounds in their products is of utmost importance for enterprises. At the moment, India does not have any regulatory frameworks or legislative measures in place that are relevant to the administration and distribution of short-chain chlorinated paraffins (SCCPs) inside the country's domestic market. According to data provided by the United Nations Environment Programme (UNEP/POPS/POPRC.6/INF/15), the total installed capacity for the production of chlorinated paraffins in India is 110,000 tons per year. This capacity is shared among 20 different manufacturers. There is a need for further advancement of analytical methodologies within the context of India.

References

Ballschmiter, K. (1994, May, 10). *Determination of short and medium chain length chlorinated paraffins in samples of water and sediment from surface water*. Department of Analytical and Environmental Chemistry, University of Ulm, Ulm, Germany.

Bayen, S., Obbard, O. J. P., & Thomas, G. O. (2006). Chlorinated paraffins: Are view of analysis and environmental occurrence. *Environment International, 32*(7), 915–929.

Bogdal, C., Alsberg, T., Diefenbacher, P. S., MacLeod, M., & Berger, U. (2015). Fast quantification of chlorinated paraffins in environmental samples by direct injection high-resolution mass spectrometry with pattern deconvolution. *Analytical Chemistry, 87*(5), 2852–2860.

Canadian Environmental Protection Act. (1999). Federal Environmental Quality Guidelines Chlorinated Alkanes, May 2018, http://www.ec.gc.ca/ese-ees/C4148C43-C35E-44EA87A7-866E5907C42C/FEQG_Chlorinated%20Alkanes_EN.pdf

Castells, P., Parera, J., Santos, F. J., & Galceran, M. T. (2008). Occurrence of polychlorinated naphthalenes, polychlorinated biphenyls and short-chain chlorinated paraffins in marine sediments from Barcelona (Spain). *Chemosphere, 70*(9), 1552–1562. https://doi.org/10.1016/j.chemosphere.2007.08.034. Epub 2007 Oct 29. PMID: 17910973.

Chaemfa, C., Xu, Y., Chakraborty, P., Syed, J. H., Malik, R. N., Wang, Y., Tian, C. G., Zhang, G., & Jones, K. C. (2014). Screening of atmospheric short- and medium-chain chlorinated paraf-

fins in India and Pakistan using polyurethane foam based passive air sampler. *Environmental Science & Technology, 48*, 4799–4808.
Chen, M. Y., Luo, X. J., Zhang, X. L., He, M. J., Chen, S. J., & Mai, B. X. (2011). Chlorinated paraffins in sediments from the Pearl River Delta, South China: Spatial and temporal distributions and implication for processes. *Environmental Science & Technology, 45*, 9936e9943.
Corden, C., Grebot, B., Kirhensteine, I., Shialis, T., & Warwick, O. (2011). *Evidence. Abatement cost curves for chemicals of concern*. The Environment Agency. Horizon House. Bristol, United Kingdom. Available from: https://www.gov.uk/government/uploads/system/uploads/attachment_data/file/290505/scho0811bucc-e-e.pdf
CSTEE (Scientific Committee on Toxicity, Ecotoxicity, and the Environment). (2003). Opinion on the results of the Risk Assessment of: Tetrasodium EDTA and Edetic Acid (EDTA). Human Health Part. European Commission. Brussels. Pp 1–7. https://citeseerx.ist.psu.edu/document?repid=rep1&type=pdf&doi=e666fa6a7f218bc9bef068cf1a427b6dde7ca6d4.
EPA. (2009). https://www.epa.gov/sites/production/files/2015-09/documents/sccps_ap_2009_1230_final.pdf
European Union Risk Assessment Report. (2000). Alkanes, C10–13, chloro (CAS No: 85535–84-8, Einecs No.: 287–476-5). Risk Assessment. Office for Official Publications of the European Communities.
Fiedler, H. (2010). Short-Chain Chlorinated Paraffins: Production, use and international regulations. In *Handb. Environ. Chem. Chlorinated Paraffins* (p. 1–40).
Filyk G., Lander L., Eggleton, M., Muir D., & Puckett K. (2003). Short-Chain Chlorinated. Paraffins (SCCPs) Substance-Final Draft II. Environment Canada. Dossier prepared for UNECE ad hoc Expert Group on POPs.
Fridén, U. E., McLachlan, M. S., & Berger, U. (2011). Chlorinated paraffins in indoor air and dust: Concentrations, congener patterns, and human exposure. *Environment International, 37*, 1169–1174. https://doi.org/10.1016/j.envint.2011.04.002
Gao, Y., Zhang, H., Su, F., Tian, Y., & Chen, J. (2012). Environmental occurrence and distribution of short chain chlorinated paraffins in sediments and soils from the Liaohe river basin, P. R. China. *Environmental Science & Technology, 46*, 3771e3778.
Hussy, I., Webster, L., Russell, M., & Moffat, C. (2012). Determination of chlorinated paraffins in sediments from the firth of Clyde by gas chromatography with electron capture negative ionisation mass spectrometry and carbon skeleton analysis by gas chromatography with flame ionisation detection. *Chemosphere, 88*, 292e299.
Hüttig, J., & Oehme, M. (2006). Congener group patterns of chloroparaffins in marine sediments obtained by chloride attachment chemical ionization and electron capture negative ionization. *Chemosphere, 64*(9), 1573–1581.
IARC (International Agency for Research on Cancer). (1990). Chlorinated paraffins of average carbon chain length C12 and average degree of chlorination approximately 60. *IARC Monographs on the Evaluation of Carcinogenic Risks to Humans, 48*, 55–72.
Iino, F., Takasuga, T., Senthilkumar, K., Nakamura, N., & Nakanishi, J. (2005). Risk assessment of short-chain chlorinated paraffins in Japan based on the first market basket study and species sensitivity distributions. *Environmental Science & Technology, 39*(3), 859–866.
IPCS. (1996). Environmental Health Criteria 181 "Chlorinated Paraffins". Available at: http://www.inchem.org/documents/ehc/ehc/ehc0181.htm
Kumari, K. (Ed.). (2021). *Persistent organic pollutants: Gaps in management and associated challenges*. ISBN 9780367821029, 288 Pages, Published December 21, 2021 by CRC Press. https://doi.org/10.1201/9781003046806
Li, Q. L., Li, J., Wang, Y., Xu, Y., Pan, X. H., Zhang, G., Luo, C. L., Kobara, Y., Nam, J.-J., & Jones, K. C. (2012). Atmospheric short-chain chlorinated paraffins in China, Japan, and South Korea. *Environmental Science & Technology, 46*, 11948–11954.
Ma, X., Chen, C., Zhang, H., Gao, Y., Wang, Z., Yao, Z., Chen, J., & Chen, J. (2014a). Congener-specific distribution and bioaccumulation of short-chain chlorinated paraffins in sediments and bivalves of the Bohai Sea, China. *Marine Pollution Bulletin, 79*, 299e304.

Ma, X., Zhang, H., Wang, Z., Yao, Z., Chen, J., & Chen, J. (2014b). Bioaccumulation and trophic transfer of short chain chlorinated paraffins in a marine food web from Liaodong bay, North China. *Environmental Science & Technology, 48*, 5964e5971.

Madeley, J. R., & Maddock, B. G. (1983). *The bioconcentration of a chlorinated paraffin in the tissues and organs of rainbow trout* (Salmo gairdneri). Imperial Chemical Industries PLC, Devon, UK (Brixham Report No. BL/B/2310).

Marvin, C. H., Painter, S., Tomy, G. T., Stern, G. A., Braekevelt, E., & Muir, D. C. G. (2003). Spatial and temporal trends in short-chain chlorinated paraffins in Lake Ontario sediments. *Environmental Science & Technology, 37*(20), 4561–4568.

Muir, D., Alaee, M., Dube, D., Lockhart, L., & Bidleman, T. (2001). Current status. *THREATS, 13*.

National Center for Biotechnology Information. (2023). PubChem Compound Summary for CID 6537497, Cereclor. Retrieved 18 Oct 2023 from https://pubchem.ncbi.nlm.nih.gov/compound/Cereclor.

Nevondo, V., & Okonkwo, O. J. (2021). Status of short-chain chlorinated paraffins in matrices and research gap priorities in Africa: A review. *Environmental Science and Pollution Research, 28*, 52844–52861.

Parera, J., Santos, F. J., & Galceran, M. T. (2004). Microwave-assisted extraction versus Soxhlet extraction for the analysis of short-chain chlorinated alkanes in sediments, *Journal of Chromatography A, 1046*(1–2), 19–26, ISSN 0021-9673, https://doi.org/10.1016/j.chroma.2004.06.064.

Pellegriti, G., Frasca, F., Regalbuto, C., Squatrito, S., & Vigneri, R. (2013). Worldwide increasing incidence of thyroid cancer: Update on epidemiology and risk factors. *Journal of Cancer Epidemiology, 2013*, 965212.

POPRC. (2007). Brazil. Submission of information for Brazil specified in Annex E to the Stockholm Convention. Technical report, 2007. URL http://chm.pops.int/Portals/0/docs/from_old_website/documents/meetings/poprc/submissions/ AnnexE_2007/Short-chainedchlorinatedparaffinsBrazil.doc.

POPRC (Persistent Organic Pollutant Review Committee). (2015). Short-chained chlorinated paraffins: Risk profile document UNEP/POPS/POPRC.11/10/Add.2. United Nations Environmental Programme Stockholm Convention on Persistent Organic Pollutants, Geneva.

POPRC (Persistent Organic Pollutant Review Committee). (2017). Eighth meeting of the conference of the Parties to the Stockholm convention. http://www.pops.int/TheConvention/ThePOPs/TheNewPOPs/tabid/2511/Default.aspx

Qiao, L., Gao, L., Zheng, M., Xia, D., Li, J., Zhang, L., et al. (2018). Mass fractions, congener group patterns, and placental transfer of short-and medium-chain chlorinated paraffins in paired maternal and cord serum. *Environmental Science & Technology, 52*(17), 10097–10103.

RPA. (2010). Evaluation of possible restrictions on short chain chlorinated paraffins (SCCPs). Technical Report July, National Institute for Public Health and the Environment (RIVM) The Netherlands, 2010. http://chm.pops.int/Default.aspx?tabid=3145

Štejnarová, P., Coelhan, M., Kostrhounová, R., Parlar, H., & Holoubek, I. (2005). Analysis of short chain chlorinated paraffins in sediment samples from The Czech Republic by short-column GC/ECNI-MS. *Chemosphere, 58*(3), 253–262.

Stern, G. A., Tomy, G. T., Muir, D. C. G., Westmore, J. B., Dewailly, E., & Rosenberg, B. (1997). Polychlorinated n-alkanes in aquatic biota and human milk. Convention on Long-range Transboundary Air Pollution. Working Group on Strategies (Twenty-first session, 16–20 June 1997). Informed in-session document No. 2.

Thomas, G. O., Farrar, D., Braekevelt, E., Stern, G., Kalantzi, O. I., Martin, F. L., & Jones, K. C. (2006). Short- and medium-chain length chlorinated paraffins in UK human milk fat. *Environment International, 32*, 34–40.

Tomy, G. T., Stern, G. A., Muir, D. C. G., Fisk, A. T., Cymbalisty, C. D., & Westmore, J. B. (1997). Quantifying C10–C13 polychloroalkanes in environmental samples by high-resolution gas chromatography/electron capture negative ion high-resolution mass spectrometry. *Analytical Chemistry, 69*, 2762–2771.

Tomy, G. T., Stern, G. A., Lockhart, W. L., & Muir, D. C. G. (1999). Occurrence of C10– C13 polychlorinated n-alkanes in Canadian midlatitude and arctic lake sediments. *Environmental science & technology, 33*(17), 2858–2863.

Tomy, G. T., Muir, D. C. G., Stern, G. A., & Westmore, J. B. (2000). Levels of C10–C13 polychloro-nalkanes in marine mammals from the Arctic and the St. Lawrence River estuary. *Environmental Science & Technology, 34*, 1615–1619.

UNEP (United Nations Environmental Programme). (2017). Draft guidance on preparing inventories of polychlorinated naphthalenes (PCNs).

UNEP (United Nations Environment Programme). (2019a). Detailed guidance on preparing inventories of short-chain chlorinated paraffins (SCCPs) Draft of 2019.

UNEP (United Nations Environment Programme). (2019b). Plastic and toxic additives, and the circular economy: the role of the Basel and Stockholm Conventions. UNEP/CHW.14/INF/29/Add.1-UNEP/POPS/COP.9/INF/28/Add.1.

UNEP (United Nations Environmental Programme). (2019c). Technical guidelines, addendum general technical guidelines on the environmentally sound management of wastes consisting of, containing or contaminated with persistent organic pollutants.

van Mourik, L. M., Leonards, P. E., Gaus, C., & de Boer, J. (2015). Recent developments in capabilities for analysing chlorinated paraffins in environmental matrices: A review. *Chemosphere, 136*, 259–272.

Wei, G. L., Liang, X. L., Li, D. Q., Zhuo, M. N., Zhang, S. Y., Huang, Q. X., Liao, Y. S., Xie, Z. Y., Guo, T. L., & Yuan, Z. J. (2016). Occurrence, fate and ecological risk of chlorinated paraffins in Asia: A review. *Environment International, 92*, 373–387. https://doi.org/10.1016/j.envint.2016.04.002

Yang, L., Wang, Z., Li, J., Ma, Y., Kong, L., Yang, H., et al. (2018). Evaluation of short chain chlorinated paraffins in human milk and their intake by infants in Hebei Province, China. *Food Additives & Contaminants: Part A, 35*(10), 2011–2021.

Zeng, L., Wang, T., Wang, P., Liu, Q., Han, S., Yuan, B., Zhu, N., Wang, Y., & Jiang, G. (2011). Distribution and trophic transfer of short-chain chlorinated paraffins in an aquatic ecosystem receiving effluents from a sewage treatment plant. *Environmental Science and Technology, 45*, 5529e5535

Zeng, L., Zhao, Z., Li, H., Wang, T., Liu, Q., Xiao, K., ... & Jiang, G. (2012). Distribution of short chain chlorinated paraffins in marine sediments of the East China Sea: influencing factors, transport and implications. *Environmental Science & Technology, 46*(18), 9898–9906.

Zeng, L., Chen, R., Zhao, Z., Wang, T., Gao, Y., Li, A., ... & Sun, L. (2013). Spatial distributions and deposition chronology of short chain chlorinated paraffins in marine sediments across the Chinese Bohai and Yellow Seas. *Environmental Science & Technology, 47*(20), 11449–11456.

Zhao, Z., Li, H., Wang, Y., Li, G., Cao, Y., Zeng, L., Lan, J., Wang, T., & Jiang, G. (2013). Source and migration of short-chain chlorinated paraffins in the coastal East China Sea using multiproxies of marine organic geochemistry. *Environmental Science & Technology, 47*, 5013e5022.

Zhou, Y., Yin, G., Du, X., Xu, M., Qiu, Y., Ahlqvist, P., Chen, Q., & Zhao, J. (2018). Short-chain chlorinated paraffins (SCCPs) in a freshwater food web from Dianshan Lake: Occurrence level, congener pattern and trophic transfer. *Science of The Total Environment, 615*, 1010–1018. https://doi.org/10.1016/j.scitotenv.2017.10.026. Epub 2017 Oct 17. PMID: 29751405.

Index

K. Kumari, *Pollutants of Global Concern*, Emerging Contaminants and Associated Treatment Technologies, https://doi.org/10.1007/978-3-031-50996-4

www.ingramcontent.com/pod-product-compliance
Ingram Content Group UK Ltd.
Pitfield, Milton Keynes, MK11 3LW, UK
UKHW022215051026
15624UKWH00001B/176
9783031509957